Solar Hybrid Systems

Solar Hybrid Systems

Design and Application

Ahmet AKTAŞ
Department of Energy Systems Engineering, Faculty of Technology,
Gazi University, Ankara, Turkey

Yağmur KIRÇİÇEK
Department of Energy Systems Engineering, Faculty of Technology,
Gazi University, Ankara, Turkey

ACADEMIC PRESS

An imprint of Elsevier

Academic Press is an imprint of Elsevier
125 London Wall, London EC2Y 5AS, United Kingdom
525 B Street, Suite 1650, San Diego, CA 92101, United States
50 Hampshire Street, 5th Floor, Cambridge, MA 02139, United States
The Boulevard, Langford Lane, Kidlington, Oxford OX5 1GB, United Kingdom

Notices
Knowledge and best practice in this field are constantly changing. As new research and experience broaden our understanding, changes in research methods, professional practices, or medical
treatment may become necessary.

Practitioners and researchers must always rely on their own experience and knowledge in evaluating and using any information, methods, compounds, or experiments described herein. In using
such information or methods they should be mindful of their own safety and the safety of others,
including parties for whom they have a professional responsibility.

To the fullest extent of the law, neither the Publisher nor the authors, contributors, or editors,
assume any liability for any injury and/or damage to persons or property as a matter of products
liability, negligence or otherwise, or from any use or operation of any methods, products, instruc-
tions, or ideas contained in the material herein.

Library of Congress Cataloging-in-Publication Data
A catalog record for this book is available from the Library of Congress

British Library Cataloguing-in-Publication Data
A catalogue record for this book is available from the British Library

ISBN: 978-0-323-88499-0

For information on all Academic Press publications
visit our website at https://www.elsevier.com/books-and-journals

Publisher: Joe Hayton
Acquisitions Editor: Lisa Reading
Editorial Project Manager: Grace Lander
Production Project Manager: Poulouse Joseph
Designer: Victoria Pearson

Typeset by Thomson Digital

Working together
to grow libraries in
developing countries

www.elsevier.com • www.bookaid.org

This book is dedicated to our daughter Su

Contents

Preface

Solar energy is widely used in heating, steam generation, and electricity generation. Solar energy is an environmentally friendly, easily accessible, free, and unlimited acceptable renewable energy source. Solar energy has an intermittent and unstable the output power characteristic due to its nature. Hybrid energy systems are used to eliminate this fluctuating output power characteristic. It is aimed to ensure the continuity of energy with solar hybrid systems. A complementary characteristic is gained by using solar energy together with other renewable energy sources and energy storage technologies. It has many applications and uses with alternative technological structures such as solar energy, concentrated solar systems, and photovoltaic panels. Energy storage technologies have become an inevitable part of solar hybrid systems. Solar hybrid systems form the infrastructure of smart grids. In addition, electric vehicles, which are developing rapidly today, take their place in smart grids.

This book primarily provides information about the power generation characteristics of solar systems. This book explains why hybrid solar systems need storage or support units with examples from the present literature and application areas. First of all, the working characteristics, advantages, and disadvantages of solar systems are explained. By simulation and experimental results, it has been shown that the solar hybrid system is a complementary factor in solar energy. The book includes hybrid solar energy generation and hybrid energy storage system design and simulation studies. What makes this book unique is that it is giving experimental results from our studies on application examples. The contribution of the hybrid solar system to energy production and quality is explained in this book. On-grid and off-grid solar hybrid systems are mentioned. The contribution and usefulness of the energy storage system is explained by supporting the experimental results. In case the solar system has more than one generation and storage unit, how the smart energy management algorithm that controls these units works is explained with application examples.

The main themes included in the book are as follows:

1. Solar hybrid energy systems operating characteristic
2. Advantages and disadvantages of solar energy systems
3. Why solar hybrid systems are needed?
4. Benefits of a solar hybrid system
5. Contributes to the power generation and end-user of the solar hybrid system
6. How to resolve the intermittent structure of renewable energy sources

7. Analysis of the fluctuating power output of renewable energy sources
8. How to design a solar hybrid energy system
9. An example; solar hybrid system layouts, design guidelines

The book focuses on academics, researchers, application engineers, technologists, and students on developments in solar hybrid systems. It will also be a sample resource for applications in solar systems and hybrid energy storage systems.

In the future, thanks to solar hybrid systems, the grid will be seen as a backup energy source in residential applications. Energy storage provides energy when it is needed just as transmission provides energy where it is needed.

We have been supported by Grace Lander, Editorial Project Manager at Elsevier, to complete the publishing process. We would like to express our deepest sense of gratitude and thanks to Grace Lander and Fisher Michelle, Acquisitions Editor at Elsevier and Poulouse Joseph, Senior Project Manager at Elsevier, for assisting and guiding us for this publication.

Ahmet AKTAŞ
Yağmur KIRÇİÇEK

Chapter 1

Solar System Characteristics, Advantages, and Disadvantages

1 Solar energy

Solar energy is the radiant energy released by the fusion process in the nucleus of the sun (the hydrogen gas turns into helium). The intensity of solar energy outside the earth's atmosphere is approximately 1370 W/m^2. An amount of 0–1100 W/m^2 of this energy reaches the earth because of its atmosphere. Even a small portion of this energy is many times more than the current energy consumption of humanity. The studies on the use of solar energy gained speed especially after the 1970s. Solar energy systems are technologically advanced, and their costs are decreasing. It has established itself as a clean energy source in terms of environmental impact. All of the solar radiation does not reach the earth, and up to 25 % is reflected back by the earth's atmosphere. Only 47 % of the solar radiation passes through the atmosphere and reaches the earth's surface. This energy rises the world's temperature, and lives on earth become possible. It causes wind movements and ocean fluctuations. Only 18 % of the radiation from the sun is kept in the atmosphere and clouds. Less than 1 % of the solar radiation coming from the earth is used by plants in the event of photosynthesis. All the solar radiation coming to the earth is eventually converted to heat and returned to space [1–3]. The flow of energy between the atmosphere and the earth of solar radiation is given in Fig. 1.1.

Solar energy is used with different technologies. The solar energy types in the indirect use method are the parabolic trough collector (PTC), linear Fresnel collector (LFC), parabolic dish collector (PDC), tower plants with central receiver power. On the basis of the previous technologies, heat is collected in a center to obtain high-rate heat. This heat energy is converted to steam, and then the electricity is obtained through the turbines. In direct use, it is converted directly into electrical energy. Solar panels are used for direct use. This technology is based on the semiconductor material base. The sunlights can be converted directly into electricity through semiconductor diodes.

Solar Hybrid Systems. http://dx.doi.org/10.1016/B978-0-323-88499-0.00001-X

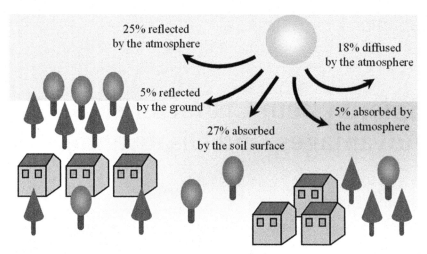

FIGURE 1.1   **The flow of energy between the atmosphere and the earth of solar radiation [1].**

2   Photovoltaic panels

Solar cells are semiconductor materials that can convert sunlight directly into their electrical energy. Solar cells, surfaces of which are square, rectangular, and circular, are generally around 100 cm^2, and their thickness is between 0.1 and 0.4 mm [4,5]. Solar cells work on the basis of the photovoltaic (PV) principle. When the light falls on them, a potential voltage occurs at their terminals. The source of the electrical energy supplied by the cell is the solar energy coming to its surface. Depending on the structure of the solar cell, PV panels can produce electrical energy in efficiency between 5 % and 30 % [6,7]. Fig. 1.2 shows the single crystal silicon and polycrystalline silicon solar panels made of different materials.

In order to increase the output power, a plurality of solar cells is connected to each other in series or in parallel. This structure is called a solar cell module or a PV module. Thus solar panels are produced at the desired current and voltage values. Depending on the power demand, from several watts to megawatts are created with the connections of the modules. Fig. 1.3 shows the sections of the PV generator process.

Solar cells can be produced using many different materials. The most widely used module technologies are thin film and crystal silicone. The a-Si, CdTe, Ci(G)S, a-Si/c-µSi, and dye cells are used in the thin-film production technology. Crystalline silicon technology is mono- and multi-crystal technology. For a 1 cm^2 cell, the highest commercially achievable cell efficiencies are Ci(G)S, with a maximum of 12 % as a thin-film cell and module [6]. The efficiency is 22 % for a mono-crystalline silicon solar cell and is 18 % for a multi-crystalline silicon solar cell. When examined as a module, total efficiency is reduced due to the interconnection between cells and internal resistances [8]. In terms of the

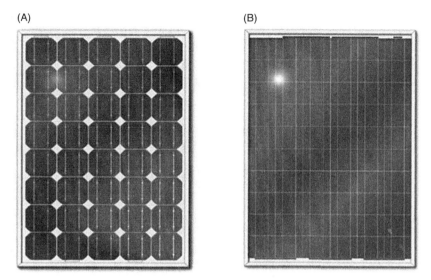

(A)   (B)

FIGURE 1.2   Mono-crystalline (A) and polycrystalline silicon (B) solar panels [7].

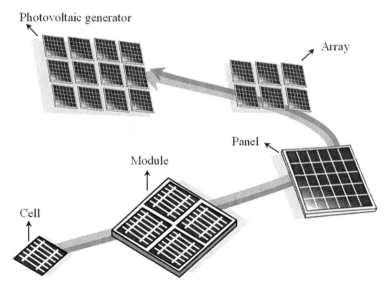

FIGURE 1.3   PV generator process.

PV plant production, it is approximately 12–15 m^2 in thin-film technology as the required space per kW. When looking at the crystalline silicon PV panel, an average area of 7–8 m^2 is needed for the same power [9]. Commercial PV technology efficiencies are given in Table 1.1.

TABLE 1.1 Commercial PV technology efficiencies [7].

PV technology	Thin film					Crystalline silicon	
	a-Si	CdTe	Ci(G)S	a-Si/μc-Si	Dyes	Mono	Multi
Cell efficiency	4 %–8 %	10 %–11 %	7 %–12 %	7 %–9 %	2 %–4 %	16 %–22 %	14 %–18 %
Module efficiency						13 %–19 %	11 %–15 %
Area required for kW	~15 m^2	~10 m^2	~10 m^2	~12 m^2		~7 m^2	~8 m^2

2.1   Structure and operation of solar cells

In today's electronic products, transistors, rectifier diodes such as solar cells, semiconductor materials are made. Among the many substances with semiconductor properties, the most suitable for making solar cells are silicon, gallium arsenide, and cadmium telluride [10].

In order for semiconductor materials to be used as solar cells, negative (N)- or positive (P)-type additives are required. Additive is made by a controlled addition of the desired additives into the pure semiconductor melt. The type of semiconductors obtained depends on the additive. In order to obtain N-type semiconductor materials from the silicon used as the most common solar cell, elements such as phosphorus are used. Phosphorus in V group of the periodic table is added to the silicon melt. Since silicon has four electrons in its outer orbit and the phosphorus has five electrons in its outer orbit, the single electron of the phosphorous gives an electron to the crystal structure. Therefore V group elements are called "transmitter" or "N-type" additives.

In order to obtain P-type silicon, an element from the III group (aluminum, indium, and boron) is added to the melt. Since there are three electrons in the outer orbit of these elements, there is a lack of electrons in the crystal. The absence of this electron is called a hole or a space and is assumed to carry a positive charge. Such substances are also called "receiver" or "P-type" additives. By adding the necessary additives into the main material of type N or P, the semiconductor junctions are formed. Electrons in the N-type semiconductor and holes in the P-type semiconductor are the main carriers. Before the N- and P-type semiconductors come together, both substances are electrically neutral. Negative energy levels and holes are equal in P-type materials, while positive energy levels and electron numbers are equal in N-type materials. When the N–P junction formed, the electrons of the N-type major carrier form a current through the P type. This event continues until the load balance occurs on both the sides. In the junction of the type N–P, a positive charge accumulates on the N side, while a negative charge on the P side.

This junction is referred to as the "transition zone" or "neutral zone." The electric field in this region is called "structural electric field." In order for the semiconductor junction to work as a solar cell, PV transformation in the junction region is required. This transformation takes place in two stages: first, by giving light to the junction region, electron–gap pairs are formed, and second, they are separated by the electric field in the region.

Energy conversion is based on the PV event. In the PV event, the light photons form free charge pairs, especially when they reach the junction region. Each negatively charged electron stimulated leaves a positively charged space behind. These load carriers form the natural internal inverse electric field (E_i) formed by the junction. The natural electric field activates the charge carriers that gain energy with photons. Thus negatively charged electrons produced by photons are collected in the N region, while positively charged carriers are

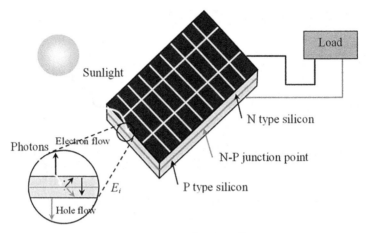

FIGURE 1.4 Process of generating electricity of a solar cell.

collected in the P region to produce a voltage [11]. The process of generating electricity of a solar cell is in principle given in Fig. 1.4.

The current–voltage (*I–V*) characteristic of a PV panel varies with the radiation intensity (W/m^2). An example of PV panel current–output graph is given in Fig. 1.5. The maximum power points that can be taken from the PV panel also vary depending on the radiation. In this case the radiation intensity directly affects the short-circuit current (I_{SC}) produced by the PV panel. The open-circuit voltage (V_{OC}) changes at a lower rate than the short-circuit current. The performance of a PV panel operating under varying radiation intensity is of great importance in PV power system designs [12,13].

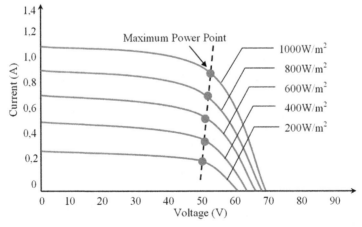

FIGURE 1.5 Current–voltage (*I–V*) characteristic of a sample PV panel in different radiations.

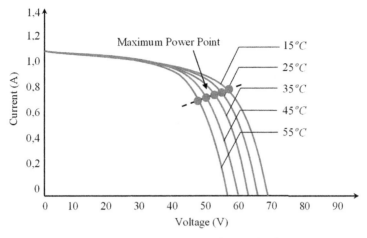

FIGURE 1.6 Current–voltage (*I–V*) characteristic of a sample PV panel at different temperatures.

In addition to the effect of radiation intensity on PV panels, another important effect is the temperature. At a given temperature, the short-circuit current of the PV panel increases in proportion to the amount of radiation. Under constant radiation, the increase of temperature causes the short-circuit current to increase and the open-circuit voltage decreases [14–16]. The effect of temperature to an example PV panel is shown in Fig. 1.6.

Fig. 1.7 shows the characteristic curve of a sample PV panel at a temperature of 25 °C and a radiation rating of 1000 W/m^2. The maximum power point

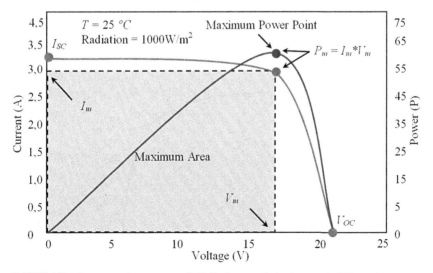

FIGURE 1.7 Current–voltage–power (*I–V–P*) characteristic of a sample PV panel.

is the maximum power that can be received from a PV panel. This point is equal to the multiplication of current and voltage values of the PV panel [17,18]. The maximum rectangular area below the *I–V* curve gives the maximum power (P_m) that the PV panel currently produces. The current in this power is indicated by I_m and the voltage is by V_m. When the PV panel terminals are open, the measured voltage is the highest voltage value of the PV panel and is indicated by V_{OC}. When the PV panel terminal starts to load, the PV panel voltage decreases and the PV panel current increases. While the current continues to rise after a point, the voltage decreases rapidly [19–21]. When the PV panel reaches V_m and I_m values, maximum power is obtained from the PV panel.

2.2   Serial and parallel connection of PV panels

In the PV panels connected to the series, the current of each passing through the system is equal. The system voltage for any current value is the sum of all PV panel voltages. The current of the PV system with serial connection is a single PV panel short-circuit current, and the voltage is a single PV panel open-circuit voltage. The total voltage of the circuit decreases considerably when there is shading in the series of connected PV panels [22–24]. Fig. 1.8 shows the *I–V* graph of the PV panels connected in series.

In a parallel connected system, a single PV panel voltage is the total voltage of the system. The current of each PV panels constitutes the total PV panel current of the system. The open-circuit voltage of the system is a single PV panel open-circuit voltage. The short-circuit current is the sum of the short-circuit currents of each PV panel. The partial shading in the parallel connected PV panel system does not change the total voltage in the system. But the total current value in the PV panel system is reduced [25–27]. Fig. 1.9 shows the *I–V* graph of the parallel connected PV panels.

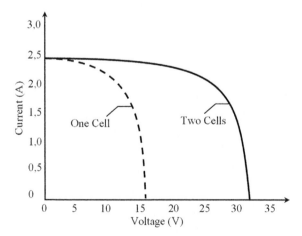

FIGURE 1.8   *I–V* graph of series connected PV panels.

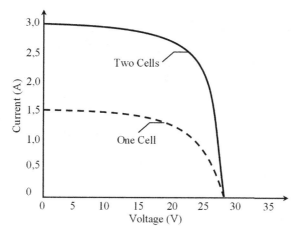

FIGURE 1.9 *I–V* **graph of parallel connected PV panels.**

2.3 Solar cell electrical model

Single-exponent equation model is commonly used for the *I–V* characteristic of a solar cell and the power definitions that can be removed from this curve. The equivalent circuit of this model is given in Fig. 1.10. The equivalent circuit acts as a source of current depending on radiation [28,29]. The relationship of current that can be obtained from a solar cell is given by the following equation:

$$I_{PV} = I_{ph} - I_d - I_p = I_{ph} - I_o \left[e^{\left(\frac{qV_d}{KFT_{PV}} \right)} - 1 \right] - \frac{V_d}{R_p} \tag{1.1}$$

where I_{ph} is current or short-circuit current produced by photons (A), I_o is reverse leakage current (A), I_d is diode current in junction area (A), I_p is current through parallel resistor (A), R_p is parallel resistor (Ω), R_s is serial resistor (Ω),

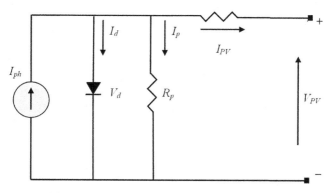

FIGURE 1.10 **Solar cell electrical equivalent circuit [30].**

q is electron charge ($1.6 \cdot 10^{-19}$ C), K is the Boltzmann constant ($1.38 \cdot 10^{-23}$ J/K), T_{PV} is the absolute temperature of the solar cell (K), F is the ideal factor of the solar cell, I_{PV} is solar cell panel current (A), V_d is diode voltage (V), and V_{PV} is solar cell panel output voltage (V).

I_{ph}, the current produced by photons, varies depending on the radiation falling on the solar cell and the solar cell temperature [31], as in the following equation:

$$I_{ph} = \left[\mu_{sc} \left(T_c - T_r \right) + I_{sc} \right] S \tag{1.2}$$

where μ_{sc} is the temperature coefficient in the short circuit current of the solar cell, T_r is the reference temperature of solar cell, I_{sc} is short-circuit current at 1 kW/m^2 and 25 °C, S is solar radiation in W/m^2.

To determine the I–V curve with a single-diode model, the specific I_o and F parameters of the respective solar cell must be known. These parameters are usually not available in the manufacturer's catalogs. The leakage current I_o can be measured with precise measurement possibilities. Also, according to the changing temperatures, the new values of I_o should be considered in the analysis. I_o changes exponentially with temperature. I_o saturation current is given in the following equation:

$$I_o = I_{o\alpha} \left[\frac{T_{PV}}{T_r} \right]^3 e \left[\frac{qV_g}{KF} \left(\frac{1}{T_r} - \frac{1}{T_c} \right) \right] \tag{1.3}$$

$$I_{o\alpha} = \frac{I_{sc}}{e^{\left(\frac{qV_{PV}}{KFT_c} \right)}} \tag{1.4}$$

where $I_{o\alpha}$ is the reverse saturation current of the solar cell at 1000 W/m^2 solar radiation and reference temperature (A), V_g is semiconductor bandgap voltage used by the solar cell (V), V_{OC} is the open-circuit voltage of solar cell (V), and F is the ideal factor of the solar cell. It depends on the solar cell technology used.

Table 1.2 shows the values of the F factor according to the solar cell technology used.

TABLE 1.2 Ideal factor (F) values depending on solar cell technology [32].

Solar cell technology	Ideal factor (F)
Si-poly	1.3
Si-mono	1.2
a-Si-triple	5
a-Si:tandem	3.3
a-Si:H	1.8

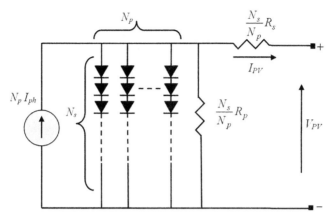

FIGURE 1.11   PV panel arrays electrical equivalent circuit.

A solar cell is usually produced as 0.5 V 2 W. In order to increase this pow-er, the solar cells can be connected in series or parallel to produce solar panels at desired voltage and current values. Fig. 1.11 shows the PV panel arrays circuit connected with the N_s series and N_p parallel.

According to the single-diode model, the total system consists of up to N_s series and up to N_p parallel connected solar cells. The $I–V$ relation of the PV panel arrays is given in the following equation:

$$I_{PV} = N_p I_{ph} - I_o \left[e^{\left(\frac{q\left(\frac{V_d}{R_s} + \frac{IR_s}{R_p} \right)}{N_s KFT_{PV}} \right)} - 1 \right] - \frac{N_p V_d / N_s}{R_p} \qquad (1.5)$$

where N_s is the number of solar cells connected in serial, and N_p is the number of solar cells connected in parallel. In fact, the efficiency of the PV panel mod-ule R_s affects very little, while R_p has a greater effect [25].

2.3.1   PV panel fill factor

A direct calculation of current and voltage values at the maximum power point of a PV panel is quite difficult due to the nonlinear $I–V$ curve. Even if the inter-nal resistance of the solar cell is neglected, the maximum power (P_m) from the PV panel is smaller than the $V_{OC} \times I_{SC}$ value due to its electrical characteristic. There is a need for rigid insertion to bring the term P_m closer to the $V_{OC} \times I_{SC}$.

The maximum utilization rate of the *I–V* curve of a PV panel is defined by the fill factor (*FF*) [33]. The FF is expressed by the following equation:

$$FF = \frac{V_m I_m}{V_{OC} I_{SC}} \tag{1.6}$$

2.3.2 PV efficiency (η)

The efficiency of a PV panel is the ratio of the electric power output (P_{out}) to the solar input (P_{in}). P_{out} is the output power of PV panel which can be accepted as P_{max}. A PV panel can be operated up to maximum power output [34]. PV panel efficiency is given in the following equation:

$$\eta = \frac{P_{out}}{P_{in}} = \frac{P_{max}}{P_{in}} \tag{1.7}$$

3 Concentrating solar power technologies

Thermal solar technologies can be divided into three applications: low-, medium-, and high temperature. The most common example of low-temperature applications is linear collector systems (LCS). Medium-temperature applications are LCS (PTCs, LFCs) and are given in Fig. 1.12. High-temperature applications include point concentrator systems [PDC and solar tower plants (STPs) with central receiver] and are given in Fig. 1.13.

Classic plate type linear solar collectors are mostly used in domestic hot water heating. Solar energy is transferred from heat to fluid by focusing. The temperature of the circulating fluid in these systems can reach up to 70 °C–80 °C.

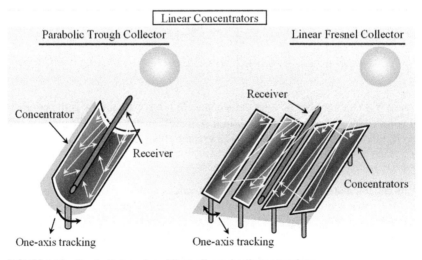

FIGURE 1.12 Parabolic trough and linear Fresnel collector topology.

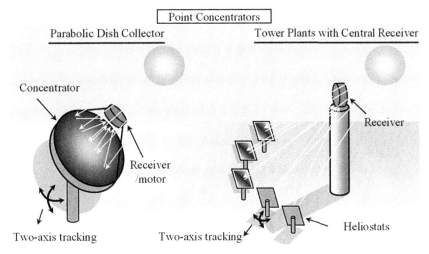

FIGURE 1.13   Parabolic dish collector and tower plants topology.

Apart from need for hot water, they are also used for hot water supply in swimming pools and small industrial facilities. In the last period the use of vacuum heat pipes has started to become widespread. The fluid temperature can reach up to 120 °C in the linear collectors [35–37].

The most common of thermal solar energy applications is water heating systems. Hot water preparation systems with solar energy vary depending on the way the water is used, the way the water is heated, and the circulation of the water in the system [38,39].

The second largest group of thermal solar energy applications is the parabolic and Fresnel trough collector systems. The temperature rises to 300 °C–400 °C in these collectors. In tower plants with central receiver and dish collectors, the temperature can reach up to 1400 °C. Concentrators are intended to use direct sunlight at the highest possible rate. For this purpose the sun collectors are equipped with a monitoring mechanism that allows the sun to be continuously monitored [40–42].

3.1   Linear concentrator systems (LCS)

3.1.1   Parabolic trough collector power system

PTCs are composed of parabolic arrays that can make linear condensation. Reflective surfaces on the inside of the trough reflect sunlight to a black absorbent pipe located at the focus of the parabolic. The heat-collecting glass tube consists of steel receiver pipes and glass–metal connectors with an absorptivity of approximately 97 % on the surface [43]. The air between the glass tube and the receiving pipe is vacuumed to reduce heat losses due to high temperature occurring on the receiving pipe. This gap pressure is about 0.1 atm. The heat-resistant

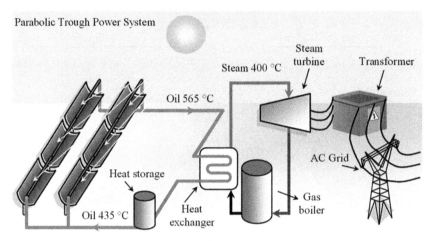

FIGURE 1.14 **Parabolic trough power system diagram.**

glass tube has a high permeability and an antireflective structure to minimize radiation losses. Glazed glass–metal binders are used to remove the effects of temperature expansions. The system has an automation-tracking control unit that allows the mirrors to monitor the sun [44].

Solar energy collectors can obtain saturated or superheated steam at medium and high temperatures. Industrial plants can be used directly for thermal purposes; it can also be used in electricity production by passing through a suitable thermodynamic cycle. The general operation principle of parabolic trough power system is shown in Fig. 1.14. In the first phase of such plants, the heat transfer fluid circulates. This fluid is usually high-temperature synthetic oil [45]. By means of a synthetic oil heat exchanger, the water contained in the electrical power generation system is converted into a steam phase. This superheated steam is generated electricity through the turbine generator. There is a heat storage unit in the cycle to ensure the continuity of electricity generation. In this heat storage unit, the salt melt is generally used. When the sunlights are insufficient, the heat storage unit is activated and plays a supporting role in the evaporation of water in the power circuit [46,47].

In thermal solar power plants, one of the most important measures implemented in recent years for continuous and regular electricity production is the use of hybrid systems. The use of hybrid energy sources plays an important role in ensuring the uninterrupted operation of plants [48].

Steam generation system consists of preheating, steam generation, and superheating sections. Steam is passed through these sections to 400 °C and 100 bar pressure. It is sent to the turbine for electricity generation [49]. After the electricity generation, the steam that is not cooled sufficiently is heated to the same temperature and sent back to the turbine without being sent to a new cycle. In this second cycle the residual cooling steam is sent to a new cycle after

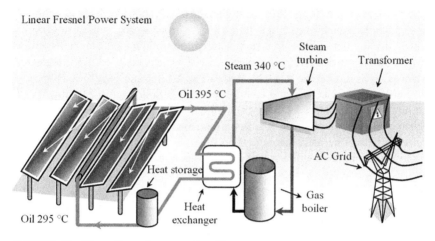

FIGURE 1.15   Linear Fresnel power system diagram.

it becomes liquid. These systems, which can produce 25–200 MW of electricity in terms of the highest radiation, are generally the lowest cost electric power plants per kW h.

3.1.2   Linear Fresnel power system

High temperatures can be achieved with linear or point concentrators in CSP technologies [50,51]. Linear Fresnel power (LFP) plants are constructed using one-dimensional planar mirrors. A heat-collecting tube is placed along the focal point of the mirrors. In linear concentrator systems, the sunlight is collected on a linear focus by the reflective surface. The fluid passing through the pipe forming the surface is heated. The focus mentioned here as a line is a narrow and long space in the form of strips [52,53]. The temperature of the fluid varies depending on the flow rate, concentration ratio, and the instantaneous solar radiation value. The highest theoretical temperature in linear concentrator systems is very close to the temperature of the sun. The control unit is available for the linear mirrors to follow the sun [54]. An LFP system diagram is shown in Fig. 1.15.

3.2   Point concentrator systems (PCS)

3.2.1   Parabolic dish collector power system

Dish collector systems consist of dish, collector, and a motor unit. The solar energy is densified to a receiving surface by a dish-shaped surface. A Stirling or a Brayton engine is used as receiving surface [55]. The receiver may be used as the direct heat energy of the radiation collected on the surface or transfer to the fluid in a Stirling engine. The Stirling engine converts heat into mechanical power. A gas of low specific gravity such as hydrogen or helium is preferred in these systems. The compressed gas expands by being heated by the sunlight.

The cylinder pistons in the Stirling engine move with this gas. This mechanical power is converted into electrical power by means of a generator. Dish systems follow the sun in two axes. It is very suitable for droughty environments where water sources are not available since the water cycle is not used in the collector systems. As it is modular, it can be used as a single or as many dishes. However, it is difficult to make a stable and economical Stirling engine. Hydrogen and helium gases, which are lightweight, should not leak for the robust operation of Stirling engines. In addition, it is not possible to store energy in dish collector systems. Therefore, it is recommended that the dish collector systems be designed in conjunction with another energy source in a hybrid structure [56,57].

3.2.2 Central tower power system

Central tower power systems consist of hundreds of heliostat mirrors spread over a wide area. Heliostats concentrate the sunlight on the tower to the collector by performing two-axis solar tracking [58,59]. A central tower power system diagram is given in Fig. 1.16. The temperature of the salt melt circulated through the collector increases with the help of this concentrated radiation. The salt melt efficiently absorbs heat and is pumped from the storage tank at 277 °C into a collector on the tower. The salt melt is heated to 777 °C in a tower and stored in a salt melt tank. When the power is needed, the salt melt is pumped into a boiler and used to produce superheated steam. In solar tower systems, the condensing rate varies between 300 and 1500, while the temperature can range from 550 °C to 1500 °C. The maximum electrical power is 10 MW in central tower power plants. It is the most efficient among the concentrating solar power (CSP) systems [60]. As the operating temperature in heliostat systems is very high compared to the others, reliability and robustness are the most important factors. The heliostats are controlled by a computer to ensure that the tower

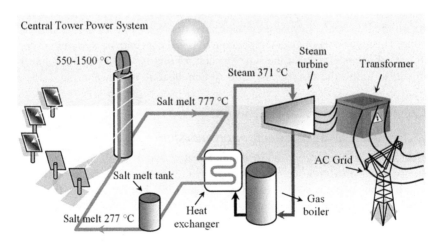

FIGURE 1.16 Central tower power system diagram.

collector is continuously receiving sunlight. As STP systems require complex structure and high technology, investment costs are high [61].

4   Advantages and disadvantages of solar systems

Renewable energy sources (RESs) are being replaced by fossil fuels to produce electricity in today's and future technologies. Solar energy is clearly more environment friendly compared to all other energy sources, so its use is increasing worldwide. RESs are inexhaustible and no harmful gases such as CO_2 leave solid wastes.

The following are the advantages of RES:

- the amount of greenhouse gases released into the environment is zero (CO_2, NO_x),
- no release of toxic gases to the environment throughout using of sources (SO_2, particles),
- making barren lands useful,
- decreasing the transmission line costs in existing electricity grids,
- allow for individual installation and use,
- provide national energy independence to countries, and
- the rapid increase in the rate of use in rural areas.

The use of solar energy is encouraged to meet the rapidly increasing energy consumption demand, to prevent global climate change and to minimize greenhouse gas emissions. International regulations are launching to ensure easy access to these resources [1]. Countries add the issue of generating electricity from RES to their national strategic objectives. Countries are trying to establish various solar energy policies within this framework [2,3]. Moreover, different areas of the world are being worked for the use of solar energy in electricity generation. Advantages and disadvantages of PV panels are given in Table 1.3.

Advantages and disadvantages of CSP systems are given in Tables 1.4 and 1.5. When the CSP systems are examined in terms of technological, economic, and area utilization and ease of operation, the differences between these systems are as follows:

Technologically
PTC: Their efficiency is low because they make solar tracking on one-axis and linear condensation. The solar tracking unit of these systems is cheap and simple. These systems can be used in low and high electrical power plant applications. The manufacture of heat-collecting pipes and mirrors used in the system requires special technologies.
LFC: Their efficiency is low because they make solar tracking on one-axis and linear condensation. The solar tracking system is cheap and very simple. It can be used in all applications with low and high power when it is used as an electrical power plant. There are no elements that require special technology.

TABLE 1.3 Advantages and disadvantages of PV panels.

PV panels	
Advantages	• PV panels do not release harmful greenhouse gas emissions during electricity generation • PV panels can be made available wherever there is sunlight • PV panels are suitable for smart grid structures with distributed power generation • PV panel costs are on the way to a rapid decline thanks to developing technology • Direct electrical energy can be achieved with PV panels • The operating and maintenance costs of PV panels are almost negligible compared to other RES • PV panels do not produce any sound during operation. Suitable for urban areas and residential applications • PV panel installations are quite easy and do not require special equipment • There are incentives for PV panels to increase use in many countries • Thanks to the financial incentives provided to PV panels, an attractive investment is becoming an alternative
Disadvantages	• As with all RESs, electrical energy generated from PV panels depends on the natural events • The electrical energy generated from PV panels is intermittent and unstable. It requires an energy storage unit to eliminate this disadvantage • The use of the energy storage unit together with the PV panels increases the cost • PV panels produce direct current (DC) electrical energy. An inverter is required for use in the alternative current (AC) electricity grid • Large powerful PV panel systems require large area for installation • PV panel efficiency levels are relatively low in today's technology (14 %–25 %) • The surfaces of PV panels must be cleaned periodically in order not to reduce system efficiency

DC: The solar tracking system is of two axes. A special motor (Stirling or Brayton engine, etc.) may be required in the focusing unit. It requires special engineering.

STP: Their efficiency is high because they make solar tracking on two-axis and point condensation. Solar tracking systems are expensive and complex. Low power and small sizes cannot be produced. The support structure carrying the heliostats is large and bulky.

Economically

PTC: Costs are high due to the special technology of heat-collecting pipes and mirrors. It can be used to generate electricity or to produce heat at medium-high temperature.

TABLE 1.4 Advantages and disadvantages of PTC and LFC.

	Linear concentrators	
	Parabolic trough collector	**Linear Fresnel collector**
Advantages	• Operating potential up to 600 °C • About 14 % net electricity production efficiency annual • Applicable investment and operating costs • Modular structure • Good land use factor • Minimum material demand • Hybrid concept applicable • Thermal storage capability	• Easy setup • Lower installation costs compared to other solar systems • Reflective flat mirrors are sufficient • Low production costs • Hybrid processing possible • 1.5–3 Times more energy per land area is produced • The most land-efficient solar energy technology
Disadvantages	• Oil-based heat transfer is used during heat cycle. It limits operating temperatures to 400 °C in oil-based system. Therefore the steam quality produced is moderate	• Low system efficiency due to low operating temperature • Implementation only in low power plants

TABLE 1.5 Advantages and disadvantages of DC and STP.

	Point concentrators	
	Dish collector	**Tower plants with central receiver**
Advantages	• Small land requirement than other solar systems • High working temperature • High thermodynamic efficiency • High sunlight collecting rate thanks to two-axis solar tracking	• High conversion efficiency in terms of system • Operating temperature potential above 1000 °C • Thermal storage at high temperatures • Hybrid concept applicable • Suitable for dry cooling concepts • Good for use in rough areas
Disadvantages	• Low thermal storage capability • High engineering requirements • High installation cost	• Large land requirement • Higher installation costs than other solar systems • Complex control mechanism

LFC: There is no special part requiring technology. Due to its simple structure, unit cost is the lowest among all systems. It can be used to generate electricity or to produce heat at medium-high temperature.

DC: Costs are high because it requires the use of special motors in the heat collection unit. These systems use a lightweight gas as a heat exchanger and, therefore, require a special installation infrastructure.

STP: It is an expensive system due to the bulky structure of towers and heliostats used for sunlight collection. It is suitable for electrical energy production purposes.

In terms of area utilization and ease of operation

PTC: In order for the mirrors to not shade each other, it is necessary to leave a certain amount of space between the mirror arrays. At least three times the active mirror area is required. The two most important elements of the system are heat-collecting pipes and special technologies mirrors. In the case of failure, original spare part supply and sometimes outsourced workmanship is required. Cleaning of mirrors or reflective surfaces is important for system efficiency. In addition, vacuum must be provided in the heat-collecting pipe.

LFC: A mirror settlement area of 1.1–1.2 times of the active mirror area is sufficient. The whole system can be produced with local materials and workmanship. Because of its simple structure, all maintenance and repairs can be carried out with the standard maintenance team.

DC: There is not much space for mirrors. In order to prevent the leakage of the gas used as a heat exchanger in the system, their maintenance should be done frequently. Due to the special motor, there may be outsourced workmanship and material requirements.

STP: A mirror settlement area of approximately 10 times the active mirror area is needed. It is possible to use a large proportion of local materials and workmanship. It is therefore relatively easy to operate and repair.

References

[1] E. Kabir, P. Kumar, S. Kumar, A.A. Adelodun, K.-H. Kim, Solar energy: potential and future prospects, Renew. Sustain. Energy Rev. 82 (2018) 894–900, doi: 10.1016/J.RSER.2017.09.094.

[2] O. López-Lapeña, R. Pallas-Areny, Solar energy radiation measurement with a low–power solar energy harvester, Comput. Electron. Agric. 151 (2018) 150–155, doi: 10.1016/J.COMPAG.2018.06.011.

[3] N. Kannan, D. Vakeesan, Solar energy for future world: a review, Renew. Sustain. Energy Rev. 62 (2016) 1092–1105, doi: 10.1016/J.RSER.2016.05.022.

[4] G. Liu, J. Liu, E. Jiaqiang, Y. Li, Z. Zhang, J. Chen, et al. Effects of different sizes and dispatch strategies of thermal energy storage on solar energy usage ability of solar thermal power plant, Appl. Therm. Eng. 156 (2019) 14–22, doi: 10.1016/J.APPLTHERMALENG.2019.04.041.

[5] G. Wang, F. Wang, Z. Chen, P. Hu, R. Cao, Experimental study and optical analyses of a multi-segment plate (MSP) concentrator for solar concentration photovoltaic (CPV) system, Renew. Energy 134 (2019) 284–291, doi: 10.1016/J.RENENE.2018.11.009.

[6] F.C.S.M. Padoan, P. Altimari, F. Pagnanelli, Recycling of end of life photovoltaic panels: a chemical prospective on process development, Sol. Energy 177 (2019) 746–761, doi: 10.1016/J.SOLENER.2018.12.003.

[7] R. Venkateswari, S. Sreejith, Factors influencing the efficiency of photovoltaic system, Renew. Sustain. Energy Rev. 101 (2019) 376–394, doi: 10.1016/J.RSER.2018.11.012.

[8] Y. Xing, P. Han, S. Wang, P. Liang, S. Lou, Y. Zhang, et al. A review of concentrator silicon solar cells, Renew. Sustain. Energy Rev. 51 (2015) 1697–1708, doi: 10.1016/j. rser.2015.07.035.

[9] M.A. Fares, L. Atik, G. Bachir, N. Delmonte, M. Lazzaroni, Photovoltaic panels characterization and experimental testing, Energy Procedia 119 (2017) 945–952, doi: 10.1016/J.EGYPRO.2017.07.127.

[10] N. Delmonte, M. Lazzaroni, Photovoltaic plant maintainability optimization and degradation detection: modelling and characterization, Microelectron. Reliab. 88–90 (2018) 1077–1082, doi: 10.1016/J.MICROREL.2018.07.021.

[11] M.A. Hasan, S.K. Parida, An overview of solar photovoltaic panel modeling based on analytical and experimental viewpoint, Renew. Sustain. Energy Rev. 60 (2016) 75–83, doi: 10.1016/J.RSER.2016.01.087.

[12] S. Mahmoudinezhad, A. Rezania, L.A. Rosendahl, Behavior of hybrid concentrated photovoltaic-thermoelectric generator under variable solar radiation, Energy Convers. Manage. 164 (2018) 443–452, doi: 10.1016/J.ENCONMAN.2018.03.025.

[13] H. Sun, N. Zhao, X. Zeng, D. Yan, Study of solar radiation prediction and modeling of relationships between solar radiation and meteorological variables, Energy Convers. Manage. 105 (2015) 880–890, doi: 10.1016/J.ENCONMAN. 2015.08.045.

[14] C. Coskun, U. Toygar, O. Sarpdag, Z. Oktay, Sensitivity analysis of implicit correlations for photovoltaic module temperature: a review, J. Cleaner Prod. 164 (2017) 1474–1485, doi: 10.1016/J.JCLEPRO. 2017.07.080.

[15] E. Klugmann-Radziemska, P. Wcisło-Kucharek, Photovoltaic module temperature stabilization with the use of phase change materials, Sol. Energy 150 (2017) 538–545, doi: 10.1016/J. SOLENER.2017.05.016.

[16] P. Sánchez-Palencia, N. Martín-Chivelet, F. Chenlo, Modeling temperature and thermal transmittance of building integrated photovoltaic modules, Sol. Energy 184 (2019) 153–161, doi: 10.1016/J.SOLENER.2019.03.096.

[17] P. Soulatiantork, L. Cristaldi, M. Faifer, C. Laurano, R. Ottoboni, S. Toscani, A tool for performance evaluation of MPPT algorithms for photovoltaic systems, Measurement 128 (2018) 537–544, doi: 10.1016/J.MEASUREMENT.2018.07.005.

[18] O. Ezinwanne, F. Zhongwen, L. Zhijun, Energy performance and cost comparison of MPPT techniques for photovoltaics and other applications, Energy Procedia 107 (2017) 297–303, doi: 10.1016/J.EGYPRO.2016.12.156.

[19] P. Kofinas, S. Doltsinis, A.I. Dounis, G.A. Vouros, A reinforcement learning approach for MPPT control method of photovoltaic sources, Renew. Energy 108 (2017) 461–473, doi: 10.1016/J.RENENE.2017.03.008.

[20] H. Abouadane, A. Fakkar, Y. Elkouari, D. Ouoba, Performance of a new MPPT method for photovoltaic systems under dynamic solar irradiation profiles, Energy Procedia 142 (2017) 538–544, doi: 10.1016/J.EGYPRO.2017.12.084.

[21] Y. Cheddadi, F. Errahimi, N. Es-sbai, Design and verification of photovoltaic MPPT algorithm as an automotive-based embedded software, Sol. Energy 171 (2018) 414–425, doi: 10.1016/J.SOLENER.2018.06.085.

[22] S. Paul Ayeng'o, H. Axelsen, D. Haberschusz, D.U. Sauer, A model for direct-coupled PV systems with batteries depending on solar radiation, temperature and number of serial connected PV cells, Sol. Energy 183 (2019) 120–131, doi: 10.1016/J.SOLENER.2019.03.010.

[23] Z. Chen, F. Han, L. Wu, J. Yu, S. Cheng, P. Lin, et al. Random forest based intelligent fault diagnosis for PV arrays using array voltage and string currents, Energy Convers. Manage. 178 (2018) 250–264, doi: 10.1016/J.ENCONMAN.2018.10.040.

[24] M. Dhimish, V. Holmes, B. Mehrdadi, M. Dales, B. Chong, L. Zhang, Seven indicators variations for multiple PV array configurations under partial shading and faulty PV conditions, Renew. Energy 113 (2017) 438–460, doi: 10.1016/J.RENENE.2017.06.014.

[25] S.R. Pendem, S. Mikkili, Modeling, simulation and performance analysis of solar PV array configurations (Series, Series–Parallel and Honey-Comb) to extract maximum power under Partial Shading Conditions, Energy Rep. 4 (2018) 274–287, doi: 10.1016/J.EGYR.2018.03.003.

[26] O. Bingöl, B. Özkaya, Analysis and comparison of different PV array configurations under partial shading conditions, Sol. Energy 160 (2018) 336–343, doi: 10.1016/J.SOLENER.2017.12.004.

[27] M. Matam, V.R. Barry, Variable size dynamic PV array for small and various DC loads, Sol. Energy 163 (2018) 581–590, doi: 10.1016/J.SOLENER.2018.01.033.

[28] X. Gao, Y. Cui, J. Hu, N. Tahir, G. Xu, Performance comparison of exponential, Lambert W function and Special Trans function based single diode solar cell models, Energy Convers. Manage. 171 (2018) 1822–1842, doi: 10.1016/J.ENCONMAN.2018.06.106.

[29] M.A. Cappelletti, G.A. Casas, A.P. Cédola, E.L. Peltzer y Blancá, B. Marí Soucase, Study of the reverse saturation current and series resistance of p-p-n perovskite solar cells using the single and double-diode models, Superlattices Microstruct. 123 (2018) 338–348, doi: 10.1016/J.SPMI.2018.09.023.

[30] F.J. Toledo, J.M. Blanes, Analytical and quasi-explicit four arbitrary point method for extraction of solar cell single-diode model parameters, Renew. Energy 92 (2016) 346–356, doi: 10.1016/J.RENENE.2016.02.012.

[31] H. Guo, Q. Sun, Y. Wu, Simulation of solar cells by delocalized recombination model and its applications, Sol. Energy 181 (2019) 83–87, doi: 10.1016/J.SOLENER.2019.01.075.

[32] S.M. Perovich, M.D. Djukanovic, T. Dlabac, D. Nikolic, M.P. Calasan, Concerning a novel mathematical approach to the solar cell junction ideality factor estimation, Appl. Math. Model. 39 (2015) 3248–3264, doi: 10.1016/J.APM.2014.11.026.

[33] D. Kiermasch, L. Gil-Escrig, H.J. Bolink, K. Tvingstedt, Effects of masking on open-circuit voltage and fill factor in solar cells, Joule 3 (2019) 16–26, doi: 10.1016/J.JOULE.2018.10.016.

[34] F. Schindler, A. Fell, R. Müller, J. Benick, A. Richter, F. Feldmann, et al. Towards the efficiency limits of multicrystalline silicon solar cells, Sol. Energy Mater. Sol. Cells 185 (2018) 198–204, doi: 10.1016/J.SOLMAT.2018.05.006.

[35] G. Xiao, G. Zheng, M. Qiu, Q. Li, D. Li, M. Ni, Thermionic energy conversion for concentrating solar power, Appl. Energy 208 (2017) 1318–1342, doi: 10.1016/j.apenergy.2017.09.021.

[36] Z. Wang, S. Sun, X. Lin, C. Liu, N. Tong, Q. Sui, et al. A remote integrated energy system based on cogeneration of a concentrating solar power plant and buildings with phase change materials, Energy Convers. Manage. 187 (2019) 472–485, doi: 10.1016/J.ENCONMAN.2019.02.094.

[37] G. Sala, Chapter IID-1 - Concentrator Systems, in: Augustin McEvoy (Ed.), Practical Handbook of Photovoltaics Fundamentals and Applications, Academic Press, 2012, pp. 837–862.

[38] J.J.C.S. Santos, J.C.E. Palacio, A.M.M. Reyes, M. Carvalho, A.J.R. Freire, M.A. Barone, Concentrating solar power, Advances in Renewable Energies and Power Technologies, Jonathan Simpson, 2018, pp. 373–402.

[39] M. Sadi, A. Arabkoohsar, Modelling and analysis of a hybrid solar concentrating-waste incineration power plant, J. Cleaner Prod. 216 (2019) 570–584, doi: 10.1016/J.JCLEPRO.2018.12.055.

[40] P. del Río, P. Mir-Artigues, Designing auctions for concentrating solar power, Energy Sustain. Dev. 48 (2019) 67–81, doi: 10.1016/J.ESD.2018.10.005.

[41] Y. Jia, G. Alva, G. Fang, Development and applications of photovoltaic–thermal systems: a review, Renew. Sustain. Energy Rev. 102 (2019) 249–265, doi: 10.1016/J.RSER.2018.12.030.

[42] C.K. Ho, A review of high-temperature particle receivers for concentrating solar power, Appl. Therm. Eng. 109 (2016) 958–969, doi: 10.1016/J.APPLTHERMALENG.2016.04.103.

[43] E. Bellos, C. Tzivanidis, Alternative designs of parabolic trough solar collectors, Prog. Energy Combust. Sci. 71 (2019) 81–117, doi: 10.1016/J.PECS.2018.11.001.

[44] R.K. Donga, S. Kumar, Thermal performance of parabolic trough collector with absorber tube misalignment and slope error, Sol. Energy 184 (2019) 249–259, doi: 10.1016/J.SOLENER.2019.04.007.

[45] G.K. Manikandan, S. Iniyan, R. Goic, Enhancing the optical and thermal efficiency of a parabolic trough collector – a review, Appl. Energy 235 (2019) 1524–1540, doi: 10.1016/J.APENERGY.2018.11.048.

[46] E. Bellos, I. Daniil, C. Tzivanidis, Multiple cylindrical inserts for parabolic trough solar collector, Appl. Therm. Eng. 143 (2018) 80–89, doi: 10.1016/J.APPLTHERMALENG.2018.07.086.

[47] A.J. Abdulhamed, N.M. Adam, M.Z.A. Ab-Kadir, A.A. Hairuddin, Review of solar parabolic-trough collector geometrical and thermal analyses, performance, and applications, Renew. Sustain. Energy Rev. 91 (2018) 822–831, doi: 10.1016/J.RSER.2018.04.085.

[48] W. Qu, R. Wang, H. Hong, J. Sun, H. Jin, Test of a solar parabolic trough collector with rotatable axis tracking, Appl. Energy 207 (2017) 7–17, doi: 10.1016/J.APENERGY.2017.05.114.

[49] B. Widyolar, L. Jiang, J. Ferry, R. Winston, D. Cygan, H. Abbasi, Experimental performance of a two-stage (50×) parabolic trough collector tested to 650°C using a suspended particulate heat transfer fluid, Appl. Energy 240 (2019) 436–445, doi: 10.1016/J.APENERGY.2019.02.073.

[50] R. Abbas, A. Sebastián, M.J. Montes, M. Valdés, Optical features of linear Fresnel collectors with different secondary reflector technologies, Appl. Energy 232 (2018) 386–397, doi: 10.1016/J.APENERGY.2018.09.224.

[51] R. Abbas, J.M. Martínez-Val, A comprehensive optical characterization of linear Fresnel collectors by means of an analytic study, Appl. Energy 185 (2017) 1136–1151, doi: 10.1016/J.APENERGY.2016.01.065.

[52] M. Cagnoli, D. Mazzei, M. Procopio, V. Russo, L. Savoldi, R. Zanino, Analysis of the performance of linear Fresnel collectors: encapsulated vs. evacuated tubes, Sol. Energy 164 (2018) 119–138, doi: 10.1016/J.SOLENER.2018.02.037.

[53] P. Boito, R. Grena, Optimization of the geometry of Fresnel linear collectors, Sol. Energy 135 (2016) 479–486, doi: 10.1016/J.SOLENER.2016.05.060.

[54] M. Alhaj, A. Mabrouk, S.G. Al-Ghamdi, Energy efficient multi-effect distillation powered by a solar linear Fresnel collector, Energy Convers. Manage. 171 (2018) 576–586, doi: 10.1016/J.ENCONMAN.2018.05.082.

[55] A.Z. Hafez, A. Soliman, K.A. El-Metwally, I.M. Ismail, Solar parabolic dish Stirling engine system design, simulation, and thermal analysis, Energy Convers. Manage. 126 (2016) 60–75, doi: 10.1016/j.enconman.2016.07.067.

[56] A. Bianchini, A. Guzzini, M. Pellegrini, C. Saccani, Performance assessment of a solar parabolic dish for domestic use based on experimental measurements, Renew. Energy 133 (2019) 382–392, doi: 10.1016/J.RENENE.2018.10.046.

[57] T. Venkatachalam, M. Cheralathan, Effect of aspect ratio on thermal performance of cavity receiver for solar parabolic dish concentrator: an experimental study, Renew. Energy 139 (2019) 573–581, doi: 10.1016/J.RENENE.2019.02.102.

[58] K. Mohammadi, J.G. McGowan, M. Saghafifar, Thermoeconomic analysis of multi-stage recuperative Brayton power cycles: Part I – Hybridization with a solar power tower system, Energy Convers. Manage. 185 (2019) 898–919, doi: 10.1016/J.ENCONMAN.2019.02.012.

[59] Q. Zhang, Z. Wang, X. Du, G. Yu, H. Wu, Dynamic simulation of steam generation system in solar tower power plant, Renew. Energy 135 (2019) 866–876, doi: 10.1016/J.RENENE.2018.12.064.

[60] Y. Luo, X. Du, D. Wen, Novel design of central dual-receiver for solar power tower, Appl. Therm. Eng. 91 (2015) 1071–1081, doi: 10.1016/J.APPLTHERMALENG.2015.08.074.

[61] M.J. Wagner, W.T. Hamilton, A. Newman, J. Dent, C. Diep, R. Braun, Optimizing dispatch for a concentrated solar power tower, Sol. Energy 174 (2018) 1198–1211, doi: 10.1016/J.SOLENER.2018.06.093.

Chapter 2

Eliminate the Disadvantages of Renewable Energy Sources

1 Disadvantages of renewable energy sources

Although renewable energy sources (RES) come from nature, there are processes in which the energy they produce is interrupted. RES are affected by natural events such as sun, wind, wave, and tide. Wind energy can only generate energy in times when the wind blows. In the same way, offshore energy sources can provide energy when there are waves, currents, and tides [1–3]. Solar energy systems produce energy depending on hourly, daily, and seasonal conditions [4,5].

In photovoltaic (PV) panels the shading effect occurs when a PV array or part of it is not fully irradiated. When PV panels have partial shading, the PV panel acts as if there is a load in the system instead of generating energy. If this reverse tendency exceeds the degradation voltage of the shaded PV cell, it may function as an open circuit and damage the entire PV module. To reduce the effects of PV panel partial shading, PV panel manufacturers add bypass diodes to the modules [6]. However, the number of diodes in the PV panel is less than the number of cells connected in series within the module. As a result, the PV panel open-circuit case is reduced. However, the total power output is reduced as the PV panel remains unstable [7].

Fig. 2.1 shows two series of PV panels in partial shading state. There is a bypass diode connected in parallel to the PV panel array in PV arrays at the first row. There is no partial shading in these PV panels. PV arrays in the second row have partial shading. The power produced by the PV panels in this sequence is lower [8].

Fig. 2.2 shows the V–P curve of the PV module in fully irradiation and partial shading conditions. In the case of partial shading of PV panels, two different power points emerge. One of these points is the actual maximum power point (MPP) and the other is only the peak power point. The peak power point is due to the bypass diode in the PV modules [9]. MPP algorithms need to work correctly to find the actual MPP in such cases.

Solar Hybrid Systems. http://dx.doi.org/10.1016/B978-0-323-88499-0.00002-1

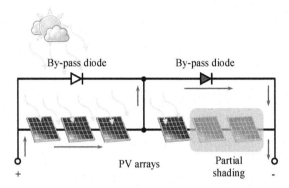

FIGURE 2.1 Power flows of PV arrays in the case of partial shading.

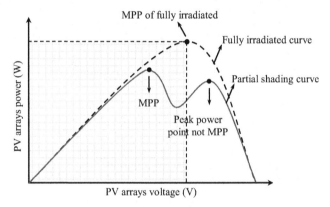

FIGURE 2.2 MPP changes of PV arrays under partial shading.

2 Maximum power point tracking techniques in solar systems

PV power generation systems have two major problems in terms of efficiency and low irradiation. The efficiency of these power conversion systems is very low (9 %–17 %) [10]. Particularly, under low irradiation conditions, the amount of electrical energy produced by the solar panels changes constantly depending on the weather conditions. Moreover, the electrical characteristics of the solar panels draw a nonlinear curve with varying temperature and radiation [11]. This peak, which is often displaced continuously on the I–V or V–P curve, is called the MPP. Maximum power point tracking (MPPT) control technique is a method that enables efficient use of PV panels. This method is a control structure that captures the highest point of the solar panel. This power point varies with variables such as radiation, temperature, PV panel slope, and PV panel aging. When the entire PV system (arrays, converters, etc.) is operated at this power point, it operates at maximum efficiency and receives maximum output power from

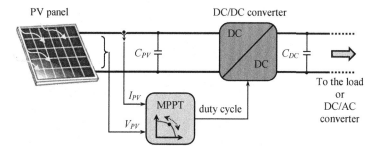

FIGURE 2.3   Implementation of MPPT technique to the PV panel.

the PV panel. MPP can only be detected by computational models or by search algorithms. In order to perform this process, the control mechanism evaluates the PV panel variables depending on the control technique and changes the reference of the power converter to reach the power point [12–15]. The more input units associated with the PV panel in the system, the closer and more stable the control system is to the MPP. However, too much input increases the number of operations in the run algorithm and may result in late response time. As a result, it may be delayed in capturing MPP in rapidly changing weather conditions, which is undesirable for the system. Therefore the use of a technique that provides the same operating efficiency with fast response time without forcing the processing algorithm with the minimum input unit becomes important in terms of cost and simplicity [16]. Fig. 2.3 has shown the use of the PV panel in a system and the topology of the MPPT method.

2.1   Maximum power point tracking operating principle

The electrical characteristic of the solar cell in the case of specific irradiation is shown in Fig. 2.4. The internal resistance of the solar cell is high on the left

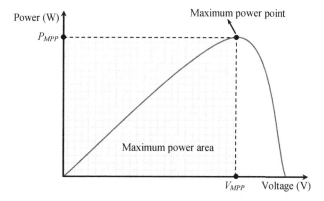

FIGURE 2.4   Electrical characteristic of PV panel for a specific irradiation condition.

side of the curve and low on the right side. The MPP is at the exact peak of this curve. According to the maximum power transfer theory, the maximum power is transferred to the load when the source and load impedances are equal [17]. When the impedance of the switched converter is equal to the impedance of the solar panel, the condition of operation at the MPP is provided.

Many conventional DC/DC converter circuits have naturally negative impedance characteristics. This behavior is due to the constant input power character and the adjustable output voltage of the power supply. If the system is operated on the high impedance side (low-voltage side) of the solar module characteristic curve, the PV panel voltage will decrease. Therefore in the case of MPPT, the solar panel must be operated in the right zone of the characteristic. Otherwise, the PV panel voltage will change according to the irradiation condition, while the PV panel will operate in the maximum duty cycle rate. Thus the system will not reach MPPT and will operate at the wrong point [18,19]. In electrical systems used as PV panel source, a nonlinear change in the output voltage occurs due to the structure of the PV panels. PV panel output voltage change is seen in Fig. 2.4.

The output voltage of the PV panels differs depending on the current drawn, as well as the intensity of the sunlight falling on the PV panels and the ambient temperature. The mentioned output power of commercial PV panels is determined as the electrical energy that the PV panel can generate in an environment where the intensity of the irradiation per unit area is 1000 W/m^2. It is assumed that said output power is the maximum power that the PV panel can deliver. As the intensity of the sunlight decreases, the current that can be drawn from the PV panel decreases is seen in the curve in Fig. 2.5.

Fig. 2.5 shows that the PV panel current is 2.3 A and the voltage is 19 V in an environment with 1000 W/m^2 irradiation (A point). In this case the power

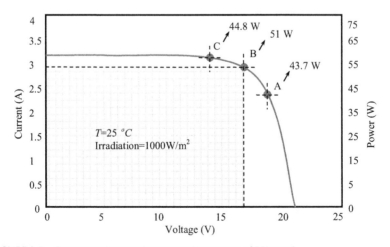

FIGURE 2.5 **Current–voltage and power–voltage curve of PV panel.**

generated at point A is 43.7 W. When the current drawn from the PV panel is operated at 3 A in the same irradiation, the curve shows that the output voltage drops to 17 V (B point). The power produced according to the new current voltage values is 51 W. In the same case, 3 A is drawn from the PV panel at point C and the output voltage is 14 V. The power generated at point C is 44.8 W.

As a result of these values, although the load increases gradually, the amount of power that can be taken from the PV panel varies in a nonlinear manner. The power generated up to a point increases depending on the load but decreases at the power generated after a point. If this situation is not kept under control, the system will operate with low efficiency. MPPT systems always try to keep this output power at its peak. The MPPT circuits include a design infrastructure that is much more than a simple current limiter [20].

Under varying temperature and irradiation conditions, the MPP of the PV panel is constantly changing so the PV system must be changed to produce maximum power at the operating point. Therefore MPPT techniques are used to enable PV panels to work in MPP. There are many MPPT techniques to capture the MPP of PV panels [21–25]. These commonly used techniques are constant voltage method, constant current method, short-circuit current pulse method, open voltage method, perturb and observe method, incremental conductance method, and temperature method.

2.2 Constant voltage method

The constant voltage algorithm is a simple MPPT control method. This algorithm tries to keep the panel close to the MPP by matching the PV panel voltage and a constant V_{ref} reference voltage. This V_{ref} value is the characteristic V_{MPP} of the PV panel. This method assumes that the temperature variations of the only PV panels are insignificant and that the constant reference voltage is a sufficient approach to the actual MPP. Therefore the process can never be processed in the full MPP, and different data must be collected for different geographical areas [26]. In the constant voltage method, it is necessary to measure the PV panel voltage of the solar panel in order to establish the duty cycle and to process the algorithm in the block diagram given in Fig. 2.6. When low irradiation conditions of the PV panel are important, it is observed that the constant voltage technique is more effective than the perturb and observe method or incremental conductance method.

The constant voltage method is a simple, inexpensive, and convenient method that does not require complex circuits. However, the need to separate the

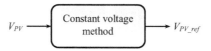

FIGURE 2.6 **Constant voltage method block diagram.**

load from the PV module in order to read the V_{PV} voltage information required to generate a reference mark is a problem for loads. The lack of utilization of solar energy during PV panel voltage measurement and the inability to sustain the actual MPPT process are negative effects [27].

2.3   Short current pulse method

The short-circuit current pulse method manages to operate in the MPP by providing the current controlled power converter I_m operating current. In fact, the optimum operating current I_m for maximum output power is proportional to Eq. (2.1) with the irradiation level S under various conditions and the short-circuit current I_{ph}.

$$I_m(S) = k \cdot I_{ph}(S) \tag{2.1}$$

where k is a proportional constant and I_m can be determined immediately when I_{ph} is detected in Eq. (2.1). Although the temperature varies from 0 °C to 60 °C, this relationship between I_m and I_{ph} is continuously proportional. This proportional parameter is estimated to be about 92 % [27].

Therefore this control algorithm requires I_{ph} measurement. To achieve this measurement, a parallel static switch must be connected to the PV panel to generate a short-circuit current. In the case of a PV panel short circuit, no power is obtained from the PV panel and power generation is stopped. For the short current pulse algorithm, V_{PV} and I_{ph} data must be processed in the block diagram given in Fig. 2.7. Although the short current pulse method is simple and does not require complex circuits, energy is lost during the measurement of the short-circuit current of PV panel. Due to the determination of the constant k and the contamination that may occur on the panel surface, deviations of the constant k are among the negative effects of this method [27,28].

2.4   Open voltage method

The open voltage method is based on the observation that the voltage at the MPP is always close to the percentage of open-circuit voltage. The temperature and solar irradiation level change the position of the MPP within the 2 % tolerance band. In general, the open voltage method uses 76 % of the open-circuit voltage of the panel as the optimum operating voltage, and maximum output power can be obtained from the PV panel. For this control algorithm

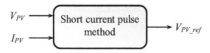

FIGURE 2.7   **Short current pulse block diagram.**

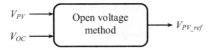

FIGURE 2.8   Open voltage method block diagram.

the open-circuit voltage V_{OC} shown in the block diagram in Fig. 2.8 must be measured. This method uses a static switch with the PV panel and must be connected in series to the PV panel to open the switch circuit [29]. When $I_{PV} = 0$, no power is generated by the PV system and as a result, the total energy from the PV panels is reduced.

2.5   Perturb and observe method

The perturb and observe method is one of the most commonly used methods. This method changes the fluctuating (increasing or decreasing) PV panel voltage or current at intervals and operates by comparing the PV output power with the previous power value. If the PV panel operating voltage changes and the power rises, the control system PV operating point changes in this direction, otherwise the operating point moves in the opposite direction. In the next cycle the algorithm continues in the same way. In the perturb and observe method, the PV output power is continuously monitored and a correlation between the movement of the control variable and the movement of the power is determined to reduce or increase the reference. It has weaknesses such as slow response to rapid change of atmospheric conditions and searching the MPP in the wrong direction for sudden changes. Since the system continuously making perturb and observe, it is not fixed here when it reaches MPP. It continuously oscillates around the MPP and there is some power loss in the system [30–32].

While the power decreases in the left region of the MPP, the PV panel voltage also decreases as shown in Fig. 2.9. In the right region of the MPP, the

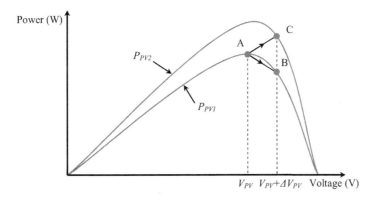

FIGURE 2.9   The change of maximum power points in the perturb and observe method.

power decreases as the PV panel voltage increases. Therefore if there is an increase in power, the next perturb should be kept the same to reach the MPP, and if there is a decrease in power perturb must be reversed. Perturb and observe algorithm can also be applied on the basis of voltage to power or current to power variation. As a result of the comparison process, the next pulse width ratio is determined to achieve the MPP. Perturb and observe method, as seen in Fig. 2.9, can work incorrectly in rapidly changing atmospheric conditions. When starting at point A, when the atmospheric conditions remain approximately constant, perturb of the PV panel voltage ΔV will move the operating point to B and the next perturb will be reversed due to a reduction in power. However, if solar irradiation increases, the power curve shifts from P_{PV1} to P_{PV2} within a sampling time, and the operating point is moved from A to C. This means an increase in power and perturb is kept the same. Therefore the operating point is separated from the MPP and if the solar irradiation increases continuously, the operating point remains separated from the MPP. The actual power point is compared with the two previous points before a perturb signal decision is made to ensure that the MPP is monitored even in the case of momentary changing solar irradiation [31].

The flow diagram of the perturb and observe algorithm is given in Fig. 2.10. In this algorithm, first the current and voltage values from the sensors are read, and the power of the PV panel is calculated. This power value is compared with the previous value. If there is an increase or decrease in power, the voltage of the PV panel is compared with the previous voltage value as in power. If there is a decrease in power and an increase in voltage value, it is determined that the MPP is perturb in the right region. To capture the MPP the reference voltage must be reduced. However, if there is a decrease both in power and voltage, it is determined that the MPP is perturb in the left region, and the reference voltage must be increase. This process continues until the MPP is reached and then the algorithm circulates around the MPP. If this sampling step is large, the system will oscillate at points remote from the MPP, reducing the total amount of energy from the PV panels. However, a smaller perturb slows the MPPT and causes late response time. As a solution, MPP should be approached using small perturbs [32]. In order to calculate the power, it is necessary to measure the voltage and current of the PV panel given in the block diagram in Fig. 2.11.

2.6 Incremental conductance method

This incremental conductance algorithm is based on the logic of finding the MPP with the conductivity of the PV panel. It has been proposed to quickly detect changes in atmospheric conditions. The incremental conductance algorithm is based on the principle of following the MPP by comparing the incremental and instantaneous conductivity of the PV panel. Sample changes in the current and voltage values produced by the PV panel may also change the conductivity gradually increasing. This method is based on Eq. (2.2). This is to see the change in

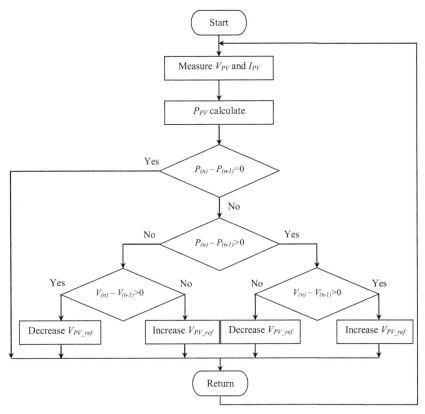

FIGURE 2.10   Flow diagram of perturb and observe method.

PV panel power due to the voltage and set the result to zero [33]. V–P characteristic curve showing the $dP_{PV}/d/V_{PV}$ change of the PV panel is given in Fig. 2.12.

$$\frac{dP_{PV}}{dV_{PV}} = I_{PV}\frac{dV_{PV}}{dV_{PV}} + V_{PV}\frac{dI_{PV}}{dV_{PV}} = I_{PV} + V_{PV}\frac{dI_{PV}}{dV_{PV}} = 0 \tag{2.2}$$

$$-\frac{I_{PV}}{V_{PV}} = \frac{dI_{PV}}{dV_{PV}} \tag{2.3}$$

The left side of Eq. (2.3) expresses the contrast of incremental conductance. The right side of Eq. (2.3) shows gradually incremental conductance and can be written as shown in the following equation:

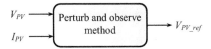

FIGURE 2.11   Perturb and observe method block diagram.

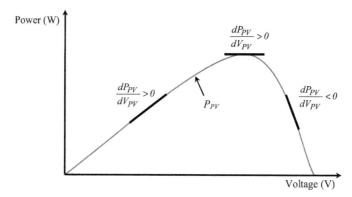

FIGURE 2.12 *V–P* characteristic curve showing the dP_{PV}/dV_{PV} change of the PV panel.

$$G = \frac{dP_{PV}}{dV_{PV}} \tag{2.4}$$

On the other hand, the incremental variables dV_{PV} and dI_{PV} are both obtained by comparing the instantaneous measurements with the previous instantaneous measurements, which are given in the following equations:

$$dV_{PV}(n) = V_{PV}(n) - V_{PV}(n-1) \tag{2.5}$$

$$dI_{PV}(n) = I_{PV}(n) - I_{PV}(n-1) \tag{2.6}$$

Thus by analyzing the derivative of the P_{PV}, it can be tested that the P_{PV} operates at or beyond the MPP. In Eq. (2.7) the operating is to the left of the MPP and the V_{PV} panel voltage is less than the voltage in the V_m MPP. The V_{PV} is equal to the operating in the V_m MPP and is given in Eq. (2.8). The V_{PV} panel voltage is more than the voltage in the V_m MPP and is given in Eq. (2.9). It is seen that the study is to the right of the MPP and the V_{PV} panel voltage is greater than the voltage at V_m MPP.

$$\frac{dP_{PV}}{dV_{PV}} > 0 \; for \; V_{PV} < V_m \tag{2.7}$$

$$\frac{dP_{PV}}{dV_{PV}} = 0 \; for \; V_{PV} = V_m \tag{2.8}$$

$$\frac{dP_{PV}}{dV_{PV}} < 0 \; for \; V_{PV} > V_m \tag{2.9}$$

According to Eq. (2.2), the instantaneous conductivity (dI_{PV}/dV_{PV}) and incremental conductance ($\Delta I_{PV}/\Delta V_{PV}$) can be compared to the MPP point as shown in the flow diagram in Fig. 2.13. At the MPP point the reference voltage

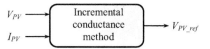

FIGURE 2.13   Flow diagram of incremental conductance method.

V_{ref} is equal to the V_m voltage in MPP. When the MPP is reached, the PV panel operating point is maintained at this point, if no change is observed in the MPP or a change in atmospheric conditions. The algorithm increases or decreases the V_{ref} reference voltage to monitor the new MPP [34,35]. In order to calculate the power in this control algorithm, it is necessary to measure the PV panel voltage and current shown in the block diagram in Fig. 2.14.

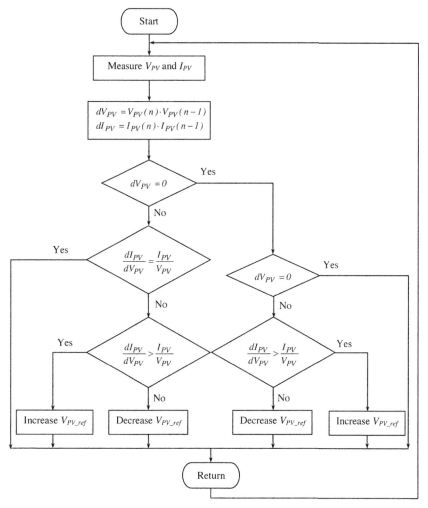

FIGURE 2.14   Incremental conductance method block diagram.

The most important advantage of the incremental conductance method is that it can adapt to rapidly changing atmospheric conditions and that the oscillation in the MPP is much less than that of the perturb and observe method. However, circuits controlled by the incremental conductance control method are somewhat more complex and expensive [35].

2.7 Temperature method

The open-circuit voltage often varies with the V_{OC} PV cell temperature but does not change in a relatively direct proportion on short-circuit current, irradiation level, and PV cell temperature changes. The following equation defines the open-circuit voltage V_{OC} of PV cell:

$$V_{OC} \cong V_{OCSTC} + \frac{dV_{OC}}{dT_{PV}} \cdot (T_{PV} - T_{STCPV}) \tag{2.10}$$

where V_{OCSTC} is the PV cell open-circuit voltage under the standard test conditions (STC), T_{STCPV} is the PV cell temperature under STC. On the other hand, the MPP voltage V_m in each operating condition is expressed by the following equation:

$$V_m \cong \left[(u + S \cdot v) - T_{PV} \cdot (w + S \cdot y) \right] \cdot V_{m\_STC} \tag{2.11}$$

where V_{m_STC} is the MPP voltage under STC. Table 2.1 shows the parameters of the optimum voltage equation with respect to the S irradiation level.

The temperature parametric equation method calculates according to Eq. (2.11) and determines the MPP voltage by measuring the T_{PV} and S values.

TABLE 2.1 Optimum voltage equation parameters [36].

S (kW/m^2)	u (S)	v (S)	w (S)	y (S)
0.1 % 0.2	0.43404	0.1621	0.00235	$-6e - 4$
0.2 % 0.3	0.45404	0.0621	0.00237	$-7e - 4$
0.3 % 0.4	0.46604	0.0221	0.00228	$-4e - 4$
0.4 % 0.5	0.46964	0.0131	0.00224	$-3e - 4$
0.5 % 0.6	0.47969	-0.0070	0.00224	$-3e - 4$
0.6 % 0.7	0.48563	-0.0169	0.00218	$-2e - 4$
0.7 % 0.8	0.49270	-0.0270	0.00239	$-5e - 4$
0.8 % 0.9	0.49190	-0.0260	0.00223	$-3e - 4$
0.9 % 1.0	0.49073	-0.0247	0.00205	$-1e - 4$

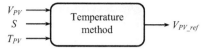

FIGURE 2.15   Temperature method block diagram.

The temperature parametric method generally also requires S solar irradiation. Fig. 2.15 shows the block diagram of the temperature method [36].

2.8   Compare of MPPT techniques

MPPT techniques vary in many aspects, including simplicity, convergence speed, hardware implementation, sensor requirement, cost, efficiency range, and measurability. In Table 2.2, MPPT techniques are listed according to their efficiency and the amount of energy they produce [27–36].

According to the data in Table 2.2, it is seen that energy production is superior to other methods and that there is a similar performance among themselves. This result is seen with the widespread use of commercial applications. The incremental conductance method, among other techniques, enables energy with high efficiency to be obtained from PV panels. The output of the incremental conductance method has the same shape as the solar irradiation input, except that it has a small transition time at the time of fast irradiation. As a result, this method adapts to atmospheric conditions faster than others and has a very small response time with very little oscillation in the MPP. Thus the total amount of energy to be taken from the PV panel increases. The same results are also obtained with a perturb and observe method. The difference between these two methods is that the number of processes used in the incremental conductance technique is slightly higher than the perturb and observe method. Each process causes the hardware and software differences to be used in the system and

TABLE 2.2 Ranking of MPPT techniques according to their efficiency [35].

MPPT techniques	Efficiency (%)	Energy production ranking
Incremental conductance	99.48	1
Perturb and observe	99.29	2
Temperature	97.01	3
Open voltage	94.56	4
Short current pulse	90.72	5
Constant voltage	79.51	6

increases the cost. The incremental conductance algorithm efficiency is the same as that of the perturb and observe method.

The open voltage and short current pulse methods require an additional static switch to the system but still provide much lower energy transfer than the perturb and observe and incremental conductance directions. Loss of energy occurs during switching. Moreover, since the open voltage and short current pulse methods do not make instant time tracking, the solar irradiation change during switching cannot be detected [28]. In fact, these techniques cannot calculate new MPP unless there is a new irradiation level. Furthermore, sampling time selection plays a very critical role for these techniques, since the shorter selection of the sampling time will increase the number of switches, which means that the power generation will be very low. If the sampling time is long, it cannot follow MPP closely in fast changing weather conditions.

Open voltage and short current pulse methods are quite simple, do not require complex circuits, and are cheap and convenient methods. However, it is problematic to remove the PV panel from the load in order to generate a reference mark. In the meantime, the solar energy cannot be utilized and the actual MPPT process that cannot be sustained has negative effects [29].

The temperature method is providing less energy than incremental conductance technique and perturb and observe technique. Unlike other techniques, irradiation and temperature values are read, but the MPP cannot be fully monitored since only the voltage of the PV panel is read.

The open voltage method is the lowest among the other methods. In fact, this technique does not follow the MPP but instead takes the optimum voltage under the STC or another constant voltage as the reference voltage, keeping it constant under any operating condition.

In terms of cost, MPPT costs are electrical components, electronic components, switchboards, controllers, and number of analog and digital sensors, software applicability and all other power components. The topology of the MPPT applications is largely based on the preference of the end users and is a good option for analog circuits with short current pulse, open voltage, or constant voltage methods. With the digital circuit that requires the use of processor, the perturb and observe, incremental conductance, and temperature methods can be easily applied. In addition, analog circuitry is less expensive than digital circuitry (processors and embedded systems). This cost comparison calculation is formulated taking into account the current spread of the MPPT methods [33].

The number of sensors required to implement the MPPT method also affects the total cost. Often, measuring voltage is easier and more reliable than measuring current, current sensors are often more expensive and cumbersome. Solar irradiation and temperature sensors are very expensive and not easily available. After all these cost assessments, Table 2.3 provides a simplified classification of sensors, processors, and additional power components [34].

TABLE 2.3 Cost classification of MPPT techniques [34].

MPPT techniques	Additional component	Sensor requirement	Processor calculation
Constant voltage	No	Low	Low
Short current pulse	Yes	Medium	Low
Open voltage	Yes	Low/medium	Low
Perturb and observe	No	Medium	Low
Incremental conductance	No	High	Medium/high
Temperature	No	High	Medium/high

3   Mechanical solar tracking systems

Solar energy is less preferred than other RES because of the high initial cost and low efficiency. Although the installation costs of solar energy systems are high, the operating and maintenance costs are much lower than other energy types. Efforts are underway to increase the efficiency of solar energy systems and reduce costs.

The main reasons for low efficiency in solar energy systems are as follows:

- inability to utilize solar energy continuously due to the solar and earth rotation movements,
- the loss of radiation as the sun rays pass through the atmosphere,
- maximum efficiency of PV panels occurs only when the solar irradiation comes upright,
- inability to obtain desired irradiation in bad weather conditions, and
- increased cost due to the need for large surface area to obtain high power.

In order to eliminate all these negativities, some efficiency increase studies are carried out. One of these is to ensure that the solar panels can receive the solar irradiation at a right angle. PV panels are designed to be movable and allow the system to follow the sun throughout its movement. A total efficiency increase of 45 % can be achieved with the mechanical solar tracking system [37–40]. Solar tracking systems are classified into two different ways as single- and two-axis according to the number of axes.

3.1 Single-axis solar tracking system

Single-axis solar tracking systems follow the solar by moving in a single axis (vertical or horizontal). Generally, the inclination angle is adjusted manually at certain intervals during the year and automatic movement is provided in the east–west direction. Single-axis systems are more cost-effective than two-axis systems but have lower yields in terms of efficiency. Single-axis solar tracking systems are moved on the vertical or horizontal axis depending on the solar trajectory and the weather condition [41,42]. Fig. 2.16A schematically illustrates the single-axis solar tracking following the azimuth angle.

3.2 Two-axis solar tracking system

In two-axis solar tracking systems, it is possible to follow the solar with two angle values indicating the position of the solar in the sky. The control system moves on the azimuth and zenith axes. PV panel efficiency can be increased by 30 %–45 % with two-axis control system. The azimuth axis is the axis in which the movement of the PV panel in the east–west direction is adjusted, while the zenith axis is the movement in the north–south direction of the PV panel. In order for the system to follow the solar, the position of the solar must be determined first. In order to realize this work, solar detection sensors are used in the system. The two-axis solar tracking system is given in Fig. 2.16B.

The solar detection sensor (pyranometer) is synchronized with the control circuit. The information received from the pyranometer is evaluated in the central control unit, and appropriate position signals are sent to the driver of the east–west and position motors to ensure the two-axis movement of the system.

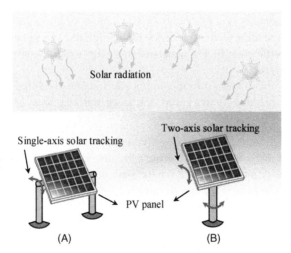

FIGURE 2.16 **Solar tracking systems.** (A) Single-axis and (B) two-axis.

Thanks to the position signals, the motor drives drive the motors and ensure the two-axis movement of the PV panel. This movement continues until the solar rays are perpendicular to the PV panels [43,44].

4 Concentrated PV panels

Concentrated solar panels are based on the principle of more intense collection of solar rays into PV cells. It keeps the peak power of the PV cell constant and ensures maximum utilization of the PV semiconductor field used. With this method, it is possible to harvest 30 times more energy than a low-concentration PV panel and 1000 times more energy than a high-concentration PV panel. Concentrated solar systems need a solar monitoring system. Single-crystal silicon PV panel normally works in 12 % efficiency, when focused; it can work in 20 % efficiency [45,46]. Concentrated PV panels produce high temperatures due to focus. This heat can be recovered using cogeneration applications. The disadvantage of concentrated PV panels is that the efficiency of the PV cells decreases due to the high temperature. In such applications, PV panels need to be cooled. Concentrated PV panels are in commercial development. Fig. 2.17 is given the concentrated PV panel structure.

5 Cylindrical PV panels

Cylindrical PV panels consist of cylindrical panels coated with 360° thin PV cells. A reflector is placed on their surface. Thus in addition to the solar irradiation directly exposed during the day, they can generate electricity from the light reflected from the reflector surface. Cylindrical PV panels can be mounted side by side horizontally for optimum energy generation. Cylindrical PV panel systems are lightweight and unlike planar panels, they are not exposed to wind

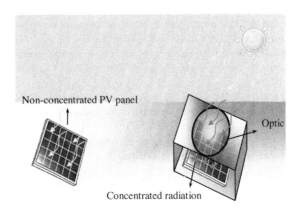

FIGURE 2.17   **Concentrated PV panel structure.**

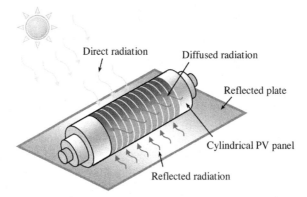

FIGURE 2.18 Cylindrical PV panel structure.

resistance. Therefore weights are not required for the installation of cylindrical panels [47–49]. In Fig. 2.18, cylindrical PV panel structure is presented.

References

[1] Q. Wang, L. Zhan, Assessing the sustainability of renewable energy: an empirical analysis of selected 18 European countries, Sci. Total Environ. 692 (2019) 529–545, doi: 10.1016/J. SCITOTENV.2019.07.170.

[2] H. Labar, M.S. Kelaiaia, Real time partial shading detection and global maximum power point tracking applied to outdoor PV panel boost converter, Energy Convers. Manag. 171 (2018) 1246–1254, doi: 10.1016/J.ENCONMAN.2018.06.038.

[3] S. Jenniches, Assessing the regional economic impacts of renewable energy sources—a literature review, Renew. Sustain. Energy Rev. 93 (2018) 35–51, doi: 10.1016/J.RSER.2018.05.008.

[4] A.R. Gupta, V.K. Rathod, Solar radiation as a renewable energy source for the biodiesel production by esterification of palm fatty acid distillate, Energy 182 (2019) 795–801, doi: 10.1016/J.ENERGY.2019.05.189.

[5] S. Guo, Q. Liu, J. Sun, H. Jin, A review on the utilization of hybrid renewable energy, Renew. Sustain. Energy Rev. 91 (2018) 1121–1147, doi: 10.1016/J.RSER.2018.04.105.

[6] Z. Dostál, L. Ladányi, Demands on energy storage for renewable power sources, J. Energy Storage 18 (2018) 250–255, doi: 10.1016/J.EST.2018.05.003.

[7] A.S. Dagoumas, N.E. Koltsaklis, Review of models for integrating renewable energy in the generation expansion planning, Appl. Energy 242 (2019) 1573–1587, doi: 10.1016/J.APEN-ERGY.2019.03.194.

[8] P. Bórawski, A. Bełdycka-Bórawska, E.J. Szyma ska, K.J. Jankowski, B. Dubis, J.W. Dunn, Development of renewable energy sources market and biofuels in The European Union, J. Clean. Prod. 228 (2019) 467–484, doi: 10.1016/J.JCLEPRO.2019.04.242.

[9] J. Benedek, T.-T. Sebestyén, B. Bartók, Evaluation of renewable energy sources in peripheral areas and renewable energy-based rural development, Renew. Sustain. Energy Rev. 90 (2018) 516–535, doi: 10.1016/J.RSER.2018.03.020.

[10] R. Stropnik, U. Stritih, Increasing the efficiency of PV panel with the use of PCM, Renew. Energy 97 (2016) 671–679, doi: 10.1016/J.RENENE.2016.06.011.

[11] A. Machniewicz, D. Knera, D. Heim, Effect of transition temperature on efficiency of PV/PCM panels, Energy Procedia 78 (2015) 1684–1689, doi: 10.1016/J.EGYPRO.2015.11.257.

[12] A.E. Yaakoubi, L. Amhaimar, K. Attari, M.H. Harrak, M.E. Halaoui, A. Asselman, Non-linear and intelligent maximum power point tracking strategies for small size wind turbines: performance analysis and comparison, Energy Rep. 5 (2019) 545–554, doi: 10.1016/J.EGYR.2019.03.001.

[13] N.A. Zainal, A.R. Yusoff, A. Apen, Integrated cooling systems and maximum power point tracking of fuzzy logic controller for improving photovoltaic performances, Measurement 131 (2019) 100–108, doi: 10.1016/J.MEASUREMENT.2018.08.056.

[14] L. Samani, R. Mirzaei, Model predictive control method to achieve maximum power point tracking without additional sensors in stand-alone renewable energy systems, Optik (Stuttg) 185 (2019) 1189–1204, doi: 10.1016/J.IJLEO.2019.04.067.

[15] J.C. Camilo, T. Guedes, D.A. Fernandes, J.D. Melo, F.F. Costa, A.J. Sguarezi Filho, A maximum power point tracking for photovoltaic systems based on Monod equation, Renew. Energy 130 (2019) 428–438, doi: 10.1016/J.RENENE.2018.06.017.

[16] M. Bahrami, R. Gavagsaz-Ghoachani, M. Zandi, M. Phattanasak, G. Maranzanaa, B. Nahid-Mobarakeh, et al. Hybrid maximum power point tracking algorithm with improved dynamic performance, Renew. Energy 130 (2019) 982–991, doi: 10.1016/J.RENENE.2018.07.020.

[17] L.-L. Li, G.-Q. Lin, M.-L. Tseng, K. Tan, M.K. Lim, A maximum power point tracking method for PV system with improved gravitational search algorithm, Appl. Soft Comput. 65 (2018) 333–348, doi: 10.1016/J.ASOC.2018.01.030.

[18] D. Kler, K.P.S. Rana, V. Kumar, A nonlinear PID controller based novel maximum power point tracker for PV systems, J. Franklin Inst. 355 (2018) 7827–7864, doi: 10.1016/J.JFRANKLIN.2018.06.003.

[19] J.P. Ram, T.S. Babu, N. Rajasekar, A comprehensive review on solar PV maximum power point tracking techniques, Renew. Sustain. Energy Rev. 67 (2017) 826–847, doi: 10.1016/J.RSER.2016.09.076.

[20] A.R. Jordehi, Maximum power point tracking in photovoltaic (PV) systems: a review of different approaches, Renew. Sustain. Energy Rev. 65 (2016) 1127–1138, doi: 10.1016/J.RSER.2016.07.053.

[21] P. Joshi, S. Arora, Maximum power point tracking methodologies for solar PV systems—a review, Renew. Sustain. Energy Rev. 70 (2017) 1154–1177, doi: 10.1016/J.RSER.2016.12.019.

[22] C.V. Vimalarani, K. Nagappan, B. Chitti Babu, Improved method of maximum power point tracking of photovoltaic (PV) array using hybrid intelligent controller, Optik (Stuttg) 168 (2018) 403–415, doi: 10.1016/J.IJLEO.2018.04.114.

[23] G. Li, Y. Jin, M.W. Akram, X. Chen, J. Ji, Application of bio-inspired algorithms in maximum power point tracking for PV systems under partial shading conditions—a review, Renew. Sustain. Energy Rev. 81 (2018) 840–873, doi: 10.1016/J.RSER.2017.08.034.

[24] A.M. Eltamaly, H.M.H. Farh, Dynamic global maximum power point tracking of the PV systems under variant partial shading using hybrid GWO-FLC, Sol. Energy 177 (2019) 306–316, doi: 10.1016/J.SOLENER.2018.11.028.

[25] F. Belhachat, C. Larbes, Comprehensive review on global maximum power point tracking techniques for PV systems subjected to partial shading conditions, Sol. Energy 183 (2019) 476–500, doi: 10.1016/J.SOLENER.2019.03.045.

[26] S. Li, A variable-weather-parameter MPPT control strategy based on MPPT constraint conditions of PV system with inverter, Energy Convers. Manag. 197 (2019) 111873, doi: 10.1016/J.ENCONMAN.2019.111873.

[27] M. Lasheen, A.K.A. Rahman, M. Abdel-Salam, S. Ookawara, Performance enhancement of constant voltage based MPPT for photovoltaic applications using genetic algorithm, Energy Procedia 100 (2016) 217–222, doi: 10.1016/J.EGYPRO.2016.10.168.

[28] A. Harrag, S. Messalti, IC-based variable step size neuro-fuzzy MPPT improving PV system performances, Energy Procedia 157 (2019) 362–374, doi: 10.1016/J.EGYPRO.2018.11.201.

[29] M. Balato, L. Costanzo, A. Lo Schiavo, M. Vitelli, Optimization of both perturb & observe and open circuit voltage MPPT techniques for resonant piezoelectric vibration harvesters feeding bridge rectifiers, Sens. Actuat. A Phys. 278 (2018) 85–97, doi: 10.1016/J.SNA.2018.05.017.

[30] E. Bianconi, J. Calvente, R. Giral, E. Mamarelis, G. Petrone, C.A. Ramos-Paja, et al. Perturb and observe MPPT algorithm with a current controller based on the sliding mode, Int. J. Electr. Power Energy Syst. 44 (2013) 346–356, doi: 10.1016/J.IJEPES.2012.07.046.

[31] Y. Yang, F.P. Zhao, Adaptive perturb and observe MPPT technique for grid-connected photovoltaic inverters, Procedia Eng. 23 (2011) 468–473, doi: 10.1016/J.PROENG.2011.11.2532.

[32] M. Kamran, M. Mudassar, M.R. Fazal, M.U. Asghar, M. Bilal, R. Asghar, Implementation of improved perturb & observe MPPT technique with confined search space for standalone photovoltaic system, J. King Saud Univ. Eng. Sci. 32 (2018) 432–441, doi: 10.1016/J.JK-SUES.2018.04.006.

[33] H. Shahid, M. Kamran, Z. Mehmood, M.Y. Saleem, M. Mudassar, K. Haider, Implementation of the novel temperature controller and incremental conductance MPPT algorithm for indoor photovoltaic system, Sol. Energy 163 (2018) 235–242, doi: 10.1016/J.SOLENER.2018.02.018.

[34] A. Loukriz, M. Haddadi, S. Messalti, Simulation and experimental design of a new advanced variable step size incremental conductance MPPT algorithm for PV systems, ISA Trans. 62 (2016) 30–38, doi: 10.1016/J.ISATRA.2015.08.006.

[35] S. Necaibia, M.S. Kelaiaia, H. Labar, A. Necaibia, E.D. Castronuovo, Enhanced auto-scaling incremental conductance MPPT method, implemented on low-cost microcontroller and SEPIC converter, Sol. Energy 180 (2019) 152–168, doi: 10.1016/J.SOLENER.2019.01.028.

[36] A. Mohapatra, B. Nayak, P. Das, K.B. Mohanty, A review on MPPT techniques of PV system under partial shading condition, Renew. Sustain. Energy Rev. 80 (2017) 854–867, doi: 10.1016/J.RSER.2017.05.083.

[37] N. AL-Rousan, N.A.M. Isa, M.K.M. Desa, Advances in solar photovoltaic tracking systems: a review, Renew. Sustain. Energy Rev. 82 (2018) 2548–2569, doi: 10.1016/J.RSER.2017.09.077.

[38] J. Zhang, Z. Yin, P. Jin, Error analysis and auto correction of hybrid solar tracking system using photo sensors and orientation algorithm, Energy 182 (2019) 585–593, doi: 10.1016/J.ENERGY.2019.06.032.

[39] A.Z. Hafez, A.M. Yousef, N.M. Harag, Solar tracking systems: technologies and trackers drive types—a review, Renew. Sustain. Energy Rev. 91 (2018) 754–782, doi: 10.1016/J.RSER.2018.03.094.

[40] M.A. Abdelghani-Idrissi, S. Khalfallaoui, D. Seguin, L. Vernières-Hassimi, S. Leveneur, Solar tracker for enhancement of the thermal efficiency of solar water heating system, Renew. Energy 119 (2018) 79–94, doi: 10.1016/J.RENENE.2017.11.072.

[41] W. Batayneh, A. Bataineh, I. Soliman, S.A. Hafees, Investigation of a single-axis discrete solar tracking system for reduced actuations and maximum energy collection, Autom. Constr. 98 (2019) 102–109, doi: 10.1016/J.AUTCON.2018.11.011.

[42] F. Sallaberry, R. Pujol-Nadal, M. Larcher, M.H. Rittmann-Frank, Direct tracking error characterization on a single-axis solar tracker, Energy Convers. Manag. 105 (2015) 1281–1290, doi: 10.1016/J.ENCONMAN.2015.08.081.

[43] L. Barker, M. Neber, H. Lee, Design of a low-profile two-axis solar tracker, Sol. Energy 97 (2013) 569–576, doi: 10.1016/J.SOLENER.2013.09.014.

[44] F.M. Hoffmann, R.F. Molz, J.V. Kothe, E.O.B. Nara, L.P.C. Tedesco, Monthly profile analysis based on a two-axis solar tracker proposal for photovoltaic panels, Renew. Energy 115 (2018) 750–759, doi: 10.1016/J.RENENE.2017.08.079.

[45] F.M. Zaihidee, S. Mekhilef, M. Seyedmahmoudian, B. Horan, Dust as an unalterable deteriorative factor affecting PV panel's efficiency: why and how, Renew. Sustain. Energy Rev. 65 (2016) 1267–1278, doi: 10.1016/J.RSER.2016.06.068.

[46] S. Zuhur, . Ceylan, A. Ergün, Energy, exergy and environmental impact analysis of concentrated PV/cooling system in Turkey, Sol. Energy 180 (2019) 567–574, doi: 10.1016/J.SOLENER.2019.01.060.

[47] M. Herrando, A.M. Pantaleo, K. Wang, C.N. Markides, Solar combined cooling, heating and power systems based on hybrid PVT, PV or solar-thermal collectors for building applications, Renew. Energy 143 (2019) 637–647, doi: 10.1016/J.RENENE.2019.05.004.

[48] H.M.S. Bahaidarah, A.A.B. Baloch, P. Gandhidasan, Uniform cooling of photovoltaic panels: a review, Renew. Sustain. Energy Rev. 57 (2016) 1520–1544, doi: 10.1016/J.RSER.2015.12.064.

[49] F. Yazdanifard, M. Ameri, Exergetic advancement of photovoltaic/thermal systems (PV/T): a review, Renew. Sustain. Energy Rev. 97 (2018) 529–553, doi: 10.1016/J.RSER.2018.08.053.

Chapter 3

Why Solar Hybrid System?

1 Solar hybrid energy systems

Solar-based renewable energy sources produce output characteristics according to the position of the sun. This characteristic differs depending on the resource type. While the temperature level of the heated liquid varies in concentrated solar systems, there is a change in output current and voltage in photovoltaic (PV) panels [1–3]. Solar systems are also renewable energy sources that generate energy depending on the solar irradiance. Solar irradiance utilization period is an average of 8–10 hours per day [4], depending on the region and season. In cases of cloudy and evening hours, the solar energy cannot be used. These problems have revealed the need for "Why solar hybrid system?". By using another renewable energy source together with solar systems, uninterrupted and continuous electricity generation can be provided. One of the most preferred sources here is wind energy. Wind energy is easy to install and access, just like solar power [5,6]. Unlike solar energy systems, wind energy systems have a moving structure. Wind formation depends on the temperature changes on the earth's surface caused by the solar. As a result, wind energy is indirectly dependent on the solar source. Unlike solar energy, wind energy can occur at any time of the day. Wind energy can be used in the evening and at night also, while solar energy is used only during the day [7,8]. By using wind energy together with solar energy, generation characteristics at different power frequencies can be combined. In this case a solar hybrid energy system is created by combining solar energy with a different renewable energy source. Thanks to the diversity of sources in solar hybrid systems, power generation possibilities can be increased. Despite all this diversity of resources, in order for the energy to be continuous, stable, and of high quality, energy storage units that have a complementary feature come into operation together with renewable energy sources. The cost of energy storage systems increases as the power values increase. Batteries are one of the most preferred methods in terms of usage and ease of purchase. Thanks to technological developments, more efficient and higher capacity energy storage units are produced [9–11]. As a result of these developments, unit prices of batteries are also falling. Nowadays, thanks to batteries, the concept of the word "mobile" has become an indispensable part of our lives. Mobile phones,

Solar Hybrid Systems. http://dx.doi.org/10.1016/B978-0-323-88499-0.00003-3

computers, tablets, smart watches, household/commercial electronic devices, and electric vehicles are just a few of the examples powered by batteries.

Solar hybrid energy system is a combination of energy sources with different characteristics and an energy storage system. In the case of off-grid applications, determining the best combination to reduce the initial investment cost (maintaining power supply reliability, reducing maintenance) in hybrid energy systems is a challenging process for many reasons.

The combination of renewable energy sources with different characteristics reduces the total energy potential effects of sources, the output characteristics of which change over time. Only solar power (PV panel, concentrating solar system) is available during the daylight, but when it comes to night, you have to store energy from the solar power during the day to ensure energy continuity. Wind energy has similar qualities but unlike solar energy, it often has a chaotic possibility. Due to the nature of renewable energy sources, the energy potential that changes over time makes it necessary to combine energy storage systems and distributed energy sources.

Energy generation systems play an important role in daily life. In our daily life, electrical energy is used in many areas and devices and becomes indispensable. Although we use electrical energy every day, unfortunately very few of us are concerned with energy saving. Although we always try to criticize the cost of electrical energy, we have more responsibilities as individuals, especially in terms of social responsibility. Fossil fuel resources are rapidly being exhausted, and many irreversible effects occur due to the burning of fossil fuels. For all these reasons, we are in a period when special attention should be paid to energy saving. The system design and installation phase are vital to obtain optimum benefit from an electrical energy system. A number of techno-economic and environmental aspects must always be considered in this challenging process. Design modeling for such renewable energy systems is often a difficult part. It is important to determine the design parameters before proceeding with system setup. At this stage, optimization of work is required for the renewable energy system and it is necessary to move away from classical methods. Today, there are many hybrid energy systems for commercial and domestic use [12–16]. Some of those—wind and PV panel energy, biomass and PV panel energy, concentrating solar and fuel cells, PV panel energy and battery, biomass and wind energy, and geothermal and PV panel energy—are a few examples.

1.1 Passive and active solar energy systems

1.1.1 Passive solar energy systems

Today, electrical energy has become a global demand. Concerns such as fossil fuel reserves and climate change have prompted researchers to search alternative energy sources. Solar energy is the most acceptable form of all renewable energy sources. Solar energy is a free resource available in every country even individually. The use of solar energy is carried out in two different ways, active and passive. In the active use method, energy is directly converted into electrical energy by using PV panels that have a nonlinear output characteristic. The efficiency of PV panel varies depending on the solar irradiance value, ambient temperature, and PV panel properties. In the passive method the heating/cooling feature is generally used.

Passive solar power system is used to heat/cool a living space. Thanks to the passive solar energy system, it is used to reduce energy costs in the field of heating and cooling. In this system a building or an apartment uses solar energy indirectly from this energy. The passive solar energy system contains a small amount of moving parts in the conversion system [17]. These systems are of low maintenance and eliminate the need for externally maintained heating/cooling systems. Passive solar system design and installation is not too complicated. In this system the knowledge of solar angle and climate change is a basic requirement. Passive solar power system can be easily integrated into any building. Different heating methods are provided with passive solar energy system. The first of these is direct gain and, in this method, solar irradiance directly penetrates the building and heat energy is stored. Another method is indirect gain, in which solar radiation is collected, stored, and distributed using a solar wall containing thermal storage materials [18–20]. Solar wall application topology for summer and winter is given in Fig. 3.1. The solar wall is a very suitable system especially for using in cool but sunny climates. This method is a combined system that heats and cools the air with solar energy. Another method is isolated gain, in which solar radiation is collected and used in a predesigned openable and closed area of the building.

The main purpose of passive solar energy systems is to maximize solar heat gain in winter and to minimize heat gain in summer. In the passive solar system the direction of the building should be positioned facing east to west. Windows, location and size of which are determined at the design stage, are used in buildings to maximize the heat gain from the sun in winter and to minimize the same

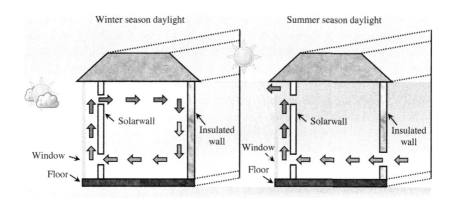

FIGURE 3.1 **Solar wall application topology for summer and winter season.**

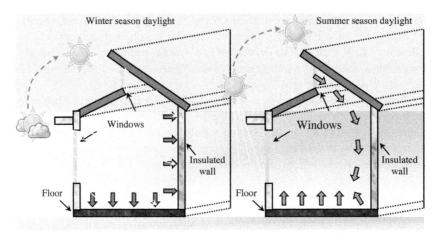

FIGURE 3.2 Roof and window overhang application topology for summer and winter season.

in summer. The roof and windows overhanging are sized to face south and heat is gained in winter. The roof and window overhang application topologies are given for summer and winter seasons in Fig. 3.2. Thermal mass heat from solar irradiation is stored in walls or floors. Thus daylight is used to provide lighting [21–23].

With this method the building is effectively cooled with natural air in summer, thanks to the roof and windows overhanging. In passive solar system design, the windows can be opened to clean the air inside at night and closed to meet the cooling requirement during the day. One of the disadvantages of passive solar system is that its activity lasts only for 16–18 hours per day. Other energy sources should be used for heating/cooling in the remaining time period. However, the passive solar-energy system used for heating/cooling saves a significant amount [24–27].

1.1.2 Active solar energy systems

Solar energy is directly converted into electrical energy by using PV panels in active solar energy system. Solar PV panels are produced with silicon-based semiconductor technology. Solar PV cells are commercially available today; these are produced with monocrystalline, polycrystalline, bar crystal silicon, and thin-film technology. The output characteristics and working principles of solar PV panels are given in detail in Chapter 1, Solar System Characteristics, Advantages, and Disadvantages.

Concentrated solar power plants have large mirror parabolic trough collector areas, a steam generation heat-exchanger system, and Rankine steam turbine generator. A thermal energy storage unit can also be added to these systems upon request. The parabolic trough power system can be used with fossil fuel power plants. The purpose of this type of hybrid system is to reduce the heat loss rate of the plant and to increase the steam cycle, power output, efficiency, and performance [28].

Linear Fresnel power systems are arranged in parallel rows and generally aligned on the north–south horizontal axis. Parabolic trough solar collectors with single-axis tracking structure are available in large modular arrays. Each solar collector has a linear parabolic-shaped reflector that focuses direct solar radiation on a linear receiver (absorber tube) located at the focal line of the parabola. The collector axis mechanism is to focus on the collectors by following the sun moving from east to west throughout the day [29]. Thus the irradiance from the sun enables a heat-transfer fluid contained in the linear receiver to be heated to approximately 400 °C (750 °F) [30].

In order for a concentrating collector to achieve high-efficiency energy harvesting, it must focus on the sunlight directly in a plane vertical to the receiver. Due to clouds, moisture, and other particles on the earth's surface, scattering occurs in the atmosphere and the solar irradiance value decreases. In the case of a linear axis trough solar energy system, effective solar irradiance is very important for high performance.

In the concentrated solar energy system, the high heat transfer fluid passes through the receivers as high pressure superheated steam or through a heat exchanger to create a thermal storage system. Superheated steam is sent to the turbine to generate electricity in a Rankine cycle system. The steam used in the turbine is returned to the heat exchangers with the steam condensation method and turned back into steam. The steam condensation method is carried out with an evaporative cooling tower or directly with an air-cooled condenser. The choice of steam condensation method will affect the total water use in the system, cycle performance, and operating cost. Dry cooling, one of the steam condensation methods, significantly reduces water use but increases investment and electricity costs. The high heat transfer fluid is reheated in the concentrated solar line after passing through the heat exchangers. Molten nitrate salts or compressed gases are used as high heat transfer fluid. The direct steam generation from the concentrated solar line is also carried out in addition to these methods [31–33].

Fig. 3.3 shows an example of solar PV panel and concentrated solar energy system realized in Karabük University Energy Systems Engineering Renewable Energy Resources Laboratory. In this study, hybrid electricity generation has been realized with both concentrated solar energy and PV panel [34].

A unique and most important feature of concentrated solar and power tower installations is their ability to store highly efficient thermal energy. Thus the thermal energy stored in the daylight is used in the evening. For example, in summer, up to 10 hours of electricity can be generated per day by thermal energy storage with solar energy plants. Another approach is to combine concentrated solar energy systems with other renewable energy sources (wind, biomass, etc.). With these hybrid renewable energy sources, solar energy system output is supported in low solar irradiation periods [35,36].

Many research and development studies, related to solar parabolic trough plants, are carried out and planned. These studies bring to light collector component improvements, power plant integration and configuration, thermal

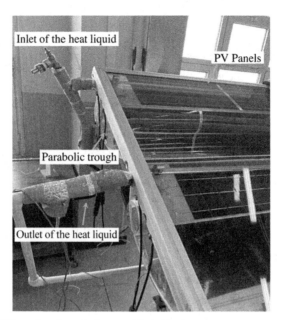

FIGURE 3.3 **PV panel and parabolic trough hybrid solar system.**

energy storage systems, cost reduction measures, and performance improvements. Innovative approaches to improve technology are being developed in these areas financed under commercial energy organizations and government programs.

In concentrated solar energy systems established with thermal storage systems, the concentrated solar line is enlarged to increase the collected energy. For example, on a sunny day, a high capacity–concentrated solar system may be sufficient to bring the thermal energy storage unit to full capacity. For a specific plant and public application, the concentrated solar system size can be optimized for grid demand model, power demand, and financial parameters.

The main difference between a concentrated solar power system and a conventional thermal power plant is the energy source that depends on natural processes. Although a thermal power plant is ready to be used in the desired period of time, a solar power plant does not have this feature. These factors lead to different methods to measure the steady-state performance of sources [37–39].

1.1.3   Power generated from concentrated solar and PV panels

The thermal power output of concentrated solar energy systems is calculated by configuring a parallel heat-exchanger array [40]. The thermal power that can be obtained from each heat-exchanger array in the concentrated solar energy system can be calculated with the next equation:

$$P_{mes} = mC_p\left(T_{in} - T_{out}\right) \tag{3.1}$$

where P_{mes} is solar thermal power, m is high heat transfer fluid mass flow rate, C_p is integral average-specific heat of high heat transfer fluid, T_{in} is high heat transfer fluid average bulk inlet temperature to solar heat-exchanger train, and T_{out} is high heat transfer fluid average bulk outlet temperature from solar heat-exchanger train.

In a large concentrated solar power plant design, all subunits of the system should be considered. In some cases the high heat transfer fluid is collected to go to the solar steam generator by being heated in a separate subunit. In this case, heat losses occurring in the entrance and exit sections of the pipeline should be taken into account. The efficiency of a concentrated solar energy system is calculated by the next equation:

$$\eta_{mes} = \frac{P_{mes}}{\cos\theta A_{ap}} \tag{3.2}$$

where η_{mes} is thermal efficiency, $\cos\theta$ is average solar irradiance angle during a test run, and A_{ap} is solar field aperture area in tracking mode during the test.

Solar PV panels are connected in series/parallel to create a PV array. The energy generated by a PV panel is given by the next equation:

$$E_{PV} = SP_f I \eta_m \eta_{PC} \tag{3.3}$$

where S is surface area of the PV panel (m^2), P_f is packing factor, I is solar irradiance (kWh/m^2), η_m is PV panel efficiency, and η_{PC} is power-conditioning efficiency.

There are usually two or more power conversion units in PV panel–based grid-connected systems. Depending on the current PV panel voltage, it may be necessary to increase or decrease this voltage level. In the first step, there is usually a direct current (DC)/DC power converter unit for making the maximum power point tracking. The alternative current (AC)/DC converter transfers energy to the grid synchronously. In such systems the other DC loads can be supplied with a DC bus [41]. Fig. 3.4 shows the topology of use of a PV panel with maximum power point tracking.

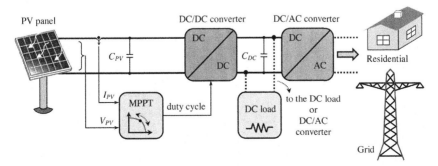

FIGURE 3.4 PV panel control and application topology.

1.2 Wind energy characteristic

Wind energy occurs as a result of the heating of the atmosphere by a small amount of 1 % of the solar energy reaching the earth. The atmosphere does not gain heat evenly due to differences in elevation, geographical shape, position, and sun angle. This causes pressure differences and air currents between regions. As it is known, the air of heat rises and is replaced by air masses at lower temperatures. In this case, wind can be called as air masses that are displaced between high- and low-pressure centers due to temperature difference. The greater the difference between these pressure centers is, the larger the airflow velocity will be. Fig. 3.5 shows the wind turbine control and application topology.

Wind is a renewable energy source that is domestic, continuous, and clean, does not pollute the environment, and has no fuel–raw material cost due to its direct usability. As a result, the most important contribution of wind turbines to the environment is that it does not generate harmful gases caused by combustion of fossil fuels and does not cause greenhouse effect and acid rain. It becomes reliable and cheaper every day. A wind turbine can be used on average for 20–30 years [42]. This means that after the investment, production can be done without interruption for a long time at a low cost. Wind turbine power plants can be built in as little as 4–5 months. This period can be of 7 years in nuclear power plants, 2–10 years in hydroelectric power plants, and 1 or 2 years in natural gas conversion plants. Considering that wind power plants can easily be transported to different regions, it is seen that these provide quite flexibility in energy production.

Wind turbines occupy less than 1 % of the agricultural land in which they are installed. Therefore wind farms do not prevent the agriculture and livestock activities in the regions where they are established. Wind farms have become economically competitive with thermal and hydraulic power plants. Wind farms enable rural areas with no economic value to be used and create opportunities for regional development by creating employment opportunities. Dismantling costs of wind turbines can be covered by the scrap value of the dismantled parts. In addition, new turbines can be built on farms and production can be continued, if desired, the used areas can easily be restored. Since wind energy can

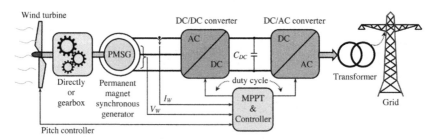

FIGURE 3.5 Wind turbine control and application topology.

be commissioned inexpensively in a short time, it provides supply flexibility. Another important advantage is that there is no need for high network investment to transport the generated power such as nuclear and thermal power plants [43–45]. Businesses in each region need to build wind farms quickly and avoid extra transport costs. As a renewable energy source without any fuel costs, wind energy is an important alternative to fossil fuels with negative environmental impacts. Under the Kyoto Protocol the use of wind power plants, which cause global warming and climate change, and which do not cause carbon emissions, is increasing in countries that are working on reducing greenhouse gases. Since wind energy has enough potential in most of the world, it will find more usage area in the coming years by using its cost advantage.

The variable wind speed is one of the biggest obstacles of wind power plants. Therefore if the wind is not sufficient at the time when energy is needed, electricity cannot be produced. It has been found that wind turbines can be harmful for some bird species. It is necessary to establish the facilities, considering the migration routes of birds. The noise caused by wind turbines adversely affects the inhabitants of that region. Although wind energy has disadvantages such as damages to the environment, its installation is increasing day by day, thanks to the developing turbine technology. Wind energy makes it more economical and attractive day by day [46–49].

1.2.1   Power generated from a wind turbine

The entire wind coming to the turbine blades does not turn into mechanical power in the rotor in wind turbines. For the expression of the mechanical power obtained from the kinetic energy of the wind, the rotor efficiency must be calculated [50]. The amount of air flowing toward the wind turbine is calculated by Eq. (3.4). The actual amount of power captured by the rotor blades is the difference in kinetic energies between the air flows at the wind channel inlet and the wind channel outlet. The amount of power that can be obtained from a wind turbine is given in Eq. (3.5).

$$m_{air} = A_1 v_{up} = A v_{blade} = A_2 v_{down} \tag{3.4}$$

$$P_{wind} = \frac{1}{2} m_{air} \left(v_{up}^2 - v_{down}^2 \right) \tag{3.5}$$

where A_1 is upwind section area, v_{up} is blow-in wind speed, A is blade section area, v_{blade} is wind speed through the turbine, A_2 is downwind section area, v_{down} is blow-out wind speed, P_{wind} is power obtained from the wind turbine, and m_{air} is mass of air.

The wind velocity change along the wind channel to which a wind turbine is exposed is given in Fig. 3.6. While the wind is moving along the channel (wind speed is v_{up} at the inlet and v_{down} at the outlet), it is not in a constant form. Thus the mass flow rate of air moving along the rotating blades (the amount of mass flowing per unit time) can be obtained by multiplying the average velocity by

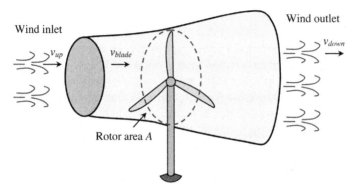

FIGURE 3.6 Wind speeds at which a wind turbine is exposed.

the air density. Also, the mass flow rate of air can be obtained by taking into account v_{down} within the rotor sweep area at the rotor plane.

The ratio of the wind mass and the generated power relationship is given in Eqs. (3.6) and (3.7). The power value to be generated from a wind turbine using Betz's law is expressed in Eq. (3.8).

$$m_{air} = \rho A \left(\frac{v_{up} + v_{down}}{2} \right) \tag{3.6}$$

$$P_{wind} = \frac{1}{2} \left(\rho A \frac{V_{up} + v_{down}}{2} \right) \left(v_{up}^2 - v_{down}^2 \right) \tag{3.7}$$

$$P_{wind} = \frac{1}{2} \rho A v_{up}^3 C_p \tag{3.8}$$

where ρ is air density, A is section area, and C_p is Betz's law.

2 Complementary feature of energy storage

Energy storage applications can provide a wide range of benefits to electricity grids and consumers. These advantages include benefits of electricity grid supply, operations and infrastructure, end consumer, and renewable energy source. In addition, energy storage benefits can be grouped as energy oriented, capacity oriented, and bulk power or distributed [51,52]. There are two general benefits of using energy storage as a resource. These are electric energy is time shifting, and electric resource capacity. When the value or price of electrical energy is cheap, storage is performed. The stored energy is used during the time period when the electricity value or price is high.

Gross benefit refers to total earnings regardless of costs. The net benefit is the difference between the gain and cost. There are three forms of cost prevention. To explain them with examples, the first one is the direct income from energy sales. If storage is located on a larger scale at the level of generation, transmission, and distribution, it may prevent additional investments that may be needed in terms of the generation, transmission, or distribution capacity of the grid. If energy storage is used as an end consumer product, the purpose may be is to reduce the electricity bill and/or use it as an uninterruptible power supply.

Sometimes it is appropriate to combine the benefits of multiple storage applications. In such cases the cost and payback period of the system may be more advantageous by combining the benefits of the systems used together. In some cases, only one system may be economically feasible. However, it is often necessary to use several systems together in order to exceed the total costs of the total benefit. In this way, in applications where different systems are used together, technical and operational characteristics should be ensured in terms of benefits [53]. In other words, systems can be used together in order to provide the intended technical benefits and to respond to the operational characteristics required to achieve these.

It is possible to face some difficulties when talking about energy storage benefits. The reasons for this may be the lack of sufficient experience and the difficulty of providing quantitative data for the assessment of the benefits mentioned. For example, the use of energy storages can contribute to reduction in emissions from electricity generation, but there are no precise cost assessments that reflect pollution values in electrical generation. The use of energy storage can also reduce fuel consumption for the end consumer (when used in place of the generator), but from the end consumers' point of view, the benefit provided here can be seen as insignificant. Nevertheless, when distributed, individual energy storage applications are considered as a whole, the effects of which cannot be clearly expressed in this way; it can be seen that significant benefits are provided on a large scale by the grid operator [54].

The transmission of power and energy in electrical grids is not just a simple task of the system. There are many auxiliary services required for reliable, stable, and efficient operation of these processes. Auxiliary services aimed at achieving the balance of energy supply and its demand at the minute and hour levels are also the basis for ensuring the frequency stability of the enterprise [55]. It can also meet the grid voltage stability and voltage quality requirements with energy storage.

In the provision of auxiliary services, energy storage applications can offer a more technically and economically efficient application compared to traditional fossil fuel power sources. There are two main reasons for this; first, the response times of the energy storage applications can be very short, and the input and output values can be controlled very quickly, thanks to the power electronics circuits. Due to these properties, energy storage systems show a characteristic suitable for use as a power source in auxiliary services. Finally, energy storage systems are more durable than conventional power systems in terms of life according to the amount of change in input and output values. While energy storage applications can more easily respond to different power demands, traditional generation units are efficient and economical only when operated with constant power at their nominal value [56,57]. If energy storage applications work outside of nominal values, an inefficient operation occurs, energy costs increase, operating life may be damaged, and negative environmental impacts may increase. In terms of auxiliary service, energy storage applications can prevent the need for additional generation, thus eliminating the burden of conventional power plants. If additional generation needs are met from energy storage units, energy costs can be kept constant and adverse environmental impacts and inefficient operating conditions are avoided.

3 Use of solar hybrid systems

The purpose of a solar hybrid renewable energy system is to ensure the continuity of energy and to provide higher energy production. A hybrid structure can be created by combining solar-based renewable energy sources (e.g., PV panel and concentrated solar plant). But the problem here is that the source of both power generating units is solar. In such a hybrid renewable power generation system, energy generation will not occur when there is no solar. In order to eliminate this shortage, it is necessary to use other renewable energy sources. Therefore, the expression of "solar hybrid systems" does not only include solar-based renewable energy generation units but also involves using it together with power generation and energy storage units such as wind, tidal, geothermal, biomass, battery, ultracapacitor, and flywheel [58–60]. The utilization topology of solar hybrid renewable energy sources is given in Fig. 3.7. Alternative energy sources can close this gap when there is no solar energy. Feeding a building or facility with 100 % renewable energy source can only be possible with hybrid renewable sources and energy storage units. The more the types of renewable energy sources are, the higher the energy harvest probability ratio is. The power generation units increase when the variety of renewable energy sources is high. This diversity of resources aims to control the entire system in an optimum way. Each power generation unit, energy storage unit, and demand unit must be controlled. Energy management algorithms are needed in such hybrid renewable power generation systems so that the end user can reach high quality and continuous energy.

Fig. 3.7 shows the hybrid renewable energy sources such as solar PV panel, concentrated solar energy, wind, and other renewable energy sources (off-shore wind, marine current, tidal, hydroelectric, geothermal, and biomass) [58–63]. There are batteries, thermal energy storage, hydrogen, and other energy storage technologies (pumped hydroelectric, flood batteries, compressed air, ultracapacitor, and flywheel) as energy storage units. As the end user in the solar hybrid renewable energy system, all power generation units are combined in the common AC line. There are different energy conversion units for each unit. The DC electrical energy taken from the solar PV panels is connected to the DC/DC

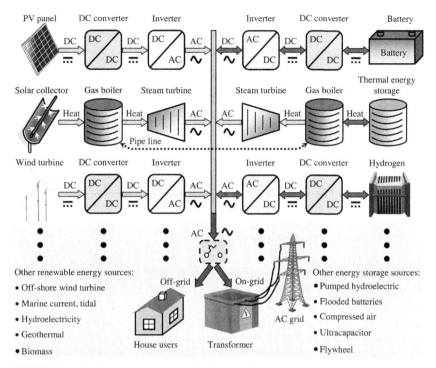

FIGURE 3.7   Topology of using solar hybrid renewable energy sources.

converter and then to the AC line after the DC/AC inverter unit. These subunits are designed to take maximum energy from the renewable energy source. For example, maximum power point tracking method is used in order to get maximum power from solar PV panels, while concentrated solar power system aims to direct solar irradiance to the collector tube located in the center of the reflective mirrors. The subunits seriously affect the total amount of energy to be taken from the whole system in hybrid power generation systems. Therefore each energy converter unit must work with maximum efficiency. In addition, energy management methods are required to control the total energy taken from the hybrid power generation system in an optimum way. In addition, energy storage systems are a complementary factor in the case of excess or lack of power produced by renewable energy sources. Energy storage units have their usage purposes depending on their different output characteristics. For example, batteries usually store the energy when there is an excess of energy in renewable energy sources. Batteries are the most suitable solution to long-term energy demand. However, ultracapacitors can meet the sudden energy demand needs. This energy storage unit preference should be chosen according to the usage purpose of the system.

Fig. 3.8 shows a PV panel output graph with an installed power of 5 kW. These power values were obtained from a hybrid renewable energy system with

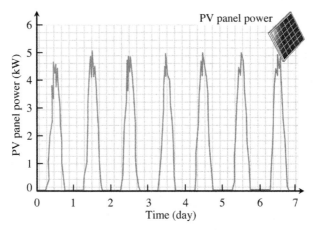

FIGURE 3.8 One-week PV panel output power.

an installed power of 5 kW at Kütahya Dumlupınar University in July 2020. When this 1-week PV panel power graph is analyzed, it is seen that daily sunlight time is close to the total installed power. The average daily output PV power is 3.5 kW (for 8–10 hours). In the power graph, it is understood that there is no daily clouding and a stable output power is obtained. The system has an average of 8–10 hours of sunlight duration in the power graph taken from PV panels. While there is energy production during the daytime, it cannot be produced at night. Generating power at the same power value every day shows that the system is steady. However, when looking at night hours, it reduces the continuity of the energy produced by the solar PV system. This situation arises due to the nature of renewable energy sources.

Fig. 3.9 shows a wind turbine output power graph with an installed power of 6 kW. These wind turbine power data were recorded at the same time periods

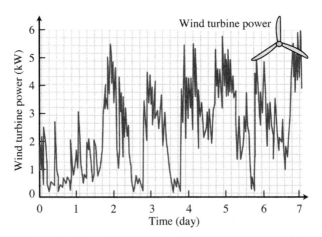

FIGURE 3.9 One-week wind turbine output power.

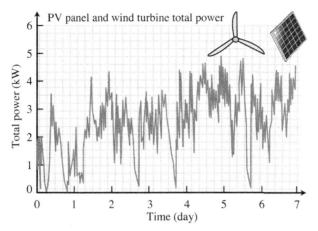

FIGURE 3.10   One-week PV panel and wind turbine output power.

with the PV panel power. Compared to the PV panel power graph, there is no steadiness and continuity in power generation. Wind turbine power data by nature have a cumulative total power value. The wind turbine power graph realizes a production in high frequency ranges according to the power of the PV panel. In the first 2 days of data acquisition, the power value of 3 kW could not be exceeded. But power graph shows that wind turbine power can be generated even at night. These rapid and sudden changes in power data allow the wind turbine power data to be interpreted over the cumulative total power. The cumulative average daily power value obtained from the wind turbine power is 2.1 kW (for 24 hours).

Fig. 3.10 shows the graph of the total power obtained from a weekly PV panel and wind turbine. When this total power graph is analyzed, a renewable power generation system with a hybrid structure is more stable than when used alone. Considering the total power value, it has the characteristics of both the sources. The oscillation of the power value at high frequency is due to the output characteristic of the wind energy. However, considering the daily average power value, there is a total generation of 3.1 kW (for 24 hours). This power graph reveals the necessity and contribution of using renewable energy sources together. Adding another renewable energy source based on load demand power optimization will increase system performance.

A structure different from the conventional grid structure occurs in hybrid solar energy production systems. Renewable energy sources in more than one building or institution and their integration are one of the important issues. These innovative grid structures are called microgrids. Microgrid is the system where energy is generated locally to meet local energy demands. While a microgrid can be applied to a small energy-demanding unit, it can also provide energy to large communities such as towns, villages, and islands [64,65].

In microgrids, only electrical energy is not produced and used. A cogeneration plant can also be established with the heat energy generated on-site in microgrids. Thus it becomes possible to meet the heating demands of the users as a by-product of electricity generation.

Microgrids are traditionally independent of the centralized grid and can be thought of as a separate entity. However, microgrids can be connected to the central grid to allow the import and export of electric energy generated by renewable energy sources. Thus while microgrids sell their energy remaining, it allows them to be seen as a highly reliable power source as they can generate energy locally whenever possible. Microgrids transfer energy to the grid when there is an excess energy and draw energy from the grid when there is a local power demand. In addition, microgrids use storage technologies such as batteries, ultracapacitors, and hot water storage systems to increase the efficiency of the energy produced [66–68]. Therefore microgrids are seen as one of the most important developments in future energy generation.

The energy management unit in microgrids is one of the most important parts of a hybrid renewable energy system. The energy management unit allows the energy produced by the resources to be used in an optimum way. The energy management unit constantly checks the energy storage units and ensures that the storage units are always full.

Many methods have been developed with the energy management unit. One of them is the advanced hybrid energy management system. The ideal item-element extension method, fuzzy-logic judgment method, analytical hierarchy process method, gray clustering method, and geographic information system methods are used in the hybrid energy management system. This method provides a convenient use with an interface in a system with two or more power generation and energy storage units at the same time. The user is informed instantly and the possible failure conditions of the units (production or storage unit) are given with this interface. In addition, this interface allows the analysis of how much power renewable energy sources generate and historical data. It enables one to make a prediction about the hybrid power generation system with these data. Thus it ensures that the energy management unit becomes even smarter and ensures optimum operation.

The renewable solar/wind energy industry is developing rapidly in developing countries. However, the difficulty in selling the generated electricity to the existing electricity grid due to instability and interruptions in electricity supply is a serious problem. In order to overcome these problems, it is especially necessary to solve the issue of generating renewable energy with unstable output power characteristic [69]. This solution is utilized to install hybrid renewable energy systems.

Improvising and applying different methods to generate and use the energy obtained in the hybrid renewable power generation system will increase the system cost. In addition, it will affect the depreciation period of the hybrid system and cause unnecessary investment. To solve this problem, it would be wise to add an energy storage unit to the system at the required power level. This will increase the reliability of the hybrid power generation system. Thus the total cost of the hybrid power generation system for energy production at the same power values can be reduced. Adding another source to the hybrid power generation system can provide a solution. The parameter to be considered here is the total demand power of the users. Choosing the right sources with power size optimization in the hybrid power generation system will reduce the capital cost.

Producing hydrogen in the hybrid power generation system created by using PV panel and wind energy together is another perspective on storage methods. The hydrogen can be stored with the excess energy in the hybrid power generation system. Later, when there is an energy demand in the system, the stored hydrogen energy is converted into electrical energy and the demand power is met.

Hydrogen is not a natural fuel but a synthetic fuel that can be produced from different raw materials such as water, fossil fuels, and biomass. During the hydrogen production phase, there are many alternative hydrogen generation technologies such as steam recovery, waste gas purification, electrolysis, photo processes, thermo chemical processes, and radiolysis. The produced hydrogen can be transported over great distances by pipelines or tankers. The absence of barriers to the transport of hydrogen will enable greater use. Hydrogen has the highest energy content of all known fuels with a value of 120.7 kJ/kg per unit. Hydrogen can be liquefied at $-252.77\ °C$, and the volume of liquid hydrogen is only 1/700 of its gaseous volume [70–75].

A hybrid power generation system, included with concentrated solar energy and wind energy, was established using hydrogen production method in an energy storage unit. This hybrid structure contributed energy to the electricity grid with the use of proton exchange membrane electrolysis in system. The energy management algorithm has enabled the flow of energy by operating the multilevel inverter. Hydrogen is produced by the hybrid power generation unit and stored in a pressurized tank. In the case of need in the hybrid system, it provides energy through the hydrogen fuel cell and the DC/DC power converter circuit. The use of hydrogen as an energy storage unit has gained a commercial dimension with the increase of such applications.

This hybrid renewable energy technology enables environment-friendly energy production for 24 hours with its clean energy generation feature. And the same systems are the most suitable solution for small electricity grids and off-grid systems. By definition, "hybrid power generation" is a system that overcomes problems when a single energy source is not sufficient. The hybrid renewable energy system is the sustainable and ecological energy generation method of the future. The significant contributions will be made to the protection of nature with these power generation methods. In addition, hybrid power generation systems will reduce transmission costs in the current grid structure.

References

[1] A.S.C. Maia, E. de, A. Culhari, V. de, F.C. Fonsêca, H.F.M. Milan, et al. Photovoltaic panels as shading resources for livestock, J. Clean. Prod. 258 (2020) 120551, doi: 10.1016/j.jclepro.2020.120551.

[2] M. Garaj, K.Y. Hong, H.S.H. Chung, A.W. lun Lo, H. Wang, Diagnostic module for series-connected photovoltaic panels, Sol. Energy 196 (2020) 243–259, doi: 10.1016/j.solener.2019.12.019.

[3] F.A. Tiano, G. Rizzo, M. Marino, A. Monetti, Evaluation of the potential of solar photovoltaic panels installed on vehicle body including temperature effect on efficiency, ETransportation 5 (2020) 100067, doi: 10.1016/j.etran.2020.100067.

[4] Q. Xuan, G. Li, Y. Lu, B. Zhao, X. Zhao, G. Pei, Daylighting characteristics and experimental validation of a novel concentrating photovoltaic/daylighting system, Sol. Energy 186 (2019) 264–276, doi: 10.1016/j.solener.2019.05.014.

[5] P.A. Adedeji, S.A. Akinlabi, N. Madushele, O.O. Olatunji, Neuro-fuzzy resource forecast in site suitability assessment for wind and solar energy: a mini review, J. Clean. Prod. 269 (2020) 122104, doi: 10.1016/j.jclepro.2020.122104.

[6] H. Liu, Y. Li, Z. Duan, C. Chen, A review on multi-objective optimization framework in wind energy forecasting techniques and applications, Energy Convers. Manag. 224 (2020) 113324, doi: 10.1016/j.enconman.2020.113324.

[7] M. Lopez-Medina, F. Hernandez-Navarro, A.I. Mtz-Enriquez, A.I. Oliva, V. Rodriguez-Gonzalez, J.P. Camarillo-Garcia, et al. Enhancing the capacity and discharge times of flexible graphene batteries by decorating their anodes with magnetic alloys NiMnMx (Mx = Ga, In, Sn), Mater. Chem. Phys. 256 (2020) 123660, doi: 10.1016/j.matchemphys.2020.123660.

[8] U.G.K. Mulleriyawage, W.X. Shen, Optimally sizing of battery energy storage capacity by operational optimization of residential PV-battery systems: an Australian household case study, Renew. Energy 160 (2020) 852–864, doi: 10.1016/j.renene.2020.07.022.

[9] L. Wei, X.Z. Fan, H.R. Jiang, K. Liu, M.C. Wu, T.S. Zhao, Enhanced cycle life of vanadium redox flow battery via a capacity and energy efficiency recovery method, J. Power Sources 478 (2020) 228725, doi: 10.1016/j.jpowsour.2020.228725.

[10] L. Liu, Z. Wang, Y. Wang, J. Wang, R. Chang, G. He, et al. Optimizing wind/solar combinations at finer scales to mitigate renewable energy variability in China, Renew. Sustain. Energy Rev. 132 (2020) 110151, doi: 10.1016/j.rser.2020.110151.

[11] A. Das, H.K. Jani, G. Nagababu, S.S. Kachhwaha, A comprehensive review of wind–solar hybrid energy policies in India: barriers and recommendations, Renew. Energy Focus 35 (2020) 108–121, doi: 10.1016/j.ref.2020.09.004.

[12] A. Herez, H. El Hage, T. Lemenand, M. Ramadan, M. Khaled, Review on photovoltaic/thermal hybrid solar collectors: classifications, applications and new systems, Sol. Energy 207 (2020) 1321–1347, doi: 10.1016/j.solener.2020.07.062.

[13] A.A. Raja, Y. Huang, Novel parabolic trough solar collector and solar photovoltaic/thermal hybrid system for multi-generational systems, Energy Convers. Manag. 211 (2020) 112750, doi: 10.1016/j.enconman.2020.112750.

[14] R. Manrique, D. Vásquez, F. Chejne, A. Pinzón, Energy analysis of a proposed hybrid solar–biomass coffee bean drying system, Energy 202 (2020) 117720, doi: 10.1016/j.energy.2020.117720.

[15] H. Yang, J. Li, Y. Huang, T.H. Kwan, J. Cao, G. Pei, Feasibility research on a hybrid solar tower system using steam and molten salt as heat transfer fluid, Energy 205 (2020) 118094, doi: 10.1016/j.energy.2020.118094.

[16] N. Lee, U. Grunwald, E. Rosenlieb, H. Mirletz, A. Aznar, R. Spencer, et al. Hybrid floating solar photovoltaics-hydropower systems: benefits and global assessment of technical potential, Renew. Energy 162 (2020) 1415–1427, doi: 10.1016/j.renene.2020.08.080.

[17] V. Saini, L. Sahota, V.K. Jain, G.N. Tiwari, Performance and cost analysis of a modified built-in-passive condenser and semitransparent photovoltaic module integrated passive solar distillation system, J. Energy Storage 24 (2019) 100809, doi: 10.1016/j.est.2019.100809.

[18] F. Signorato, M. Morciano, L. Bergamasco, M. Fasano, P. Asinari, Exergy analysis of solar desalination systems based on passive multi-effect membrane distillation, Energy Rep. 6 (2020) 445–454, doi: 10.1016/j.egyr.2020.02.005.

[19] L. Gourdo, H. Fatnassi, R. Bouharroud, K. Ezzaeri, A. Bazgaou, A. Wifaya, et al. Heating canarian greenhouse with a passive solar water–sleeve system: effect on microclimate and tomato crop yield, Sol. Energy 188 (2019) 1349–1359, doi: 10.1016/j.solener.2019.07.004.

[20] M. Vaseghi, M. Fazel, A. Ekhlassi, Numerical investigation of solar radiation effect on passive and active heating and cooling system of a concept museum building, Therm. Sci. Eng. Prog. 19 (2020) 100582, doi: 10.1016/j.tsep.2020.100582.

[21] M. Rabani, Performance analysis of a passive cooling system equipped with a new designed solar chimney and a water spraying system in an underground channel, Sustain. Energy Technol. Assess. 35 (2019) 204–219, doi: 10.1016/j.seta.2019.07.005.

[22] A.K. Singh, R.K. Yadav, D. Mishra, R. Prasad, L.K. Gupta, P. Kumar, Active solar distillation technology: a wide overview, Desalination 493 (2020) 114652, doi: 10.1016/j.desal.2020.114652.

[23] A. Savvides, C. Vassiliades, A. Michael, S. Kalogirou, Siting and building-massing considerations for the urban integration of active solar energy systems, Renew. Energy 135 (2019) 963–974, doi: 10.1016/j.renene.2018.12.017.

[24] X. Kong, L. Wang, H. Li, G. Yuan, C. Yao, Experimental study on a novel hybrid system of active composite PCM wall and solar thermal system for clean heating supply in winter, Sol. Energy 195 (2020) 259–270, doi: 10.1016/j.solener.2019.11.081.

[25] G.N. Tiwari, A.K. Mishra, M. Meraj, A. Ahmad, M.E. Khan, Effect of shape of condensing cover on energy and exergy analysis of a PVT-CPC active solar distillation system, Sol. Energy 205 (2020) 113–125, doi: 10.1016/j.solener.2020.04.084.

[26] H.A. Nasef, S.A. Nada, H. Hassan, Integrative passive and active cooling system using PCM and nanofluid for thermal regulation of concentrated photovoltaic solar cells, Energy Convers. Manag. 199 (2019) 112065, doi: 10.1016/j.enconman.2019.112065.

[27] J. Chen, K. Ge, B. Chen, J. Guo, L. Yang, Y. Wu, et al. Establishment of a novel functional group passivation system for the surface engineering of c-Si solar cells, Sol. Energy Mater. Sol. Cells 195 (2019) 99–105, doi: 10.1016/j.solmat.2019.02.039.

[28] S. Hong, B. Zhang, C. Dang, E. Hihara, Development of two-phase flow microchannel heat sink applied to solar-tracking high-concentration photovoltaic thermal hybrid system, Energy 212 (2020) 118739, doi: 10.1016/j.energy.2020.118739.

[29] X. Chen, Q. Liu, J. Xu, Y. Chen, W. Li, Z. Yuan, et al. Thermodynamic study of a hybrid PEMFC-solar energy multi-generation system combined with SOEC and dual Rankine cycle, Energy Convers. Manag. 226 (2020) 113512, doi: 10.1016/j.enconman.2020.113512.

[30] M.M. Aboelmaaref, M.E. Zayed, J. Zhao, W. Li, A.A. Askalany, M. Salem Ahmed, et al. Hybrid solar desalination systems driven by parabolic trough and parabolic dish CSP technologies: technology categorization, thermodynamic performance and economical assessment, Energy Convers. Manag. 220 (2020) 113103, doi: 10.1016/j.enconman.2020.113103.

[31] Y. Xu, Z. Li, H. Chen, S. Lv, Assessment and optimization of solar absorption-subcooled compression hybrid cooling system for cold storage, Appl. Therm. Eng. 180 (2020) 115886, doi: 10.1016/j.applthermaleng.2020.115886.

[32] Z. Song, T. Liu, Q. Lin, Multi-objective optimization of a solar hybrid CCHP system based on different operation modes, Energy 206 (2020) 118125, doi: 10.1016/j.energy.2020.118125.

[33] J. Gómez-Hernández, P.A. González-Gómez, J.V. Briongos, D. Santana, Technical feasibility analysis of a linear particle solar receiver, Sol. Energy 195 (2020) 102–113, doi: 10.1016/j.solener.2019.11.052.

[34] M.J. Wagner, W.T. Hamilton, A. Newman, J. Dent, C. Diep, R. Braun, Optimizing dispatch for a concentrated solar power tower, Sol. Energy 174 (2018) 1198–1211, doi: 10.1016/j.solener.2018.06.093.

[35] D. Fähsing, C. Oskay, T.M. Meißner, M.C. Galetz, Corrosion testing of diffusion-coated steel in molten salt for concentrated solar power tower systems, Surf. Coat. Tech. 354 (2018) 46–55, doi: 10.1016/j.surfcoat.2018.08.097.

[36] S. Shafiei Kaleibari, Z. Yanping, S. Abanades, Solar-driven high temperature hydrogen production via integrated spectrally split concentrated photovoltaics (SSCPV) and solar power tower, Int. J. Hydrog. Energy 44 (2019) 2519–2532, doi: 10.1016/j.ijhydene.2018.12.039.

[37] J.E. Rea, C.J. Oshman, M.L. Olsen, C.L. Hardin, G.C. Glatzmaier, N.P. Siegel, et al. Performance modeling and techno-economic analysis of a modular concentrated solar power tower with latent heat storage, Appl. Energy 217 (2018) 143–152, doi: 10.1016/j.apenergy.2018.02.067.

[38] T. Galiullin, B. Gobereit, D. Naumenko, R. Buck, L. Amsbeck, M. Neises-von Puttkamer, et al. High temperature oxidation and erosion of candidate materials for particle receivers of concentrated solar power tower systems, Sol. Energy 188 (2019) 883–889, doi: 10.1016/j.solener.2019.06.057.

[39] M. Kaya, A.E. Gürel, Ü. Ag˘bulut, . Ceylan, S. Çelik, A. Ergün, et al. Performance analysis of using CuO-methanol nanofluid in a hybrid system with concentrated air collector and vacuum tube heat pipe, Energy Convers. Manag. 199 (2019) 111936, doi: 10.1016/j.enconman.2019.111936.

[40] B. Dou, M. Guala, P. Zeng, L. Lei, Experimental investigation of the power performance of a minimal wind turbine array in an atmospheric boundary layer wind tunnel, Energy Convers. Manag. 196 (2019) 906–919, doi: 10.1016/j.enconman.2019.06.056.

[41] Z. Li, B. Wen, X. Dong, Z. Peng, Y. Qu, W. Zhang, Aerodynamic and aeroelastic characteristics of flexible wind turbine blades under periodic unsteady inflows, J. Wind Eng. Ind. Aerodyn. 197 (2020) 104057, doi: 10.1016/j.jweia.2019.104057.

[42] Y. Zhao, J. Pan, Z. Huang, Y. Miao, J. Jiang, Z. Wang, Analysis of vibration monitoring data of an onshore wind turbine under different operational conditions, Eng. Struct. 205 (2020) 110071, doi: 10.1016/j.engstruct.2019.110071.

[43] B. Sri Revathi, P. Mahalingam, F. Gonzalez-Longatt, Interleaved high gain DC-DC converter for integrating solar PV source to DC bus, Sol. Energy 188 (2019) 924–934, doi: 10.1016/j.solener.2019.06.072.

[44] M.E. Zayed, J. Zhao, A.H. Elsheikh, W. Li, S. Sadek, M.M. Aboelmaaref, A comprehensive review on Dish/Stirling concentrated solar power systems: design, optical and geometrical analyses, thermal performance assessment, and applications, J. Clean. Prod. 283 (2020) 124664, doi: 10.1016/j.jclepro.2020.124664.

[45] C. Kan, Y. Devrim, S. Eryilmaz, On the theoretical distribution of the wind farm power when there is a correlation between wind speed and wind turbine availability, Reliab. Eng. Syst. Saf. 203 (2020) 107115, doi: 10.1016/j.ress.2020.107115.

[46] H. Sun, C. Qiu, L. Lu, X. Gao, J. Chen, H. Yang, Wind turbine power modelling and optimization using artificial neural network with wind field experimental data, Appl. Energy 280 (2020) 115880, doi: 10.1016/j.apenergy.2020.115880.

[47] Q. Yao, Y. Hu, H. Deng, Z. Luo, J. Liu, Two-degree-of-freedom active power control of megawatt wind turbine considering fatigue load optimization, Renew. Energy 162 (2020) 2096–2112, doi: 10.1016/j.renene.2020.09.137.

[48] E. Leelakrishnan, M. Sunil Kumar, S. David Selvaraj, N.S. Vignesh, T.S.A. Raja, Numerical evaluation of optimum tip speed ratio for Darrieus type vertical axis wind turbine, Mater. Today Proc. 33 (2020) 4719–4722, doi: 10.1016/j.matpr.2020.08.352.

[49] S. Sang, H. Wen, A.X. Cao, X.R. Du, X. Zhu, Q. Shi, et al. Dynamic modification method for BEM of wind turbine considering the joint action of installation angle and structural pendulum motion, Ocean Eng. 215 (2020) 107528, doi: 10.1016/j.oceaneng.2020.107528.

[50] S. Najafi-Shad, S.M. Barakati, A. Yazdani, An effective hybrid wind-photovoltaic system including battery energy storage with reducing control loops and omitting PV converter, J. Energy Storage 27 (2020) 101088, doi: 10.1016/j.est.2019.101088.

[51] M.K. Kiptoo, M.E. Lotfy, O.B. Adewuyi, A. Conteh, A.M. Howlader, T. Senjyu, Integrated approach for optimal techno-economic planning for high renewable energy-based isolated microgrid considering cost of energy storage and demand response strategies, Energy Convers. Manag. 215 (2020) 112917, doi: 10.1016/j.enconman.2020.112917.

[52] A. Bartolini, F. Carducci, C.B. Muñoz, G. Comodi, Energy storage and multi energy systems in local energy communities with high renewable energy penetration, Renew. Energy 159 (2020) 595–609, doi: 10.1016/j.renene.2020.05.131.

[53] A. Trivedi, H. Chong Aih, D. Srinivasan, A stochastic cost–benefit analysis framework for allocating energy storage system in distribution network for load leveling, Appl. Energy 280 (2020) 115944, doi: 10.1016/j.apenergy.2020.115944.

[54] A.O. Gbadegesin, Y. Sun, N.I. Nwulu, Techno-economic analysis of storage degradation effect on levelised cost of hybrid energy storage systems, Sustain. Energy Technol. Assess. 36 (2019) 100536, doi: 10.1016/j.seta.2019.100536.

[55] S. Li, L. Caracoglia, Experimental error examination and its effects on the aerodynamic properties of wind turbine blades, J. Wind Eng. Ind. Aerodyn. 206 (2020) 104357, doi: 10.1016/j.jweia.2020.104357.

[56] P.S. Pravin, S. Misra, S. Bhartiya, R.D. Gudi, A reactive scheduling and control framework for integration of renewable energy sources with a reformer-based fuel cell system and an energy storage device, J. Process Control 87 (2020) 147–165, doi: 10.1016/j.jprocont.2020.01.005.

[57] A. Mayyas, M. Wei, G. Levis, Hydrogen as a long-term, large-scale energy storage solution when coupled with renewable energy sources or grids with dynamic electricity pricing schemes, Int. J. Hydrog. Energy 45 (2020) 16311–16325, doi: 10.1016/j.ijhydene.2020.04.163.

[58] T.B. Peffley, J.M. Pearce, The potential for grid defection of small and medium sized enterprises using solar photovoltaic, battery and generator hybrid systems, Renew. Energy 148 (2020) 193–204, doi: 10.1016/j.renene.2019.12.039.

[59] K. Li, C. Liu, S. Jiang, Y. Chen, Review on hybrid geothermal and solar power systems, J. Clean. Prod. 250 (2020) 119481, doi: 10.1016/j.jclepro.2019.119481.

[60] R. Li, S. Guo, Y. Yang, D. Liu, Optimal sizing of wind/ concentrated solar plant/ electric heater hybrid renewable energy system based on two-stage stochastic programming, Energy 209 (2020) 118472, doi: 10.1016/j.energy.2020.118472.

[61] T. Salameh, M.A. Abdelkareem, A.G. Olabi, E.T. Sayed, M. Al-Chaderchi, H. Rezk, Integrated standalone hybrid solar PV, fuel cell and diesel generator power system for battery or supercapacitor storage systems in Khorfakkan, United Arab Emirates, Int. J. Hydrog. Energy 46 (2020) 6014–6027, doi: 10.1016/j.ijhydene.2020.08.153.

[62] M. Jafari, D. Armaghan, S.M. Seyed Mahmoudi, A. Chitsaz, Thermoeconomic analysis of a standalone solar hydrogen system with hybrid energy storage, Int. J. Hydrog. Energy 44 (2019) 19614–19627, doi: 10.1016/j.ijhydene.2019.05.195.

[63] M.S. Saleem, N. Abas, A.R. Kalair, S. Rauf, A. Haider, M.S. Tahir, et al. Design and optimization of hybrid solar-hydrogen generation system using TRNSYS, Int. J. Hydrog. Energy 45 (2020) 15814–15830, doi: 10.1016/j.ijhydene.2019.05.188.

[64] W. Liu, N. Li, Z. Jiang, Z. Chen, S. Wang, J. Han, et al. Smart micro-grid system with wind/ PV/battery, Energy Procedia 152 (2018) 1212–1217, doi: 10.1016/j.egypro.2018.09.171.

[65] Y. Zhou, Z. Li, X. Tao, Urban mixed use and its impact on energy performance of micro gird system, Energy Procedia 103 (2016) 339–344, doi: 10.1016/j.egypro.2016.11.296.

[66] A.H. Fathima, K. Palanisamy, Optimization in microgrids with hybrid energy systems—a review, Renew. Sustain. Energy Rev. 45 (2015) 431–446, doi: 10.1016/j.rser.2015.01.059.

[67] S. Angalaeswari, K. Jamuna, Design and implementation of a robust iterative learning controller for voltage and frequency stabilization of hybrid microgrids, Comput. Electr. Eng. 84 (2020) 106631, doi: 10.1016/j.compeleceng.2020.106631.

[68] S. Li, F. Sun, H. He, Y. Chen, Optimization for a grid-connected hybrid PV-wind-retired HEV battery microgrid system, Energy Procedia 105 (2017) 1634–1643, doi: 10.1016/j.egypro.2017.03.532.

[69] N Endo, K Goshome, M Tetsuhiko, Y Segawa, E Shimoda, T Nozu, Thermal management and power saving operations for improved energy efficiency within a renewable hydrogen energy system utilizing metal hydride hydrogen storage, Int. J. Hydrog. Energy 46 (2020) 262–271, doi: 10.1016/J.IJHYDENE.2020.10.002.

[70] R. Singh, M. Singh, S. Gautam, Hydrogen economy, energy, and liquid organic carriers for its mobility, Mater. Today Proc. (2020)doi: 10.1016/j.matpr.2020.09.065.

[71] H. Mehrjerdi, Modeling and optimization of an island water-energy nexus powered by a hybrid solar-wind renewable system, Energy 197 (2020) 117217, doi: 10.1016/j.energy.2020.117217.

[72] A. Khouya, Levelized costs of energy and hydrogen of wind farms and concentrated photovoltaic thermal systems. A case study in Morocco, Int. J. Hydrog. Energy 45 (2020) 31632–31650, doi: 10.1016/j.ijhydene.2020.08.240.

[73] A. Baldinelli, L. Barelli, G. Bidini, Progress in renewable power exploitation: reversible solid oxide cells-flywheel hybrid storage systems to enhance flexibility in micro-grids management, J. Energy Storage 23 (2019) 202–219, doi: 10.1016/j.est.2019.03.018.

[74] S.A. Mirnezami, A. Zahedi, A. Shayan Nejad, Thermal optimization of a novel solar/hydro/biomass hybrid renewable system for production of low-cost, high-yield, and environmental-friendly biodiesel, Energy 202 (2020) 117562, doi: 10.1016/j.energy.2020.117562.

[75] Z. Cabrane, D. Batool, J. Kim, K. Yoo, Design and simulation studies of battery-supercapacitor hybrid energy storage system for improved performances of traction system of solar vehicle, J. Energy Storage 32 (2020) 101943, doi: 10.1016/j.est.2020.101943.

Chapter 4

Hybrid Renewable Generation Systems

1 Hybrid renewable energy systems

It is observed that the interest in renewable energy sources has been increasing recently. Renewable energy sources are nonpolluting and easy to access. Such advantages make them attractive for many applications. Today, hybrid renewable energy sources are created by combining more than one renewable energy source. The most widely used wind turbine and photovoltaic (PV) systems are used in various applications such as pumping water, lighting, telecommunications, and providing the electricity needs of remote areas. Hybrid renewable energy sources are preferred for systems that need to be located in remote areas such as wireless telecommunications, satellites, and ground stations. As the demand for renewable energy generating systems increases, research and development studies with these sources continue rapidly [1–4].

Solar PV panels or wind energy can be considered as a suitable solution for regions far from the city center. Battery energy storage units are generally used with such systems. These alternative solutions are generally classified as hybrid renewable energy systems. These systems are widely used and preferred in areas where there is no grid structure, because it is much lower than the cost of withdrawing a grid line. The use of more than one source and especially with energy storage units increases the reliability of such systems [5]. Thus the necessary electrical energy of a house can be met by these hybrid renewable energy sources. With the increasing applications of such hybrid systems, system topologies also change. Hybrid systems can be classified as serial hybrid energy systems, switched hybrid energy systems, and parallel hybrid energy systems [6].

1.1 Serial hybrid renewable energy system topology

All energy sources feed a battery group in the structure of serial hybrid renewable energy system topology. Each renewable energy source has a separate battery charge control unit. In this topology structure, all units are collected in a single direct current (DC) bus. For this reason the voltage and current parameters of all sources should be sized according to the common DC bus. Power

Solar Hybrid Systems. http://dx.doi.org/10.1016/B978-0-323-88499-0.00004-5

69

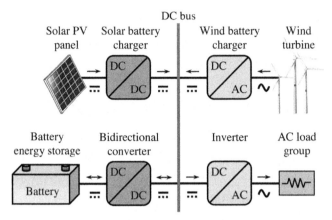

FIGURE 4.1 Serial hybrid renewable energy system topology.

converters in renewable energy source types have the feature of working at the maximum power point. In this topology structure the battery group is always active. For this reason the deep of discharge and cycle life of the battery group should be high. The DC voltage taken from the DC bus is converted into alternative current (AC) voltage with an inverter. In this topology structure the system efficiency is low since the power in all sources is covert to AC voltage with a high power inverter. The serial hybrid topology sizing is determined according to the load power requested by the system [7]. Serial hybrid renewable energy system topology is given in Fig. 4.1. Since there is more than one power generation unit in this topology, the battery group is not exposed to high discharge currents while feeding the load group. Thanks to the common DC bus, different renewable energy sources can be integrated into this topology. In addition to all these, there is a single inverter in the system to feed the load group. The inverter must have the capacity to meet the peak load demand in the system. Since the battery group is exposed to a high cycle number, its life is shortened. In addition, the battery group must have a deep of discharge feature.

1.2 Switched hybrid renewable energy system topology

Although the switched hybrid renewable energy system topology has application difficulties, it is one of the most preferred structures. In this structure, it is not possible to operate the energy sources connected to the AC bus in parallel. Wind turbine and solar PV panels charge the battery group. In this topology structure, the load group connected to the AC bus can be fed, thus reducing the number of converter units in the system. This structure increases system efficiency. In this structure the units on the AC bus and the battery group can also be charged. This structure enables DC loads to be operated in the system. As in the serial topology, it can meet all the powers in the inverter system [8]. System

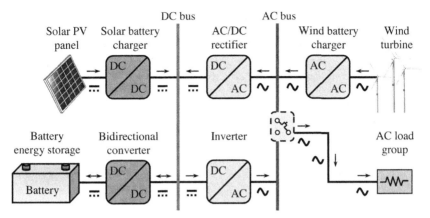

FIGURE 4.2   Switched hybrid renewable energy system topology.

efficiency increases as power conversion units decrease between energy generation units. In this topology, suddenly power cuts may occur during transitions
and this is an undesirable feature. Switched hybrid renewable energy system
topology is depicted in Fig. 4.2.

1.3   Parallel hybrid renewable energy system topology

Parallel hybrid renewable energy system topology includes both DC bus and
AC bus. Renewable energy sources generating DC power are collected in a
single DC bus. Units with AC output characteristics are collected in the AC bus.
Bidirectional inverter is used between these two buses. This bidirectional inverter operates in AC/DC rectifier or DC/AC inverter mode according to battery
and load demand power. If the energy produced by the AC power generation
units is equal to and greater than the energy demanded, the load group is fed by
the AC bus. The battery group is charged with the excess energy bidirectional
inverter in this power generation unit. If AC power generation systems cannot
meet the load demand, the battery group is activated. In this case the bidirectional inverter meets the energy requirement of the load group. The role of the
bidirectional inverter is great in this topology structure. The power capacity of
this bidirectional inverter must be high. If the bidirectional inverter faults, the
energy units on the DC bus side are canceled. Since the bidirectional inverter
operates in both rectifier and inverter mode, it reduces converter units. The reduction in power converter units increases system efficiency while reducing
cost [9].

Parallel hybrid renewable energy system topology has many advantages
compared to other topology structures. In this topology structure, each power
converter can be controlled from a single unit. This hybrid energy management
algorithm ensures optimum use of energy. Although these control structures

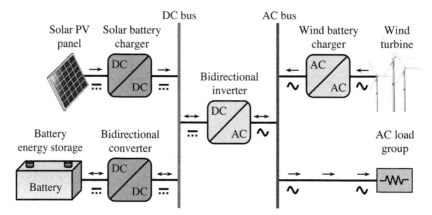

FIGURE 4.3 **Parallel hybrid renewable energy system topology.**

have a complex feature, they are indispensable for system efficiency. In this to-pology structure the energy demand of the load group is provided at an optimum level. A hybrid control algorithm is essential for the system to operate reliably in this topology. The inverter used in the system must have pure sine output fea-ture. Parallel hybrid renewable energy system topology is depicted in Fig. 4.3.

1.4   Control of hybrid renewable energy systems

The efficiency, cost, and control unit selection are important criteria in the siz-ing stage of hybrid energy systems. Before proceeding with such hybrid power applications, simulation programs will help one to make the right decision. The purpose of energy management algorithms is to provide optimum operation for the whole system. These control algorithms also ensure that the energy obtained from renewable energy sources is at the maximum level. The hybrid energy control unit aims to ensure that the cycle life of the elements in the system ex-ists as long as possible. The energy management algorithms control the charge/discharge rates, deep of discharge current and cycle life numbers of the battery group. The more parameters related to the battery group are evaluated, the lon-ger the battery will be used. On the other hand, this situation makes the con-troller more complex [10]. The complex structure of the control unit increases system reliability. The fluctuating output power characteristics of renewable energy sources are detected by the control system and the battery group is ac-tivated. The hybrid energy control algorithm must continuously respond to changing conditions by the system.

   If the demand load power in the hybrid system is supplied by energy sourc-es, the most efficient operation situation is achieved. Thus the charge/discharge cycle of the energy storage unit is reduced. The most efficient control method selection depends on system components, environmental conditions, mainte-nance period, and operating modes [11,12]. Grid-connected hybrid energy sys-tem topology is given in Fig. 4.4.

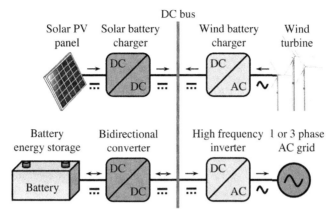

FIGURE 4.4 Grid-connected hybrid energy system topology.

Using more than one renewable energy source and energy storage unit to-gether increases the reliability of energy. Using these systems synchronously with the grid ensures that the energy is used uninterruptedly. Grid-connected hybrid renewable energy systems are usually connected to the grid with low- or medium voltage. The inverter in the hybrid power generation system is used by connecting to the grid synchronously. High power applications are usually connected to the medium voltage grid by a transformer. Since this transformer has high power, its dimensions, weight, and cost are also high. Thanks to ad-vances in power electronics, inverters can be directly connected to the grid. Isolation between the hybrid system and the grid is also provided by a trans-former. However, this transformer has very low dimensions and weight thanks to its high-frequency structure. While transformers working with low frequency (50–60 Hz) in the classical method, they work with 10–20 kHz in new-gener-ation inverters. Electrical isolation is provided in both usage methods. In such applications, renewable energy sources are combined in a single DC bus. By using high-frequency inverter between the hybrid system and the grid, both grid integration and electrical isolation are provided [13–16].

2 Hybrid system consisting of solar PV panel, wind, and battery

When renewable energy source types are used for distributed energy genera-tion, each source has some disadvantages of its own nature. As a popular so-lution to eliminate these disadvantages, hybrid systems are often created. For example, in wind energy, the power is generated during the wind blows; in solar PV panel energy, the power is generated during the daytime and a sunny time. PV systems have a modular structure. It allows for a new solar PV panel to be added to the system if more power generation is demanded from solar system. Although the level of solar irradiation and insolation varies according to the region, season, and time of day, sunlight is one of the energy sources that

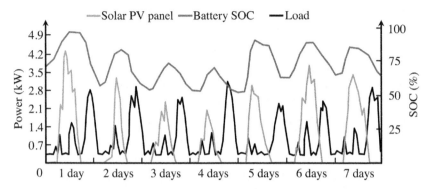

FIGURE 4.5 The power generation graph of solar PV panel system with energy storage.

can be found almost everywhere on the planet. However, the onset of winter or cloudy periods in many places limits the availability of sunlight. When the time interval needed by the load and the energy generated from the solar system do not match, the excess energy generated during intensive use is usually met by an energy storage system [17,18]. The need for a large energy storage capacity can significantly increase the cost of the system. Usually, a backup electric generator or an external grid is required to ensure continuous availability, which means higher costs. A 20-kWh capacity energy storage unit has been connected to the solar PV panel system with a power capacity of 5 kWp. Fig. 4.5 shows the battery group state of charge, load group power, and solar PV panel power on a weekly basis.

Fig. 4.6 shows the data of a 5-kWp solar PV panel power output graph for 2019 in Kütahya Dumlupınar University Electrical and Electronics Engineering Department. A total of 6.2 MWh of electricity was generated throughout the year of 2019 with this solar PV panel system. As can be seen in the power generation graph, power generation peak values vary depending on the seasons.

2.1 Wind data analysis

The wind data were analyzed for the same area where the solar PV panel is installed. Coordinates of the location where the wind data are analyzed; latitude 39.11° and longitude 29.01°. Wind data have been obtained over the National Energy and Climate Plans (NECPs) at hourly intervals since 1979 until today. These data include wind speeds, wind directions, air pressure, and air temperatures. Wind data results have been obtained for the region below 10 m.

The coordinates determined in the Navionics Boating program were used in the NECP program, and the requested data were made ready for processing. After the transfer process the data divided on a monthly basis were processed separately and as all data. First of all, all 40 years of data were processed and the wind rose profile at the selected point was created. As a result of this process,

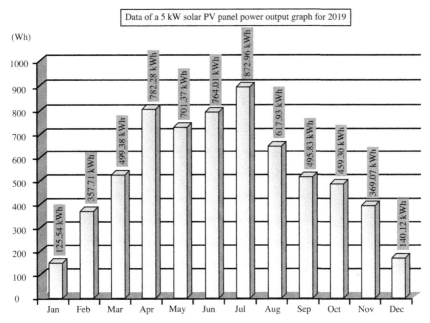

FIGURE 4.6   Data of a 5-kW solar PV panel power output graph for 2019.

the expected profile has been obtained. It has been observed that the northeastern winds are dominant. All 40 years of data are divided into 16 directions in the design of wind rose. The wind speeds by months are given in Fig. 4.7. The all speed data of the wind rose are given in Fig. 4.8.

The process of obtaining wind rose was repeated by filtering 40-year data for all months on a monthly basis. As a result of this process, it was obtained both a wind rose consisting of all data and separate wind roses for 12 months. When the monthly average wind speeds are analyzed, it is clearly seen that the northeastern and southwestern winds are dominant in the winter season, while the northeastern wind is dominant in the other seasons. Energy generation times can be determined by processing the minimum and maximum wind speeds required to generate wind energy from the graphics obtained.

2.1.1   Wind energy conversion efficiency

The energy continuity curve has been calculated in two separate ways for all wind data and speeds between 2 and 15 m/s. Unfiltered energy continuity curve obtained for all data is given in Fig. 4.9. The energy continuity curve obtained for filtered data is given in Fig. 4.10. The wind roses on a monthly basis show that the winter season is formed on the southeastern axis, and the other seasons are predominantly on the northeastern axis.

For the wind energy conversion efficiency, the cut-in and cut out speeds of the horizontal axis three-bladed turbine were calculated by filtering. Applying

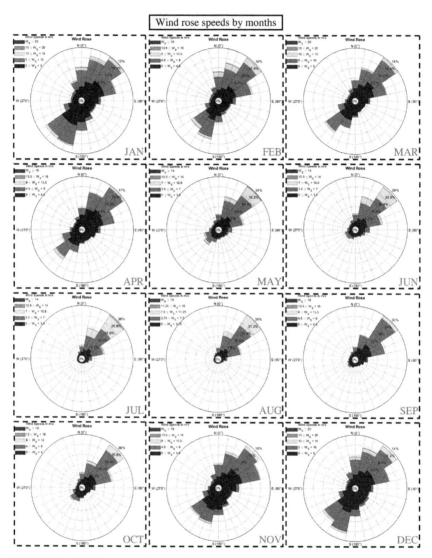

FIGURE 4.7   Wind rose speeds by months.

that the three-bladed turbine can generate energy at a minimum of 2 m/s and a maximum of 15 m/s, all speeds outside this range are taken into account as 0. Wind data have been analyzed for the date of July 18, 2019. The power chart of a 1-day 5-kWp capacity turbine is given in Fig. 4.11A. Typical year-round hourly, daily, and monthly power analysis was performed for 2019. Typical year-round hourly, daily, and monthly power data graphs are given in Fig. 4.11B, C, and D, respectively. Wind speed values are grouped and averaged separately for each month. A typical monthly average efficiency column graph is given with

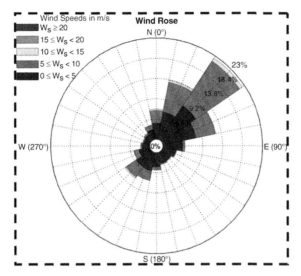

FIGURE 4.8 Wind rose speeds (all data).

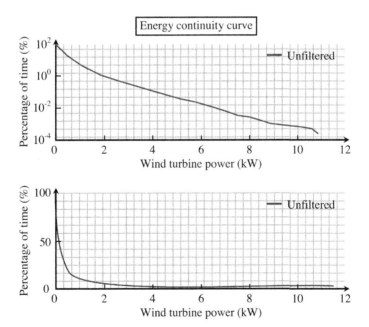

FIGURE 4.9 Unfiltered energy continuity curve.

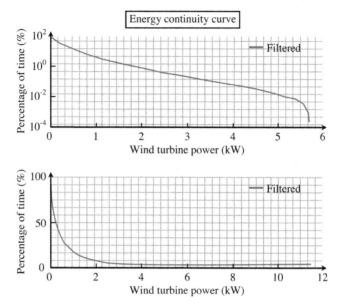

FIGURE 4.10   Filtered energy continuity curve.

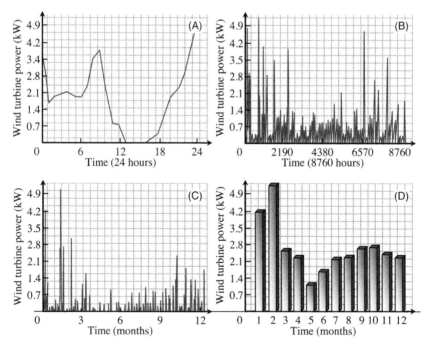

FIGURE 4.11   Graph of 5-kW wind turbine output power. (A) For 24 hours, (B) typical year-round hourly efficiency, (C) typical year-round daily efficiency, and (D) monthly average efficiency.

these average speed values. In the typical monthly average efficiency calculation, it is necessary to use the average of the speed values cubed rather than substituting the averaged speeds in the power equation.

2.2   Solar PV panel and wind energy integration

Like sunlight, wind is a free and nonpolluting source of energy. However, noise pollution should be considered in larger turbine models. Although it varies dramatically with wind speed, region, and other environmental characteristics, it can be found everywhere at a certain capacity. However, off-grid wind systems experience similar challenges to solar PV panel systems. Wind speeds head to fluctuate significantly from 1 hour to the next and from one season to the next. Unexpected delays may occur in the energy produced. Renewable energy sources cannot continuously generate energy throughout the year and cannot meet constant load demands. As with solar PV panel power, an energy storage or backup power supply is required to provide uninterrupted power [19–21].

When both solar and wind energy are integrated into a hybrid generation system, they have significant advantages over their counterparts alone. Most importantly, a hybrid use of wind and solar energy can create less fluctuations in the energy produced by the system. This increases overall energy output and reliability and reduces energy saving requirements and associated costs [22,23].

A weekly power output graph created by using solar PV panel and wind energy system together is given in Fig. 4.12. Solar PV panel and wind energy production are very low at night. During these time intervals, it seems that en-

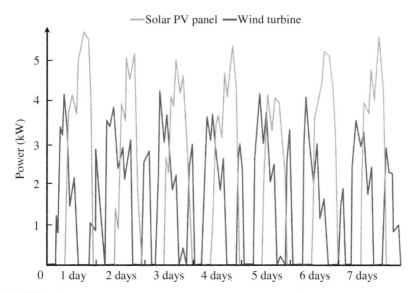

FIGURE 4.12   Weekly solar PV panel and wind turbine power generation graph.

ergy storage power is needed when there is a load demand. A lot of energy is produced from both solar PV panel and wind energy in the morning and noon hours. If this excess amount of energy is more than the amount demanded by the load, the remaining energy can be stored in a battery unit. In the evening the amount of energy produced from solar PV panel and wind energy gradually decreases. These time intervals are the point where the load demands peaks. In this case, since the power generated from the sources decreases, the energy in the storage unit can be used at noon.

On an annual basis the amount of energy obtained from wind energy reaches its maximum level in autumn and winter. Solar PV panel energy is observed to be at its maximum level in spring and summer months. This seasonal change shows that wind energy is suitable for heating and lighting applications in winter, while solar PV panel energy can be met air conditioning and irrigation needs in summer. In this way, wind and solar PV panel energy together form a complementary system that extends the daily and annual total peak production periods.

The role of the battery in a hybrid renewable energy source is basically to ensure the storage of the excess energy generated and to provide energy to the system when low or no production is produced. If the system is connected to the grid and excess energy is generated in the system, excess energy can be sold to the grid to provide economic incentives. The energy storage unit can help the energy needs during periods of time when housing needs are at their peak by storing energy when low energy is demanded. Thus the user saves money by using the energy storage unit during this high demand for several hours.

The hybrid energy generation system established with solar PV panel and wind energy has a characteristic that complements or supports each other. This complementary of sources enables the energy storage requirement capacity to be reduced. Because when no power is generated from any source, the situations where the user has to take the advantage of the energy storage unit will increase. In addition, as the batteries deep of discharges in the energy storage unit decrease, their service life will increase.

The hybrid system is ideal for villages in small isolated areas, because there is no electricity grid in this region. The needed electrical energy can be provided from the renewable energy sources. Remote areas are not the only places where the hybrid renewable energy system can be used effectively. The development of hybrid energy systems for urban environments will create an additional power supply to the grid. In urban areas where lands are costly and people often oppose the construction of new power plants, renewable energy hybrid systems can be created in small-scale housing [24–26]. Thus a distributed electricity generation system can be integrated into existing grid infrastructure to meet an increasing load demand. Such a distributed electricity generation system can reduce system losses.

The small-scale hybrid renewable energy system has been classified as the "trend of the future" in a complex urbanization environment. Hybrid renewable energy systems are growing and becoming widespread. Hybrid energy systems are beginning to be seen as the most pragmatic approach to distributed energy generation in a variety of settings. This idea not only minimizes environmental pollution but also benefits to reduce losses in transmission and distribution equipment of existing electrical systems. Hybrid renewable energy systems are a positive trend toward distributed energy generation, helping to reduce business losses on the demand side. Hybrid renewable energy systems are becoming less dependent on the electricity distribution line. Consequently, using hybrid renewable energy systems is an excellent way to advance distributed energy generation. Wind, solar, and current energy sources extend the total peak energy generation times. They complement each other by preserving the battery life in the energy storage system and reducing the total cost of the system. Hybrid renewable energy systems can be adapted to a wide variety of environments, villages, seaside settlements, and urban forests. They are rapidly becoming the preferred system for distributed renewable energy generation [27,28].

2.3   Load balancing with hybrid renewable energy sources

Renewable energy sources produce daily energy depending on the periodic characteristics. Due to this irregular output characteristic, the energy generated from these renewable energy sources must be stored. The amount of energy from these sources varies according to the time of the year (seasonal) and weather conditions. Renewable energy sources also can work synchronously with the grid. According to the system status, the excess energy generated is transferred to the grid. In this case the grid actually functions as a buffer or energy storage [29,30]. Electric load demand in residential applications varies in different time periods during the day. Generally, the demand for electricity usage increases in the evening hours. Fig. 4.13 shows 1-day sample electric power consumption

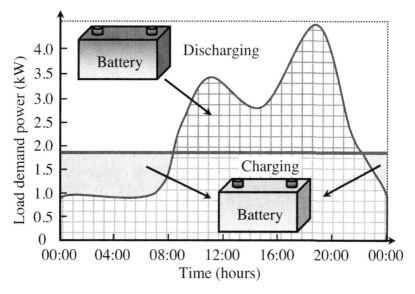

FIGURE 4.13   One-day sample electric power consumption graph for residential.

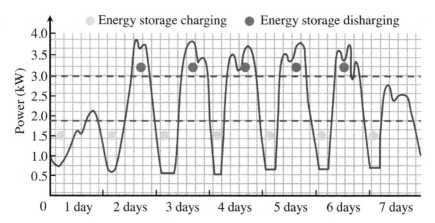

FIGURE 4.14   One-week sample electric power consumption graph for residential.

graph for residential. The devices such as oven, air conditioners, and water heaters demand high energy in residences. Residential electricity needs vary according to the seasons. While heating is used in residences in winter, air conditioners are used in summer.

Electricity use of energy in residences also varies on a weekly basis. While electricity usage for the weekdays is low, it is higher due to the weekend holiday. In order to understand the electricity needs of a residence, the daily energy model must be analyzed correctly. Fig. 4.14 shows 1-week sample electric power consumption graph for residential. Daily and weekly power demand patterns of residences are important for electricity distribution companies. According to this load model, electricity distribution companies will join an additional power plant to the existing grid. Therefore model estimation of grid-connected loads is very important.

The institution that provides electricity distribution service puts power plants operating with different sources into operation in order to meet the load demands. Thus the load demand on the grid is balanced. The power plants that are commissioned are generally coal-, natural gas–, and oil-based facilities. The electricity distribution company prefers the fastest and cheapest method. These power plants usually take 30–60 minutes to activate [31]. If there is a need for a sudden load in the grid, this is a very long time. A grid with a hybrid structure is required to overcome this problem. This is possible with hybrid renewable energy sources and energy storage units. Sudden loads on the grid can be met quickly with energy storage units. Thus voltage sags and power fluctuations are prevented in the grid. In addition, regional energy storage systems can be activated very quickly. Energy storage systems are charged when the load demand is low. The charging process of the energy storage unit can be done with renewable energy sources. Thus there will be no more energy demand from the grid and the energy storage unit will not be charged with fossil fuels.

2.4 Complementarity of hybrid renewable energy sources

Power output characteristics of solar PV panels and wind energy may change over time during the day. This power generation graph may not overlap with the load power graph demanded during the day. If the solar PV panel and wind energy are to be used separately, the sizing should generally be made with high capacity. In this case, system setup cost increases. Power generation interruptions are also high in systems installed separately. Combining solar PV panels and wind energy reduces the fluctuation in output power. Thus system performance and reliability increase and costs decrease. The energy storage sizing to be used in the hybrid system is also reduced [32]. The sizing stage of the hybrid system is more complex as the number of resources increases. Attention should be paid to the parameters in order to produce energy at the optimum level from the hybrid system. These parameters for wind energy are turbine type, turbine-rated power, and generator output voltage/current.

Solar PV panels and wind energy are widely used due to their ease of installation and cost effectiveness. The power generation characteristics of these two renewable energy sources are highly variable. Power generation capacity can be estimated for the location to be installed, thanks to the insolation and wind speed data. This is very important for system sizing. According to these estimated data, the number of solar PV panels and wind turbine generator power will be selected. Most importantly, the capacity of the energy storage unit to be used in the hybrid system will be determined. If the capacity of the energy storage unit is selected high, the energy storage unit will never be charged at full capacity. This will only increase the cost of the system. If the capacity of the energy storage unit is selected low, then the load group in the system cannot be provided energy for the calculated time [33].

The power generation characteristics of solar PV panels and wind energy are highly variable. Solar energy does not change rapidly. However, wind energy can change hourly. There are variations between the power generation characteristics of solar and wind energy and the time interval of the load demand power. Wind and solar energy sources are complementary. The total power generation graph is tried to be softened with a hybrid system created consisting of solar PV panels and wind turbine. The decision to use both sources in a hybrid system depends entirely on the load and energy storage unit [34,35]. The aim in the hybrid system is to keep consumption and production equal. It is also to ensure that the energy storage unit is always full. Generally, two power generation sources are expected to complement each other. However, this is not always possible due to natural conditions.

In general, it will not be correct to come to a definite conclusion about the complementarity of these two renewable energy sources, because the source of these two energy generation units is nature. It is possible to make predictions with solar and wind data. Another variable factor is the load profile. The load profile varies greatly over time. Hybrid power generation systems are generally used in areas far from the grid. In the hybrid system the power generation units must be able to meet the total load requirement in the system. The hybrid system must be cost-effective.

References

[1] J.A. Ferraz de Andrade Santos, P. de Jong, C. Alves da Costa, E.A. Torres, Combining wind and solar energy sources: potential for hybrid power generation in Brazil, Util. Policy 67 (2020) 101084, doi: 10.1016/j.jup.2020.101084.

[2] K. Elmaadawy, K.M. Kotb, M.R. Elkadeem, S.W. Sharshir, A. Dán, A. Moawad, et al. Optimal sizing and techno-enviro-economic feasibility assessment of large-scale reverse osmosis desalination powered with hybrid renewable energy sources, Energy Convers. Manag. 224 (2020) 113377, doi: 10.1016/j.enconman.2020.113377.

[3] A. Kafetzis, C. Ziogou, K.D. Panopoulos, S. Papadopoulou, P. Seferlis, S. Voutetakis, Energy management strategies based on hybrid automata for islanded microgrids with renewable sources, batteries and hydrogen, Renew. Sustain. Energy Rev. 134 (2020) 110118, doi: 10.1016/j.rser.2020.110118.

[4] J.C. Alberizzi, J.M. Frigola, M. Rossi, M. Renzi, Optimal sizing of a hybrid renewable energy system: importance of data selection with highly variable renewable energy sources, Energy Convers. Manag. 223 (2020) 113303, doi: 10.1016/j.enconman.2020.113303.

[5] E. Noghreian, H.R. Koofigar, Power control of hybrid energy systems with renewable sources (wind-photovoltaic) using switched systems strategy, Sustain. Energy Grids Networks 21 (2020) 100280, doi: 10.1016/j.segan.2019.100280.

[6] X. Zhang, W. Huang, S. Chen, D. Xie, D. Liu, G. Ma, Grid–source coordinated dispatching based on heterogeneous energy hybrid power generation, Energy 205 (2020) 117908, doi: 10.1016/j.energy.2020.117908.

[7] Y. Wang, L. Wang, M. Li, Z. Chen, A review of key issues for control and management in battery and ultra-capacitor hybrid energy storage systems, ETransportation 4 (2020) 100064, doi: 10.1016/j.etran.2020.100064.

[8] B. Benlahbib, N. Bouarroudj, S. Mekhilef, D. Abdeldjalil, T. Abdelkrim, F. Bouchafaa, et al. Experimental investigation of power management and control of a PV/wind/fuel cell/battery hybrid energy system microgrid, Int. J. Hydrogen Energy 45 (2020) 29110–29122, doi: 10.1016/j.ijhydene.2020.07.251.

[9] P. Singh, J.S. Lather, Dynamic power management and control for low voltage DC microgrid with hybrid energy storage system using hybrid bat search algorithm and artificial neural network, J. Energy Storage 32 (2020) 101974, doi: 10.1016/j.est.2020.101974.

[10] D. Rezzak, N. Boudjerda, Robust energy management strategy based on non-linear cascade control of fuel cells-super capacitors hybrid power system, Int. J. Hydrogen Energy 45 (2020) 23254–23274, doi: 10.1016/j.ijhydene.2020.05.250.

[11] G. Human, G. van Schoor, K.R. Uren, Genetic fuzzy rule extraction for optimised sizing and control of hybrid renewable energy hydrogen systems, Int. J. Hydrogen Energy 46 (5) (2020) 3576–3594, doi: 10.1016/j.ijhydene.2020.10.238.

[12] Q. Zhang, L. Wang, G. Li, Y. Liu, A real-time energy management control strategy for battery and supercapacitor hybrid energy storage systems of pure electric vehicles, J. Energy Storage 31 (2020) 101721, doi: 10.1016/j.est.2020.101721.

[13] A. Kadri, H. Marzougui, A. Aouiti, F. Bacha, Energy management and control strategy for a DFIG wind turbine/fuel cell hybrid system with super capacitor storage system, Energy 192 (2020) 116518, doi: 10.1016/j.energy.2019.116518.

[14] Y. Yuan, J. Wang, X. Yan, B. Shen, T. Long, A review of multi-energy hybrid power system for ships, Renew. Sustain. Energy Rev. 132 (2020) 110081, doi: 10.1016/j.rser.2020.110081.

[15] G. Debastiani, C.E. Camargo Nogueira, J.M. Acorci, V.F. Silveira, J.A. Cruz Siqueira, L.C. Baron, Assessment of the energy efficiency of a hybrid wind-photovoltaic system for Cascavel, PR, Renew. Sustain. Energy Rev. 131 (2020) 110013, doi: 10.1016/j.rser.2020.110013.

[16] J. Du, X. Zhang, T. Wang, Z. Song, X. Yang, H. Wang, et al. Battery degradation minimization oriented energy management strategy for plug-in hybrid electric bus with multi-energy storage system, Energy 165 (2018) 153–163, doi: 10.1016/j.energy.2018.09.084.

[17] I. Komušanac, B. Ćosić, N. Duić, Impact of high penetration of wind and solar PV generation on the country power system load: the case study of Croatia, Appl. Energy 184 (2016) 1470–1482, doi: 10.1016/j.apenergy.2016.06.099.

[18] K. Rajesh, A.D. Kulkarni, T. Ananthapadmanabha, Modeling and simulation of solar PV and DFIG based wind hybrid system, Procedia Technol. 21 (2015) 667–675, doi: 10.1016/j.protcy.2015.10.080.

[19] B.V. Ermolenko, G.V. Ermolenko, Y.A. Fetisova, L.N. Proskuryakova, Wind and solar PV technical potentials: measurement methodology and assessments for Russia, Energy 137 (2017) 1001–1012, doi: 10.1016/j.energy.2017.02.050.

[20] Y. Jiang, L. Lu, A.R. Ferro, G. Ahmadi, Analyzing wind cleaning process on the accumulated dust on solar photovoltaic (PV) modules on flat surfaces, Sol. Energy 159 (2018) 1031–1036, doi: 10.1016/j.solener.2017.08.083.

[21] T. Sarkar, A. Bhattacharjee, H. Samanta, K. Bhattacharya, H. Saha, Optimal design and implementation of solar PV-wind-biogas-VRFB storage integrated smart hybrid microgrid for ensuring zero loss of power supply probability, Energy Convers. Manag. 191 (2019) 102–118, doi: 10.1016/j.enconman.2019.04.025.

[22] A.W. Frazier, C. Marcy, W. Cole, Wind and solar PV deployment after tax credits expire: a view from the standard scenarios and the annual energy outlook, Electr. J. 32 (2019) 106637, doi: 10.1016/j.tej.2019.106637.

[23] A. Vasel, F. Iakovidis, The effect of wind direction on the performance of solar PV plants, Energy Convers. Manag. 153 (2017) 455–461, doi: 10.1016/j.enconman.2017.09.077.

[24] E. Nuño, P. Maule, A. Hahmann, N. Cutululis, P. Sørensen, I. Karagali, Simulation of transcontinental wind and solar PV generation time series, Renew. Energy 118 (2018) 425–436, doi: 10.1016/j.renene.2017.11.039.

[25] S. Weida, S. Kumar, R. Madlener, Financial viability of grid-connected solar PV and wind power systems in Germany, Energy Procedia 106 (2016) 35–45, doi: 10.1016/j.egypro.2016.12.103.

[26] O. Abdalla, H. Rezk, E.M. Ahmed, Wind driven optimization algorithm based global MPPT for PV system under non-uniform solar irradiance, Sol. Energy 180 (2019) 429–444, doi: 10.1016/j.solener.2019.01.056.

[27] A. Thomas, P. Racherla, Constructing statutory energy goal compliant wind and solar PV infrastructure pathways, Renew. Energy 161 (2020) 1–19, doi: 10.1016/j.renene.2020.06.141.

[28] D.B. Carvalho, E.C. Guardia, J.W. Marangon Lima, Technical-economic analysis of the insertion of PV power into a wind-solar hybrid system, Sol. Energy 191 (2019) 530–539, doi: 10.1016/j.solener.2019.06.070.

[29] C. Zomer, I. Custódio, S. Goulart, S. Mantelli, G. Martins, R. Campos, et al. Energy balance and performance assessment of PV systems installed at a positive-energy building (PEB) solar energy research centre, Sol. Energy 212 (2020) 258–274, doi: 10.1016/j.solener.2020.10.080.

[30] E. O'Shaughnessy, D. Cutler, K. Ardani, R. Margolis, Margolis, solar plus: a review of the end-user economics of solar PV integration with storage and load control in residential buildings, Appl. Energy 228 (2018) 2165–2175, doi: 10.1016/j.apenergy.2018.07.048.

[31] M.R. Sheibani, G.R. Yousefi, M.A. Latify, Economics of energy storage options to support a conventional power plant: a stochastic approach for optimal energy storage sizing, J. Energy Storage 33 (2020) 101892, doi: 10.1016/j.est.2020.101892.

[32] Y. Zhang, J. Lian, C. Ma, Y. Yang, X. Pang, L. Wang, Optimal sizing of the grid-connected hybrid system integrating hydropower, photovoltaic, and wind considering cascade reservoir connection and photovoltaic-wind complementarity, J. Clean. Prod. 274 (2020) 123100, doi: 10.1016/j.jclepro.2020.123100.

[33] F. Weschenfelder, G. de Novaes Pires Leite, A.C. Araújo da Costa, O. de Castro Vilela, C.M. Ribeiro, A.A. Villa Ochoa, et al. A review on the complementarity between grid-connected solar and wind power systems, J. Clean. Prod. 257 (2020) 120617, doi: 10.1016/j.jclepro.2020.120617.

[34] M. D'Isidoro, G. Briganti, L. Vitali, G. Righini, M. Adani, G. Guarnieri, et al. Estimation of solar and wind energy resources over Lesotho and their complementarity by means of WRF yearly simulation at high resolution, Renew. Energy 158 (2020) 114–129, doi: 10.1016/j.renene.2020.05.106.

[35] P.B.L. Neto, O.R. Saavedra, D.Q. Oliveira, The effect of complementarity between solar, wind and tidal energy in isolated hybrid microgrids, Renew. Energy 147 (2020) 339–355, doi: 10.1016/j.renene.2019.08.134.

Chapter 5

Solar Hybrid Systems and Energy Storage Systems

1 Energy storage systems using in solar hybrid systems

Batteries are chemical circuits that can store the energy resulting from chemical reactions as electrical energy. It consists of unit cells and these cells contain chemical energy that can be converted into electrical energy. One or more of these electrolytic cells are connected in series to form the battery. The grouped cells are added together to form the battery module. The battery pack consists of a combination of battery modules connected in series or parallel combinations.

The four main components of the battery cell are shown in Fig. 5.1.

1. Positive electrode: It is an active substance that provides the creation of electrical current. It is an electrode consisting of oxide, sulfite, or any compound that can decrease during discharge. While this electrode battery is discharged, it consumes electrons from the external circuit. Lead oxide and nickel oxide hydroxide are examples of positive electrodes. Electrode materials are solid.

2. Negative electrode: It is an active substance that provides the creation of electrical current. It consists of metal or alloy that can be oxidized during discharge. While this electrode battery is discharged, it produces electrons in the external circuit. Lead and cadmium are examples to negative electrodes. Negative electrode materials are also present in the battery cell in a solid state. Negative electrodes of some lithium-ion batteries are made of aluminum to prevent oxidation.

3. Electrolyte: It is a solution consisting of charged particles that can transmit and move electrical current. The electrolyte should have high and selective conductivity for the ions involved in the electrode reactions but should not conduct electrons to prevent discharge. Electrolytes can be liquids, gels, or solids, as well as acidic or alkaline depending on the type of battery. Liquid electrolytes are used in conventional batteries such as lead–acid or nickel–cadmium. In lead–acid batteries the electrolyte is in the liquid form of sulfuric acid. More developed batteries such as nickel–metal hydride and lithium ion use electrolytes in the form of gel, paste, or resin. Lithium polymer batteries use solid-state electrolytes.

Solar Hybrid Systems. http://dx.doi.org/10.1016/B978-0-323-88499-0.00005-7

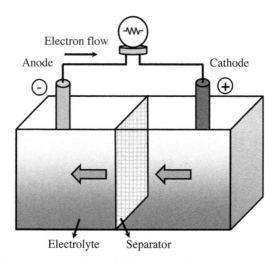

Electron flow

Anode

Cathode

Electrolyte   Separator

FIGURE 5.1   Operating principle of the battery.

4. Separator: It is an electrically insulating layer that physically separates electrodes at different poles. Separators must pass the ions of the electrolytes and have a function that can store the electrolyte and keep it immobile. Separators used today are made of synthetic polymers.

The energy stored in the battery is the difference of chemical energy components between when the battery is charged and discharged. Chemical energy in a cell is converted into electrical energy depending on demand; the unit listed earlier performs this process using the basic elements in the cell. The electrochemically active ones of the positive or negative electrodes are called active substances. The chemical oxidation event takes place in the electrodes due to the electrons that are connected and broken. Electrodes must be conductive in an electronic context, separated by a splitter, and the contacts made on them must be very strong. While the battery is working, chemical reactions in the electrodes cause electrons to flow from one electrode to another. In addition, if the electrons formed during the chemical reaction can flow in an electrical circuit that connects the two electrodes externally, the electron in the cell shows its flow resistance. The connection points between the electrodes and the external electrical circuit are called battery terminals [1–3].

1.1   Primary (nonrechargeable) batteries

Electrochemical batteries can be examined in two classes as primary (nonrechargeable) and secondary (rechargeable) batteries according to their ability to charge electrically. Since the primary-type batteries are nonrechargeable, they cannot be used once again after discharging. Primary cells in the electrolyte containing absorbent and release agent are called dry cells. Although primary batteries are generally inexpensive, they can be used as a lightweight power pack for portable electronics and electrical devices, in lighting, photographic devices, toys, backup memories, and many other areas. The benefits of primary batteries are as follows: it is easy to reach, simple to use, needs little maintenance if necessary, and can be shaped in connection with the area of use. Other general advantages should have good shelf life, reasonable energy and power densities, reliable and acceptable cost.

At first, the capacity obtained with zinc carbon cells was less than 50 Wh/kg, and now the capacity obtained with lithium cells exceeds 500 Wh/kg. The shelf life of the cells was limited to about 1 year when the batteries were stored at average temperatures during the first commercial years of the batteries. Today, this period varies between 2 and 5 years. With the ability to store even at temperatures as high as 70 °C, the shelf life of the latest lithium batteries is as long as 10 years. Working at low temperature was reduced from 0 °C to −40 °C, −55 °C and the power density was increased. The shelf life of special slow flow batteries using a solid electrolyte exceeds 20 years. Thanks to these advanced features, the usage areas of primary batteries have also expanded. Battery weight and size are also reduced with high energy density. Thus considering the developments in electronic technology, new mobile phone, practical communication, and electronic devices could also be made. The higher power density made it possible to design devices used in computers, mobile phones, military surveillance systems, and other high-power applications. Currently, many primary batteries are used in medical electronic devices, as well as in application areas such as spare memories, as they have a long shelf life [4–6].

1.2 Secondary (rechargeable) batteries

The use of secondary or rechargeable batteries is quite wide. The most known ones are starter, lighting and ignition systems, and emergency and backup power systems in the automotive industry. Small secondary batteries are also used in portable power supplies such as toys, lighting, photography, radio, and electronic devices such as computers and mobile phones.

In applications used as an energy storage device, it is generally electrically connected and charged by this source. The automotive and aircraft systems, emergency and backup power supplies, and stationary energy storage systems for load balancing are example to these applications. In other applications of secondary batteries, the battery is used like a primary battery, but it is recharged after use. For example, in terms of economy, portable electronic and electrical devices (thus, the old one can be used instead of a new battery) and some stationary battery applications. In addition to being rechargeable, the secondary batteries can be characterized by high power density, high discharge rate, full discharge profiles, and good performance at low temperature. However, compared to primary batteries, their energy density is generally lower and their charge-holding ability is weaker. Rechargeable batteries have higher energy densities, better charge-holding abilities, and other performance-enhancing properties of this high energy material, using lithium as a negative active material [7–9].

1.3 Lead–acid (Pb) batteries

Lead–acid batteries have the oldest and most mature technology. It is a low-cost first-action battery created by placing negative lead and positive lead dioxide electrodes into the sulfuric acid electrolyte. Depending on the purpose of use, the number of plates in the electrodes is increased or decreased or their sizes are changed. Parts and general appearance of a typical lead–acid battery are given in Fig. 5.2. The lead–acid battery has a low energy density of 25–35 Wh/kg, although it has passed through many years of development. However, the power density is as high as 150 W/kg. Lead–acid batteries are affected by their low temperatures. There is a serious decrease in both energy and power density at ambient temperatures below 10 °C [10,11].

Fig. 5.3 shows the graph of the terminal voltage depending on the battery charge rate of a typical lead–acid battery cell discharged. In a normally accepted

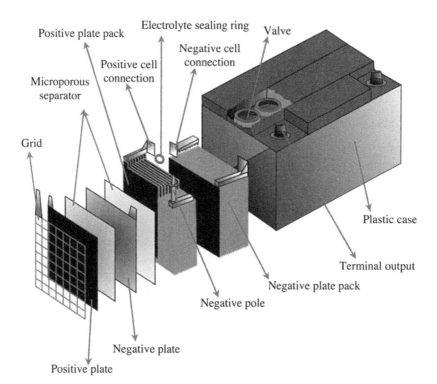

FIGURE 5.2 Parts and general view of a typical lead–acid battery.

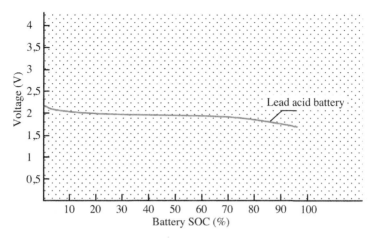

FIGURE 5.3   Discharge voltage graph of lead–acid battery cell.

2 V lead–acid cell, the actual voltage is between 2.05 and 2.15 V/cell in relation to the electrolyte density. The voltage value per cell is calculated by taking 2 V [11].

1.4   Nickel–iron (NiFe) batteries

It is formed by putting the nickel positive and iron negative electrodes into the potassium hydroxide electrolyte. These batteries are reliable and long-life but costly. These batteries were developed in order to make electric vehicles travel longer. The maximum power density is 100 W/kg and they can maintain their performance down to −20 °C. These batteries have a lifetime of up to 2000 deep discharge cycles corresponding to about 6 years [12]. Parts and general appearance of a typical nickel–iron battery are given in Fig. 5.4.

Fig. 5.5 shows the graph of the terminal voltage depending on the battery charge rate of a typical nickel–iron battery cell discharged. The open-circuit voltage of the nickel–iron battery is 1.4 V. The battery nominal voltage is 1.2 V, the maximum charging voltage is usually between 1.7 and 1.8 V. The capacity of the nickel–iron battery depends on the capacity of the positive electrode, so the length and number of each positive plate determines the capacity of the battery [12].

1.5   Nickel–zinc (NiZn) batteries

Although the energy density of these rechargeable batteries is 70 Wh/kg and the power density is 150 W/kg, it is disadvantageous to have a lifetime of 300. The use of these batteries has not become widespread, since dendrites on the zinc plate shorten the life of the battery during charging. The battery has a wide

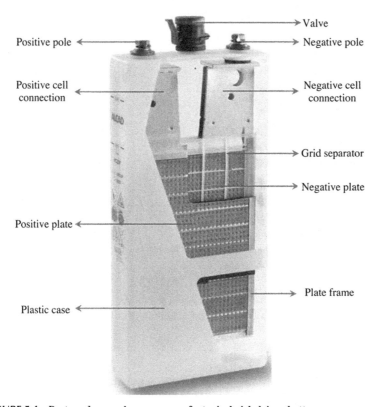

FIGURE 5.4 Parts and general appearance of a typical nickel–iron battery.

operating temperature tolerance ($-39\ ^\circ$C to $+81\ ^\circ$C). After 30 days of use, there is a 60 % reduction in charge. A single nickel–zinc cell can be combined with various designs into multiple cells in traditional ways to be transformed into a monoblock-type battery [13]. Parts and general appearance of a typical nickel–zinc battery are given in Fig. 5.6.

In practice, the working voltages of nickel–zinc batteries under load are between 1.55 and 1.65 V and approximate energy densities are 70 Wh/kg depending on special designs. A single nickel–zinc cell provides 1.65 V. Nickel–zinc batteries produced in monoblock-type are usually turned into a 12-V typical battery, containing seven or eight cell modules [14]. Fig. 5.7 shows the graph of the terminal voltage depending on the battery charge rate of a typical nickel–zinc battery cell discharged.

1.6 Nickel–cadmium (NiCd) batteries

It is formed by placing the sintered positive nickel electrode and negative cadmium electrode in the potassium hydroxide aqueous solution. In recent years, it

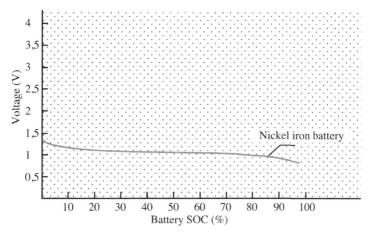

FIGURE 5.5 Discharge voltage graph of nickel–iron battery cell.

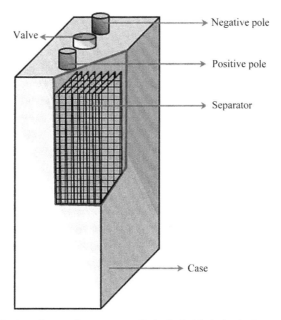

FIGURE 5.6 Parts and general appearance of a typical nickel–zinc battery.

is considered as a battery that provides good balance in terms of specific energy, specific power, cycle life, and reliability. Because cadmium is toxic and environmentally hazardous, recovery of nickel–cadmium batteries is very important and complex. Their use has been discontinued due to the damage to the environment. These batteries have a high charge/discharge rate and the number of deep discharge cycles is around 2000. The internal resistance of nickel–cadmium

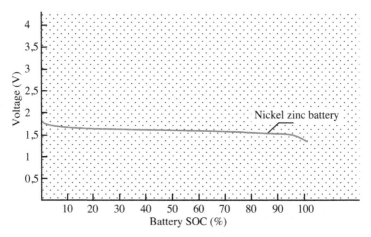

FIGURE 5.7 Discharge voltage graph of nickel–zinc battery cell.

batteries is generally very low. A typical direct current (DC) resistance value is 0.4, 1, and 4 mΩ, respectively, high-, medium-, and low charge rate for the 100 Ah charge value. The decrease in temperature and battery charge will cause an increase in internal resistance. Depending on the working conditions and battery design, the total working life varies between 8 and 25 years. Parts and general appearance of a typical nickel–cadmium battery are given in Fig. 5.8 [15].

The energy density of a typical nickel–cadmium cell is 20 Wh/kg and 40 Wh/L. The nominal voltage of the nickel–cadmium battery cell is 1.2 V. Although the battery discharge rate and battery temperature are an important variable for chemical batteries, these parameters have little effect in nickel–cadmium batteries compared to lead–acid batteries. Therefore nickel–cadmium batteries can be used at high discharge rates without losing their nominal capacity. Nickel–cadmium batteries also have a wide range of operating temperatures. A standard nickel–cadmium battery cell can operate between −20 °C and +50 °C [16]. Fig. 5.9 shows the graph of the terminal voltage depending on the battery charge rate of a typical nickel–cadmium battery cell discharged.

1.7 Nickel–metal hydride (NiMH) batteries

Nickel–metal hydride batteries have recently been used in many electric car applications since they do not have oxide properties and have better performance. Nickel–metal hydride batteries store more energy than nickel–cadmium batteries. The negative electrode, which is a metal hydride mixture, consists of the potassium hydroxide electrolyte and the positive electrode, the active material of which is nickel hydroxide. With an energy density of more than 70 Wh/kg and a power density of more than 200 W/kg, these batteries are about five times more expensive than lead–acid batteries. The battery has over 600 full charge/discharge cycles in the case of 80 % deep discharge and it is about 35 minutes

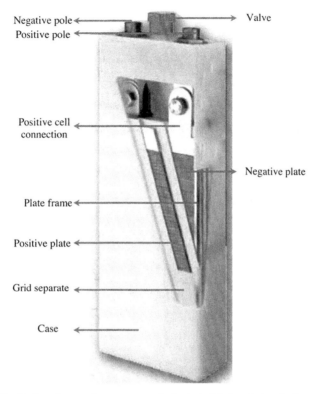

FIGURE 5.8   **Parts and general appearance of a typical nickel–cadmium battery.**

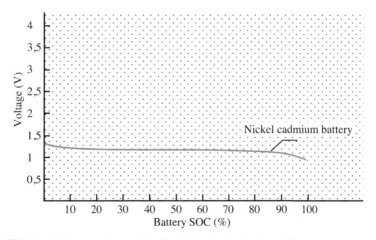

FIGURE 5.9   **Discharge voltage graph of nickel–cadmium battery cell.**

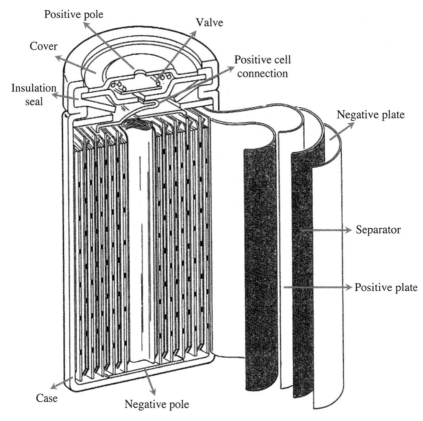

FIGURE 5.10 **Parts and general appearance of a typical nickel–metal hydride battery.**

to quickly recharge 80 %. Sealed nickel–metal hydride cells and batteries are produced in cylindrical, button, and prismatic ways [17–20]. Parts and general view of a typical nickel–metal hydride battery are given in Fig. 5.10.

The nominal voltage of a nickel–metal hydride cell is 1.2 V, and the open circuit voltage is between 1.25 and 1.35 V. Battery discharge cutoff voltage or final voltage is 1 V. The operating temperature of a standard nickel–metal hydride battery cell is between 0 °C and +40 °C. Operation of nickel–metal hydride batteries at high temperatures affects the performance characteristics of the batteries. The performance of the batteries at high temperatures is further reduced compared to the low discharge temperature due to the increased internal resistance. Similarly, temperature change effects are more pronounced at high discharge rates. Fig. 5.11 shows the graph of the terminal voltage depending on the battery charge rate of a typical nickel–cadmium battery cell discharged. At

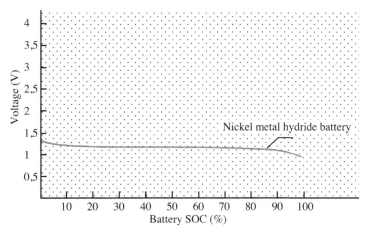

FIGURE 5.11   Discharge voltage graph of nickel–metal hydride battery cell.

low temperatures, especially when the discharge rate increases, the capacity of the battery decreases [20].

1.8   Sodium–sulfur (NaS) batteries

The sodium–sulfur battery is formed by combining the liquid states of the negative sodium and positive sulfur electrodes. Both electrode components are in a liquid state. It has been developed to provide the energy and power density required for electric vehicles. The optimum operating temperature is 350 °C and stops operating below 2000 °C. The reaction slows down as the sodium electrolyte freezes below this temperature. Freezing of sodium slows down the reaction and damages the battery due to mechanical stresses. The battery heater is needed to determine the operating temperature. It is necessary to be well insulated for easy maintenance and its temperature does not pose a danger. The charging time of the sodium–sulfur battery is 4–5 hours. Their lifespan is longer than the life of the lead–acid battery. The substances used in the structure of this battery are harmful to health. Sodium–sulfur batteries provide high energy density of 110 Wh/kg and power density of 150 W/kg [21]. Parts and general appearance of a typical sodium–sulfur battery are given in Fig. 5.12.

1.9   Sodium–nickel chloride (NaNiCl) batteries

Nickel chloride is positive electrode and sodium is negative electrode. Instead of sodium salt electrolyte, sodium chloride electrode is located. This electrode has a lower freezing point and it is around 160 °C. This type of battery operates at

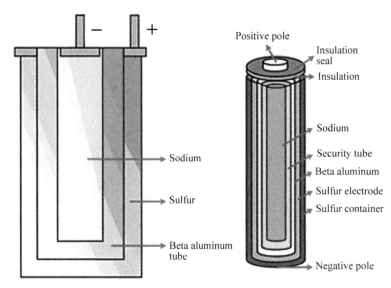

FIGURE 5.12 Parts and general appearance of sodium–sulfur battery.

somewhat low temperatures (300 °C) compared to the sodium sulfide battery and has similar energy (100 Wh/kg) and maximum power density (150 W/kg) [22].

1.10 Aluminum–air (Al–air) and zinc–air (Zn–air) batteries

Alkaline water–based electrolyte such as thin gas permeable cathode and potassium hydroxide are used in all air–metal batteries. Zinc and aluminum are the most commonly used metal electrodes in such applications. The maximum energy density of the aluminum–air battery is 220 Wh/kg, and the zinc–air battery is 200 Wh/kg. However, the rate of exchange between air and electrolyte determines the power density and this speed is very low [23].

1.11 Lithium-ion (Li-ion) batteries

Negative "hosts" such as graphite or tin oxide are used as negative electrodes in lithium-ion cells. During discharge, lithium ions change from negative "host" to organic "electrode" to manganese, cobalt or nickel oxide positive "host." The reverse process takes place during charging, and lithium ions act like a pendulum between the cathode and the anode. It has an energy density of approximately 150 Wh/kg and a deep discharge cycle of 1000 cycles. These batteries can be recharged in less than 1 hour to the 80 % charge state. Due to these features, lithium has become one of the most preferred elements of rechargeable battery technologies. Lithium metal oxide and lithium metal phosphate–based "cathode" (positive electrode) materials are especially preferred in mobile

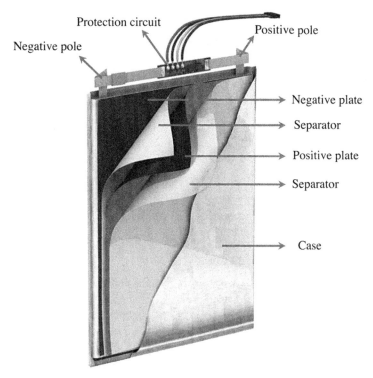

FIGURE 5.13   Parts and general appearance of lithium-ion battery.

equipment and electric vehicles [24]. They can provide very high energy storage density as a rechargeable battery. Parts and general appearance of a typical lithium-ion battery are given in Fig. 5.13.

The operating temperature of lithium-ion batteries during charging is between −20 °C and +60 °C, and when discharging the temperature is between −40 °C and +65 °C. The cell voltage of a typical lithium-ion battery is 2.5–4.2 V, and this voltage value is approximately three times the nickel–cadmium and nickel–metal hydride battery cell voltage. Fig. 5.14 shows the graph of the terminal voltage depending on the battery charge rate of a typical lithium-ion battery cell discharged [24].

Lithium-ion batteries, which are in the development phase, vary according to the chemicals used. For example, when lithium–manganese oxide ($LiMn_2O_4$) is used instead of lithium–cobalt oxide ($LiCoO_2$), the risk of explosion by cobalt is eliminated. However, the performance decreases dramatically when these types of batteries exceed 50 °C. On the other hand, when using manganese, the energy density decreases by 20 %. However, manganese is a much safer and environment-friendly element than cobalt. However, cobalt-based lithium-ion batteries are preferred due to their energy density. In the use of nickel oxide

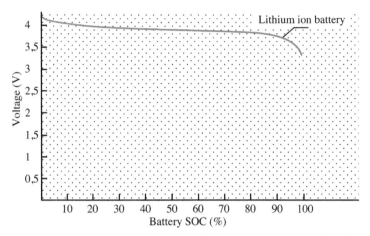

FIGURE 5.14   **Discharge voltage graph of lithium-ion battery cell.**

metal (LiNiO$_2$), the energy density of the lithium-ion battery is increased by 30 % compared to cobalt. However, because of the exothermic reaction that produces much higher heat, cooling problems occur in nickel-based batteries. In industry, three-metal Li(NiCoMn)O$_2$ type is also tried and some cheap and safe parameters can be obtained [25–27].

Some of the commercially produced lithium-ion battery types are [25–27] as follows:

- lithium–cobalt oxide (LiCoO$_2$),
- lithium–manganese oxide (LiMn$_2$O$_4$),
- lithium–nickel oxide (LiNiO$_2$),
- lithium–iron phosphate (LiFePO$_4$),
- lithium–titanate (Li$_2$TiO$_3$),
- lithium–titanate spinel (Li$_4$Ti$_5$O$_{12}$),
- lithium sulfur (Li$_2$S),
- lithium–nickel cobalt oxide (Li(NiCo)O$_2$),
- lithium–nickel cobalt magnesium oxide (Li(NiCoMn)O$_2$), and
- lithium–nickel cobalt aluminum oxide (Li(NiCoAl)O$_2$).

1.12   Lithium-ion polymer batteries

Unlike lithium-ion batteries, lithium salt is kept in the solid polymer mixture instead of organic solution. This design has many advantages, such as the solid polymer being nonflammable. Lithium-ion batteries are in prismatic or cylindrical metal sheaths and have a burning feature if exposed to sunlight. However, different shape designs can be realized by using flexible polymers in lithium-ion polymer batteries. In addition, one of the differences between commercial lithium-ion polymer and lithium-ion cells is that in lithium-ion cells, the metal

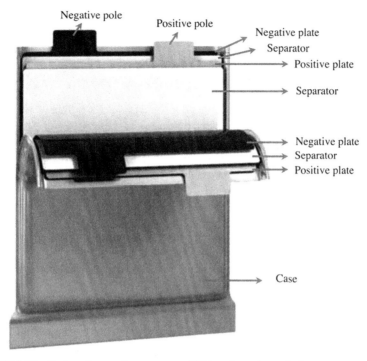

Negative pole

Positive pole

Negative plate

Separator

Positive plate

Separator

Negative plate

Separator

Positive plate

Case

FIGURE 5.15 Parts and general appearance of lithium-ion polymer battery.

sheath must press the electrodes and separators, whereas lithium-ion polymer batteries do not. As they do not need metal sheaths, they can be shaped lightly and easily. Thus 20 % more cells of equal weight can be placed than lithium-ion cells and even three times more energy density than nickel–cadmium or nickel–metal hydride batteries. The polymer cell voltage ranges from 2.7 (empty) to 4.23 V (fully charged) and must be protected against charging more than 4235 V for counter-connected cells. Excessive charging can cause an explosion or burns. Again, for the cell connected in series, the load should be removed when the discharge voltage drops below 3 V, otherwise the battery will not accept a longer charge [28–30]. Parts and general appearance of a typical lithium-ion polymer battery are given in Fig. 5.15.

1.13 Lithium–iron phosphate (LiFePO$_4$) batteries

The cathode material is made of lithium metal phosphate material instead of lithium metal oxide, which is another type of lithium-ion batteries and briefly called lithium iron or lithium ferrite in the market. As metal, iron, cobalt, manganese, or titanium are used. Lithium–iron phosphate battery technology was scientifically reported by Akshaya Padhi of the University of Texas in 1996. Lithium–iron phosphate batteries, one of the most suitable in terms of performance and production, started mass production commercially. Lithium–iron phosphate batteries have a high energy density of 220 Wh/L and 100–140 Wh/kg, and also the battery charge efficiency is greater than 90 %. The cycle life is approximately 2000 at a deep discharge rate of 80 %. The operating temperatures of lithium–iron phosphate batteries that perform well at high operating temperatures are between $-20\,°C$ and $+70\,°C$ [31].

1.14   Comparison of lithium-ion batteries

The comparison of terminal voltage and energy density of lithium–cobalt oxide ($LiCoO_2$), lithium–nickel cobalt aluminum oxide ($Li(NiCoAl)O_2$), lithium–nickel cobalt magnesium oxide ($Li(NiCoAl)O_2$), lithium–manganese oxide ($LiMn_2O_4$), and lithium–iron phosphate ($LiFePO_4$) battery cells, which are lithium-ion battery types, with numerical data is given in Table 5.1 [32]. Depending on the different chemical materials used in lithium-ion batteries, terminal voltages and energy densities of the batteries vary.

Although the use of different materials in lithium-ion batteries changes gravimetric or volumetric energy density in some battery types, it shows positive or negative effects on many issues such as cost and safety battery life. Comparison of lithium–cobalt oxide ($LiCoO_2$), lithium–manganese oxide ($LiMn_2O_4$), lithium–iron phosphate ($LiFePO_4$), lithium–nickel cobalt magnesium oxide ($Li(NiCoMn)O_2$), lithium–nickel cobalt aluminum oxide ($Li(NiCoAl)O_2$), and lithium–titanate spinel ($Li_4Ti_5O_{12}$) batteries, which are lithium-ion battery types, by scaling specific energy, specific power, safety, performance, life, and cost issues can be seen in Fig. 5.16 [33].

While different chemical materials used in lithium-ion batteries increase their specific energy in one type, performance effect is seen in another. It can be seen from the scaled comparison that the studies required to get the most out of these six subjects are still continuing.

TABLE 5.1 Voltage and energy densities of lithium-ion battery types.

Battery types	Voltage (V)	Energy densities	
		Gravimetric (Wh/kg)	Volumetric (Wh/kg)
$LiCoO_2$	3.7	195	560
$Li(NiCoAl)O_2$	3.6	220	600
$Li(NiCoMn)O_2$	3.6	205	580
$LiMn_2O_4$	3.9	150	420
$LiFePO_4$	3.2	90–130	333

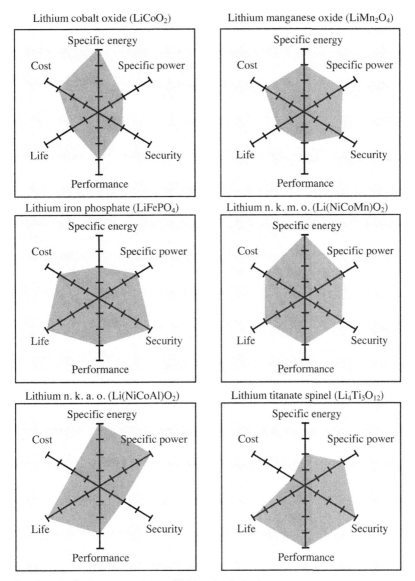

FIGURE 5.16 Scaling comparisons of lithium-ion batteries.

Lithium-ion batteries offer the optimum solution in terms of weight and volume for portable devices and electric vehicles. With today's technology the highest voltage per unit cell and the highest energy density per unit mass (Wh/kg), in other words, the highest specific energy can be provided by lithium-ion batteries. They are also efficient because they are not affected by the memory effect problem, and interrupting the charging/discharging process does not constitute a negative situation.

However, since lithium is a substance that can react very quickly, lithium-containing batteries have a risk of burning and explosion. New electrode designs are being developed and materials used for electrolyte are constantly being improved, and safer batteries are being introduced day by day. During the charging process, constant voltage method should be used and after the battery voltage reaches a certain value, this voltage should be kept constant and the charging current should be reduced. Otherwise, an overcharged battery may explode. Discharge should be stopped when a certain cutting voltage is reached during the discharge process. Damage occurs in batteries that are tried to be discharged below this cutting voltage.

1.15 Comparison of rechargeable battery types

Comparison of rechargeable batteries with different chemical structure in terms of energy density is given in Fig. 5.17 [34]. The developed battery technology has an increasingly less volume and less weight feature. In a comparison in Wh, lithium-ion technology appears to be in top condition compared to other batteries.

In Table 5.2, energy and power density, efficiency, self-discharge, charge/discharge cycle, and nominal voltage comparison of rechargeable batteries are given. As a charge/discharge cycle, lithium-ion batteries are quite high compared to lead–acid batteries and are approximately 1200 cycles. When examined in terms of efficiency, lithium-ion batteries are the most efficient battery with a rate of 99 %. In addition, when examined in terms of power and energy density, lithium-ion battery technology is at the forefront with very high values.

When examined in terms of nominal voltage, lithium-ion battery technology appears to be the highest battery terminal with 4 V. The discharge voltage graphs of the rechargeable battery cells are given in detail in Fig. 5.18 depending on the battery charge status. It is a nickel–iron battery with the lowest

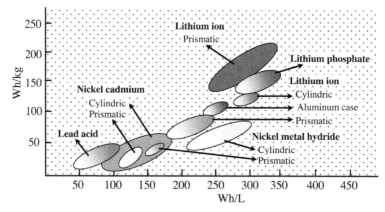

FIGURE 5.17 Comparison of secondary batteries in terms of energy density.

TABLE 5.2 Performance characteristics of rechargeable batteries.

Batteries	Energy density (Wh/kg)	Power density (Wh/kg)	Efficiency (%)	Self-discharge (%/month)	Charge/ discharge cycle	Nominal voltage (V)
Lead–acid	30–40	180	70–92	3–4	500–800	2.0
Nickel–cadmium	40–60	150	70–90	20	1500	1.2
Nickel–metal hydride	30–80	250–1000	66	20	1000	1.2
Nickel–iron	50	100	65	20–40	2000–4000	1.2
Nickel–zinc	60	900	70	<20	100–500	1.65
Zinc–silver oxide	130	>1000	70–90	5	50–100	1.5
Cadmium–silver oxide	70	<1000	90	5	300–800	1.2
Nickel–hydrogen	75	<1000	60	High	20,000	1.3
Lithium–ion	160	1800	99	5–10	1200	4.0
Lithium–polymer	130–200	>3000	99	5–10	500–1000	3.7

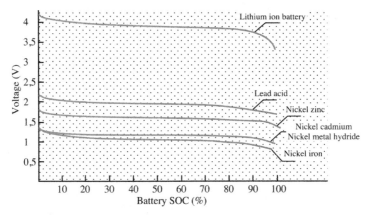

FIGURE 5.18   **Discharge voltage graphs of rechargeable battery cells.**

operating voltage below 1.2 V. Nickel–cadmium and nickel–metal hydride coil types draw approximately the same discharge curve as 1.2 V [35–37].

2   Ultracapacitors

Capacitors are devices that store energy by decomposition of positive and negative electrostatic charges. Capacitors consist of two conductive plates and insulators called dielectrics that separate them. The dielectric material prevents arcing between the two plates, helping to charge more. Classical capacitors have very high power densities (approximately 1012 W/m^3). However, their energy intensity is very low (approximately 5 Wh/m^3). Classical capacitors are commonly referred to as electrolytic capacitors. Ultracapacitors are the developed ones of classical capacitors. The power densities of these capacitors are 106 W/m^3 and the energy densities are 104 Wh/m^3. Energy intensities are low but discharge times are fast and cycle life is higher. However, capacitors actually have size problems. There is a linear connection between the capacitors' capacitance and the dielectric material. Therefore when large capacity is required, the dielectric material must also be large. Thanks to the ultracapacitors, very large capacity values provided high energy storage with very small capacitors [38–40]. Fig. 5.19 shows the parts and general appearance of a typical ultracapacitor.

Ultracapacitors are energy storage systems that act basically like capacitors but separate from capacitors in terms of high energy and power density. As a result, these new types of energy storage systems known as electrochemical capacitors or ultracapacitors are a very intensive research topic. The use of ultracapacitors as an alternative for applications that require high energy and power density where batteries and capacitors cannot be used increases their importance. Since there is no phase change in the electrochemical process in the ultracapacitors, such capacitors can be filled and discharged more than 100,000 times [40].

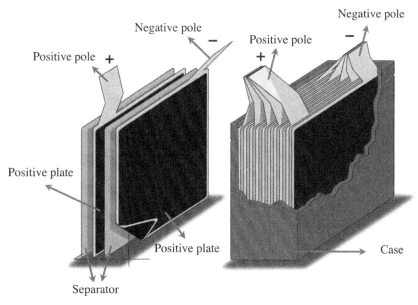

FIGURE 5.19   Ultracapacitor parts and general appearance.

Fig. 5.20 shows the typical Ragone graph that gives energy intensity versus power intensity [39]. The Ragone graph is a chart used to compare the power densities of various energy storage systems against the energy densities. In the Ragone graph the energy density (vertical axis) defines how much energy can

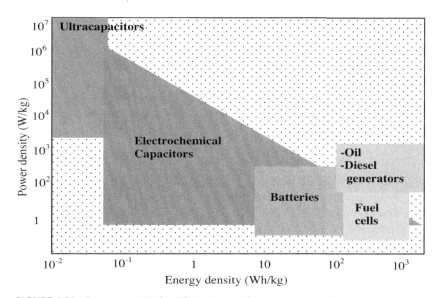

FIGURE 5.20   Ragone graphic for different types of energy storage systems.

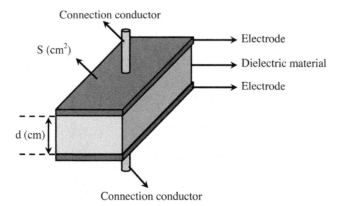

FIGURE 5.21   **Structure of conventional electrostatic capacitor.**

be stored, and the power density (vertical axis) defines how long the stored energy can be transferred. According to this graph, ultracapacitors have high power density and low energy density, while fuel cells and batteries have high energy but limited power density. The slow power transfer of lithium-ion batteries and advanced secondary batteries has increased the need for systems with higher energy and power density in many new applications. In the power density and energy density scale, ultracapacitors fill the gap between the batteries and electrolytic capacitors. Since the working principle of electrochemical capacitors contains similarities with traditional capacitors, the study of the working mechanism of traditional capacitors is an important step to illuminate the working mechanism of ultracapacitors.

Conventional capacitors are devices created by placing a dielectric material between two conductive materials, as shown in Fig. 5.21. Load accumulation on the conductive sheet is achieved by the potential application to the conductive sheet, and the sheets remain loaded when the potential applied to the sheets is removed.

The V is capacitor voltage, and the Q is load on the plates; Q/V ratio is called capacitance (C). In the following equation the capacitance formula is given:

$$C = \frac{Q}{V} \tag{5.1}$$

Capacitance is usually denoted by C and its unit is Farad (F). Since F is very large, it is usually measured in microfarad (μF) or picofarad (pF). The structure of conventional electrochemical capacitor is shown in Fig. 5.22.

Ultracapacitors must have high ionic electrolyte conductivity, high ionic separator conductivity, high electronic separator resistance, high electrode electronic conductivity, high electrode surface area, low separator, and electrode thickness [41]. The internal structure of a typical ultracapacitor is given in Fig. 5.23.

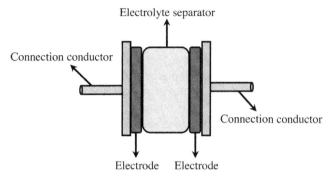

FIGURE 5.22   Structure of conventional electrochemical capacitor.

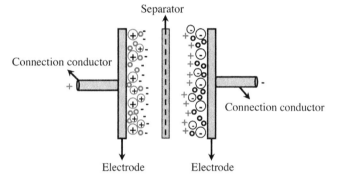

FIGURE 5.23   Internal structure of traditional ultracapacitor.

Ultracapacitors consist of the following elements:

Electrodes: The reason for the high capacitances obtained in ultracapacitors is the electrodes with high surface area that provide good contact between the electrode and the electrolyte. Electrodes must have a high surface area and be electrochemically inert (inactive substance). For this purpose, highly reactive metal oxides, carbon, graphite, conductive polymers, and their composites can be used. Capacitance can reach 250 F/g in ultracapacitors using carbon.

Electrolyte: Electrolytes used in ultracapacitors can be in solid, organic, or aqueous form. The electrolyte used must have high conductivity and electrochemical stability that allows the capacitor to operate at the highest possible voltage.

Separator: Different separators can be used depending on the type of electrolyte used. While polymer or paper separators are more suitable for organic electrolytes, glass fiber or ceramic separators are more suitable for aqueous electrolytes.

Although the working principle of electrochemical capacitors is similar to the working principle of conventional capacitors, the charge accumulation of electrochemical capacitors does not occur on conductive sheets separated by an insulator, unlike conventional capacitors. Conceptual structure of electrochemical capacitors is shown in Fig. 5.23; the charge accumulation on the electrodes depends on the change in the intersection between the electrolytic solution and the electrode surface. At this point, as with electrical double-layer capacitors, charge accumulation occurs in the called electrical double layer (electrode–electrolyte solution interface). The ionic separation distance depends on the size of the ions and the concentration of the electrolyte solution. Ultracapacitors have higher specific capacitance values than electrostatic capacitors because they are based on electrochemical double layer basis [41].

2.1 Usage areas and applications of ultracapacitors

The first examples of ultracapacitors were patented by General Electric in 1957 but were not offered for use. The first example was patented in 1978; it had 5 m^3 dimensions, 5 V voltage, and higher capacitance than 1 F. This capacitor is used to prevent data loss in computers. Although ultracapacitors did not show much improvement until the early 1990s, designers have focused on the cell design of capacitors. They also investigated the functions of working principles related to direct and alternating current.

In the early 1990s a company in Moscow developed a structure called asymmetrical design in capacitors and used it as the power supplier of an all-electric bus. The charging time was ahead of its time with a usage area limited to half an hour and 15 km in distance. The ultracapacitors used could store 8 kWh of energy. A similar solution was introduced in Shanghai in 2010. These buses are constantly charging between the stops and the charging time is only 20 seconds.

Ultracapacitors are used for the operation of camera flashes that require instant energy release. The ultracapacitor connected to the flash is first charged by the battery, then it is activated at the time of shooting and the stored high energy is transmitted to the flash at once. Thus high light is obtained instantaneously.

Increasing carbon dioxide emissions in the world have pushed many car manufacturers to develop hybrid or electric vehicles. Volvo, one of the car manufacturers, used ultracapacitor as an energy-storage element in electric vehicle application. The fact that these carbon fiber and polymer-based ultracapacitors can also be used as body elements has eliminated the need for a battery compartment that causes additional weight.

Rory Handel and Maxx Bricklin, who set off for a wind-powered car design, designed the RORMAx Formula AE, which uses alternative energy, the Formula vehicle. The concept car is intended to cause less harm to nature by using technologies such as ultracapacitor, solar panels, and airflow converter. The vehicle, which is activated with the energy provided by the ultracapacitors, can then increase its charge up to 25 % by using the wind passing over it. It is thought that this car, which is expected to complete 0–100 km acceleration under 4 seconds with the calculations made on paper, can reach 230 km final speed and can go 300 km with 90-minute charge. The prototype was completed in August 2009.

Electric buses in Shanghai can operate without the need for an electric line, thanks to ultracapacitors. Thanks to the fast charging feature of the ultracapacitors, they can only be transported by charging at the stops.

Ultracapacitors have created a solution, for electronic devices, memory of which can be erased in the case of power cuts. Examples of such devices are digital wrist watches, some computer parts, and mobile phones. The ultracapacitors on digital watches and mobile phones are activated when the battery is exhausted, and it prevents the loss of watches and some important information, in particular. When the batteries of some mobile phones are removed and inserted back for a few seconds, it reminds the clock at the opening with the help of ultracapacitors.

It is used for the purpose of creating artificial lightning in the laboratory environment by utilizing the ability of ultracapacitors to discharge suddenly. An artificial lightning is created by bringing together a large number of ultracapacitor blocks. The stored energy is short circuited at a time and directed to a point, thus creating artificial lightning [42–44].

3   Battery terms

The main features that determine the performance of the batteries differ according to the chemical and physical properties of the batteries. Depending on the area of use, these features should be taken into consideration and battery selection should be made accordingly in order to get the most performance from the battery. Battery performance characteristics are examined in the following subsections under the headings of battery capacity, energy, density, power density, efficiency, self-discharge, and battery cycle life.

3.1   Battery capacity

Battery capacity is defined as the total amount of electricity generated due to electrochemical reactions in the battery and is expressed in ampere hours. For example, a constant discharge current of 1 C (5 A) can be drawn from a 5 Ah battery for 1 hour. For the same battery a discharge current of 0.1 C (500 mA) can be withdrawn from the battery for 10 hours. For a given cell type the behavior of cells of different capacities with the same C ratio value is similar. The energy that a battery can deliver in the discharge process is called the capacity of the battery. The unit of the capacity is "ampere hour" and is briefly expressed by the letters "Ah." The label value of the battery is called rated capacity. The capacity of a battery depends on the following factors:

To the number and size of plates in a cell: Essentially, the number of plates or the large size means that the amount of active substance that stores energy increases. The battery's ability to store or deliver energy will increase if the active ingredient in the battery plates is excessive.

To the density of the electrolyte: If a high-density electrolyte is put in a battery, the capacity will increase to a certain extent. However, increasing density also means that the battery life is shorter. Therefore the electrolyte density cannot be increased as desired.

The two factors described earlier are related to the structure of the battery and are assigned to a battery that has been manufactured. Also, the capacity of a battery depends on its age. As the battery is used, the capacity decreases to a certain extent as a result of pouring the active substance from the plates, aging and wearing out of the elements forming the battery.

Electrolyte temperature: The capacity of a battery varies depending on the electrolyte temperature. Capacity increases as the temperature rises. Excessive heat causes wear on lead grates for lead–acid batteries. The worn grill rods come out and break. Therefore despite the capacity-increasing effect, batteries should not be exposed to excessive heat. Fig. 5.24 shows the relationships between the discharge voltage of a battery, discharge current, and discharge capacity [45].

The top curves in Peukert curve indicate the voltage per cell when a battery is discharged within a certain current and time. The capacity curve shows how much of the rated capacity of the battery discharged with a certain current and time should yield. The discharge current curve explains how many amperes it must be discharged to achieve a certain capacity in a given time. The values apply to a new 100-Ah battery cell with full capacity at rated temperature. For a battery group the cell voltage values on the left should be multiplied by the number of cells, and for the batteries other than 100 Ah, the current and capacity values on the right should be taken into account.

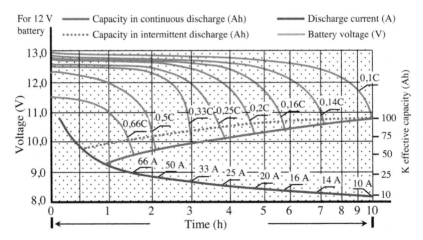

FIGURE 5.24   Peukert curve based on discharge graph.

The higher the batteries are discharged, the lower the battery capacity. For this reason the charge and discharge curves of the batteries are exponential, not linear, as seen in the graphic. The next equation is given in the Peukert law.

$$C_p = I_B{}^k t \tag{5.2}$$

C_p (Ah) is capacity according to Peukert, I_B (A) is battery discharge current, t is discharge time, and k is Peukert constant. Battery manufacturers can give their capacity values in the sales catalogs, and the k value is approximately between 1.1 and 1.3.

According to the Peukert curve, this curve is used for a battery that is less than 10 hours discharge time. In the case of discharges over 10 hours, the discharge capacity is determined by dividing the battery capacity Ah by the current value drawn. For example, if a 100-Ah battery is discharged with 10 A, its voltage drops to 13 V immediately at the start of discharge, and after supplying energy for 10 hours, its voltage becomes 11.3 V.

3.2   Battery state of charge (SOC)

It is important for the user and the health of the battery to learn how much energy is left in a battery. The capacity of the battery is defined by the state of charge (SOC) as a percentage of reference. The amount of remaining energy in the batteries is usually set as SOC instead of giving in Coulomb, kWh, or Ah for easier understanding. The preferred SOC reference should be the SOC reference of the actual battery cell, not the capacity of a new battery cell. The capacity of a battery cell gradually decreases as the battery ages. For example, when a battery cell reaches the end of its life, its actual usable capacity is approaching 80 %. In this case, even if the battery is fully charged, its nominal SOC capacity will be around 80 %. Temperature and deep discharge currents further reduce the effective capacity of the battery. Reference SOC points are of great importance for the user to learn the battery SOC correctly. Battery SOC calculation is given by the following equation:

$$SOC = 1 - \frac{1}{C_B} \int I_B \, d\tau \cdot 100 \tag{5.3}$$

SOC must be between $0 \le SOC(t) \le 100$. C_B is nominal capacity of the battery, I_B is the current drawn from the battery. The battery should be operated within a certain SOC area in order for a battery to be charged/discharged properly, and the number of cycle life increases. The critical SOC operating band range of a battery is given in Fig. 5.25. When the battery is overcharged, it becomes disturbing for battery plates and can cause overheating. This causes the battery cell plates to wear and reduce their cycle life, such as not recharging. Therefore the battery should be operated in a safe SOC area in order to work properly and prolong its life. This safe working area varies according to the

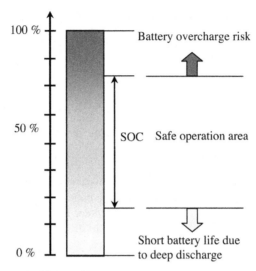

FIGURE 5.25   **Battery's SOC operating range.**

battery type, according to the technical specifications of the battery. Controlled charge and discharge equipment is required to hold the battery in this safe working area. By measuring the current and voltage of the battery real time, the battery operates in this specified SOC area.

Fig. 5.26 gives the working band range according to the lithium-ion battery, and cell voltage is given for the charge and discharge conditions of the SOC safe working area. If the battery leaves the safe working area while it is charging, it causes the cell voltage to increase, to temperature up, and to decrease its

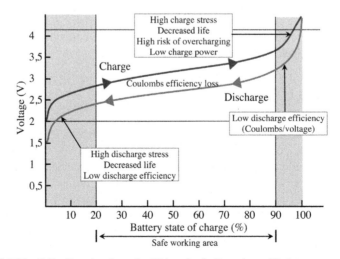

FIGURE 5.26   **Cell voltage band gap for lithium-ion battery charge/discharge.**

life. If the battery leaves the safe working area while discharging, it causes the cell voltage to gradually decrease and damage the plates. In this case the battery cannot hold its first SOC in the next charge [46].

3.3 Effect of temperature on battery

The value of the ambient temperature during discharge is a factor affecting the battery performance. As the ambient temperature decreases, losses in the output voltage increase and capacity decreases. The chemical reaction inside the battery occurs by the battery voltage or temperature. The chemical reaction will be faster in a high-temperature battery. The high temperature can accelerate the chemical reaction and the providing high performance in the battery. But at the same time, undesirable chemical reactions can occur and reduce battery life. The battery's shelf life and charge-keeping feature depend on the self-discharge of the battery. The self-discharge of the battery is an undesirable chemical reaction. Likewise, electrodes reduce the cycle life of battery as a result of negative chemical reactions such as corrosion and gasification. Therefore the temperature affects the battery's shelf life, cycle life, and the battery's charge-keeping ability. Fig. 5.27 gives SOC change in different discharge situations depending on the ambient temperature of a lead–acid battery [47].

3.4 Depth of discharge (DoD)

The amount of current that a battery can deliver during discharge is expressed by the depth of discharge (DoD). Fig. 5.28 gives the graph of the relationship between the number of cycles and DoD logarithmically. In other words, the battery cycle number is the exponential function of DoD. Fig. 5.28 gives the

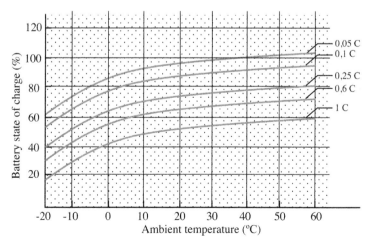

FIGURE 5.27 Change of lead–acid battery capacity by temperature.

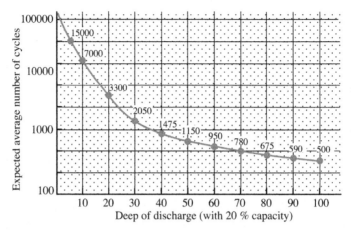

FIGURE 5.28   Cycle number and depth of discharge for lead–acid battery.

relationship between the number of cycles and the DoD for the lead–acid battery type. But it is a typical model for all battery cell chemicals, including lithium-ion batteries with different scaling. This is because the battery life depends on the total energy efficiency that active chemicals can provide. The total energy flow is constant, when the 100 % DoD cycle is calculated as 1 cycle, 50 % DoD as 2 cycles, 10 % DoD as 10 cycles, and 1 % DoD as 100 cycles equivalent. In addition, because the cycle life and the performance of active chemicals decrease due to battery aging, the performance of the battery cells decreases [48].

3.5   Energy density

Energy intensity of a battery is determined as the amount of energy per unit weight or unit volume of the battery. Energy intensity unit is shown in Wh/kg or Wh/L. The energy density of a battery can be calculated in two different ways: theoretical and practical energy density. In theoretical energy density, the calculation is made taking into account the weight or volume of the anode, cathode and electrolyte, which are included only in the electrochemical reaction of the battery. In calculating the practical energy density, the total weight or volume of the battery is taken into account. In this case, depending on the battery type and manufacturing process, there can be up to five times the difference between theoretical and practical energy densities [48].

3.6   Power density

Another feature that characterizes battery performance is battery power density. Power density is determined as the amount of power per unit weight or volume of the battery. Its unit is W/kg or W/L. The power output of a battery depends not only on its chemical properties but also on its design and physical structure.

Therefore there is usually a trade-off between the power density and the energy density during the battery manufacturing phase. For a battery with a high power density, low internal resistance and low electrode polarization are required. In order to achieve this, the amount of materials that do not react electrochemically should increase. This, in turn, increases the volume and weight of the battery, resulting in a reduction in energy density [48].

3.7 Efficiency

The efficiency of a rechargeable battery is calculated as the ratio of the total energy given by the battery to discharge to the total energy received by the battery. This rate may decrease up to 0.6 due to energy losses in the battery. Energy losses in the battery are caused by corrosion in the battery material, gas leakage in the electrolyte, battery internal resistance, and polarization losses in the electrodes. Battery charge numbers and battery life also affect battery efficiency [48].

3.8 Self-discharge

Another feature that characterizes battery performance is the amount of battery discharge when the battery is not used. Depending on the battery type and the ambient temperature, there are differences in the rate of unloaded discharge of the batteries. The battery can be fully discharged due to long-term use, and rechargeable batteries can be completely out of use. Therefore rechargeable batteries should be recharged at certain time intervals [49].

3.9 Cycle life

In rechargeable batteries, gas accumulation and leaks occur in each charge/discharge process, battery electrodes due to corrosion and corrosion, and chemical reactions in the electrolyte. As the number of battery charge/discharge increases, the internal resistance of the battery increases due to such structural defects and losses. Due to polarization losses, large drops in the output voltage occur while the battery is charged. As a result, the efficiency of the battery decreases. The battery becomes unusable after a certain number of charge/discharge cycles. There are differences in the number of battery charge/discharge depending on the battery types [49].

3.10 Lead–acid and LiFePO₄ battery DoD and temperature alarm graphics

The internal resistance of a typical lead–acid and lithium-ion battery is several mΩ (cylindrical-type lithium-ion battery internal resistance 10–50 mΩ, prismatic-type lithium-ion battery internal resistance 0.5–5 mΩ). This resistance value varies due to the chemical process taking place in the battery. Cell voltage and current decrease due to battery resistance change. The resistance value mentioned here is not the $R = V/I$ resistance value we measure in ohms. This resistance for the battery cell is dynamic resistance. Dynamic resistance is defined as $R = \Delta V/\Delta I$. This resistance value, SOC, depends on temperature, battery charge/discharge current, and lifetime. In Fig. 5.29, internal resistance change of the cell according to current, temperature, SOC, and number of cycles is given.

In Fig. 5.29A, the internal resistance is given which is high when the battery charge/discharge current is high. When looking at the change in temperature, it can be seen in Fig. 5.29B that the internal resistance of the battery is high in the low-temperature environment and very low in the high-temperature environment. However, the limit temperature should never be exceeded. Otherwise, it may cause the battery to burst or burn from excessive heat. According to SOC, the internal resistance of the battery is high when the battery is fully charged and fully discharged. The graph showing low internal resistance of the battery in the active working area is given in Fig. 5.29C. Finally, depending on the cycle life of the battery, the internal resistance change is given in Fig. 5.29D. As the battery cell nears the end of its cycle life, it is understood that the internal resistance values are high and the life of the battery has expired [50].

Due to the internal resistance of the lithium-ion cell, the loss of I^2R in the cell is spent as heat. For example, when loaded with 1 C (2.3 A) for the M1 26500 LiFePO$_4$ cell model, its internal resistance is 10 mΩ. While the battery gives energy of $P = 2.3$ A $\times$ 3.2 V $= 7.6$ W, the lost power spent in the cell becomes $P = 2.3^2 \times 10$ m$\Omega = 53$ mW. Battery cell efficiency is between 99.3 % and 98.6 % while the battery is charging and discharging. Current and voltage limits are given in the safe working area of the M1 26500 LiFePO$_4$ in Fig. 5.30 and the Vision 6FM80 lead–acid battery in Fig. 5.31. While the battery cell is charging and discharging, the limit voltage of the cell should not be exceeded for the health of the electrodes [51].

Working temperature limits of M1 26500 LiFePO$_4$ in Fig. 5.32 and the Vision 6FM80 lead–acid battery in Fig. 5.33 are given in the safe working area. While the battery is charging and discharging, the operating temperature should

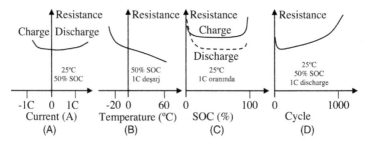

FIGURE 5.29 (A) Current, (B) temperature, (C) SOC, and (D) change of cell internal resistance according to the cycle parameters.

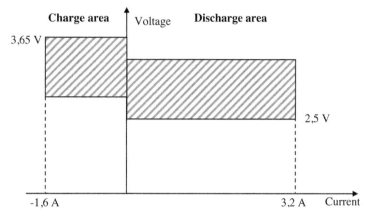

FIGURE 5.30   Safe charge/discharge working zone of LiFePO$_4$ battery.

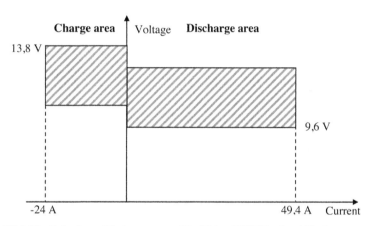

FIGURE 5.31   Safe charge/discharge zone of the Vision 6FM80 lead–acid battery.

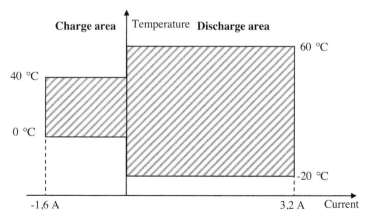

FIGURE 5.32   Safe temperature working zone of LiFePO$_4$ battery.

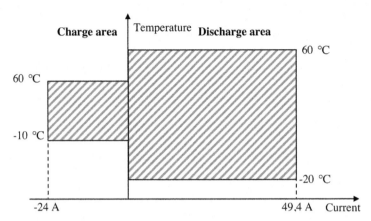

FIGURE 5.33 Safe temperature zone of the Vision 6FM80 lead–acid battery.

be within the limit range to avoid damaging the battery, dangerous situations such as explosion and fire.

4 Ultracapacitor terms

Ultracapacitors have some electrical parameters, just like batteries. These parameters vary depending on the chemical structure. Ultracapacitors with various chemical and physical structures, such as electrical double layer capacitors, pseudo capacitors, and hybrid capacitors, are common terms such as energy, each of which stores internal resistance. The capacity of an ultracapacitor is represented in F. The amount of energy it stores is shown in E_d and is expressed in Wh. The energy stored by an ultracapacitor is expressed as in the following equation [52]:

$$E_d = \frac{1/2\,C_{UC}V_{UC}{}^2}{3600} \tag{5.4}$$

C_{UC} is the capacity of the ultracapacitor, V_{UC} is terminal voltage. The amount of energy stored for the Maxwell brand BMOD0006 E160 B02 model is calculated by the next equation:

$$E_d = \frac{1/2 \cdot 6\,\text{F} \cdot 160^2}{3600} = 21.3\,\text{Wh} \tag{5.5}$$

4.1 Equivalent series resistance (ESR)

The charge and discharge of an ultracapacitor depend on the movement of ions flowing through the separator from the electrodes to the electrolyte. Losses during this movement are measured as DC resistance.

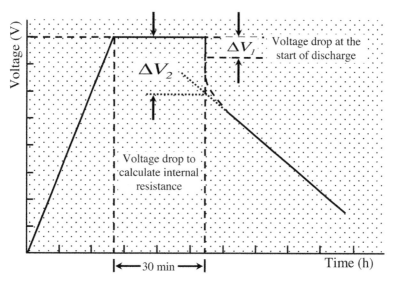

FIGURE 5.34 ΔV_2 voltage changes while calculating the ultracapacitor internal resistance.

The steps in the electrode pores can be shown with *RC* (resistance/capacitor) elements connected in series by making electrical circuit simulation. The increase of load-bearing layers in the pores causes an increase in internal resistance. DC internal resistance increases during charging/discharging depending on time. The internal resistance R_{ESR} can be calculated as in Eq. (5.6) with ΔV_2 voltage drop during discharge and the constant starting current I_{UC} ultracapacitor discharge current [40]. ΔV_2 voltage drop variation is given while calculating the ultracapacitor internal resistance in Fig. 5.34.

$$R_{ESR} = \frac{\Delta V_2}{I_{UC}} \tag{5.6}$$

4.2 Effect of temperature on ultracapacitor

Temperature affects the internal resistance and capacity in the ultracapacitor, as in the batteries. Decreasing the temperature reduces the capacity of the ultracapacitor and the efficiency of the ultracapacitor, because the chemical reaction rate in the separator part is reduced. The increase in temperature does not affect the capacity change of the ultracapacitor to a large extent, but after a certain temperature value it damages the ultracapacitor and separator plates. The effect of temperature on the internal resistance of the ultracapacitor is much more than the capacity change. While the internal resistance does not change at zero degrees and above, the ultracapacitor internal resistance increases at almost twice

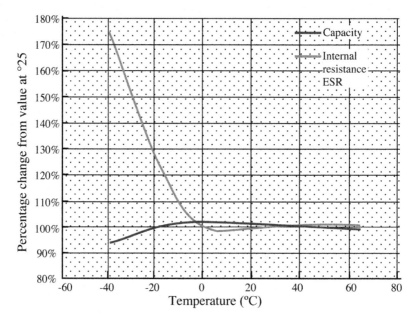

FIGURE 5.35 Internal resistance and capacity change of ultracapacitor with temperature.

at the temperature below zero. Fig. 5.35 is given the variation of the internal resistance and capacity of the ultracapacitor in terms of (%) [53].

References

[1] K. Varshney, P.K. Varshney, K. Gautam, M. Tanwar, M. Chaudhary, Current trends and future perspectives in the recycling of spent lead acid batteries in India, Mater. Today Proc. 26 (2020) 592–602, doi: 10.1016/J.MATPR.2019.12.168.

[2] P.T. Moseley, D.A.J. Rand, A. Davidson, B. Monahov, Understanding the functions of carbon in the negative active-mass of the lead-acid battery: a review of progress, J. Energy Storage 19 (2018) 272–290, doi: 10.1016/J.EST.2018.08.003.

[3] M. Li, J. Yang, S. Liang, H. Hou, J. Hu, B. Liu, et al. Review on clean recovery of discarded/spent lead-acid battery and trends of recycled products, J. Power Sources 436 (2019) 226853, doi: 10.1016/J.JPOWSOUR.2019.226853.

[4] A. Ponrouch, J. Bitenc, R. Dominko, N. Lindahl, P. Johansson, M.R. Palacin, Multivalent rechargeable batteries, Energy Storage Mater. 20 (2019) 253–262, doi: 10.1016/J.ENSM.2019.04.012.

[5] Z.J. Zhang, W. Fang, R. Ma, Brief review of batteries for XEV applications, eTransportation 2 (2019) 100032, doi: 10.1016/J.ETRAN.2019.100032.

[6] P. Goel, D. Dobhal, R.C. Sharma, Aluminum-air batteries: a viability review, J. Energy Storage 28 (2020) 101287, doi: 10.1016/J.EST.2020.101287.

[7] N. Alias, A.A. Mohamad, Advances of aqueous rechargeable lithium-ion battery: a review, J. Power Sources 274 (2015) 237–251, doi: 10.1016/J.JPOWSOUR.2014.10.009.

[8] Q. Liu, H. Wang, C. Jiang, Y. Tang, Multi-ion strategies towards emerging rechargeable batteries with high performance, Energy Storage Mater. 23 (2019) 566–586, doi: 10.1016/J. ENSM.2019.03.028.

[9] Q. Li, Y. Liu, S. Guo, H. Zhou, Solar energy storage in the rechargeable batteries, Nano Today 16 (2017) 46–60, doi: 10.1016/J.NANTOD.2017.08.007.

[10] V. Esfahanian, A.A. Shahbazi, F. Torabi, A real-time battery engine simulation tool (BEST) based on lumped model and reduced-order modes: application to lead-acid battery, J. Energy Storage 24 (2019) 100780, doi: 10.1016/J.EST.2019.100780.

[11] S.U.-D. Khan, Z.A. Almutairi, O.S. Al-Zaid, S.U.-D. Khan, Development of low concentrated solar photovoltaic system with lead acid battery as storage device, Curr. Appl. Phys. 20 (2020) 582–588, doi: 10.1016/J.CAP.2020.02.005.

[12] C.X. Guo, C.M. Li, Molecule-confined FeO_x nanocrystals mounted on carbon as stable anode material for high energy density nickel-iron batteries, Nano Energy 42 (2017) 166–172, doi: 10.1016/J.NANOEN.2017.10.052.

[13] S. Li, K. Li, E. Xiao, R. Xiong, J. Zhang, P. Fischer, A novel model predictive control scheme based observer for working conditions and reconditioning monitoring of Zinc-Nickel single flow batteries, J. Power Sources 445 (2020) 227282, doi: 10.1016/J.JPOW-SOUR.2019.227282.

[14] H. Jia, Z. Wang, B. Tawiah, Y. Wang, C.-Y. Chan, B. Fei, et al. Recent advances in zinc anodes for high-performance aqueous Zn-ion batteries, Nano Energy 70 (2020) 104523, doi: 10.1016/J.NANOEN.2020.104523.

[15] V. Innocenzi, N.M. Ippolito, I. De Michelis, M. Prisciandaro, F. Medici, F. Vegliò, A review of the processes and lab-scale techniques for the treatment of spent rechargeable NiMH batteries, J. Power Sources 362 (2017) 202–218, doi: 10.1016/J.JPOWSOUR.2017.07.034.

[16] Q. Cheng, D. Sun, X. Yu, Metal hydrides for lithium-ion battery application: a review, J. Alloys Compd. 769 (2018) 167–185, doi: 10.1016/J.JALLCOM.2018.07.320.

[17] Y. Xia, L. Xiao, J. Tian, Z. Li, L. Zeng, Recovery of rare earths from acid leach solutions of spent nickel-metal hydride batteries using solvent extraction, J. Rare Earths 33 (2015) 1348–1354, doi: 10.1016/S1002-0721(14)60568-8.

[18] T. Tirronen, D. Sukhomlinov, H. O'Brien, P. Taskinen, M. Lundström, Distributions of lithium-ion and nickel-metal hydride battery elements in copper converting, J. Cleaner Prod. 168 (2017) 399–409, doi: 10.1016/J.JCLEPRO.2017.09.051.

[19] A. Khazaeli, D.P.J. Barz, Fabrication and characterization of a coplanar nickel-metal hydride microbattery equipped with a gel electrolyte, J. Power Sources 414 (2019) 141–149, doi: 10.1016/J.JPOWSOUR.2018.12.074.

[20] M. Galeotti, C. Giammanco, L. Cinà, S. Cordiner, A. Di Carlo, Synthetic methods for the evaluation of the state of health (SOH) of nickel-metal hydride (NiMH) batteries, Energy Convers. Manage. 92 (2015) 1–9, doi: 10.1016/J.ENCONMAN.2014.12.040.

[21] D. Kumar, S.K. Rajouria, S.B. Kuhar, D.K. Kanchan, Progress and prospects of sodium-sulfur batteries: a review, Solid State Ionics 312 (2017) 8–16, doi: 10.1016/J.SSI.2017.10.004.

[22] C. Capasso, D. Lauria, O. Veneri, Experimental evaluation of model-based control strategies of sodium-nickel chloride battery plus supercapacitor hybrid storage systems for urban electric vehicles, Appl. Energy 228 (2018) 2478–2489, doi: 10.1016/J.APENERGY.2018.05.049.

[23] B. Liu, Y. Jia, C. Yuan, L. Wang, X. Gao, S. Yin, et al. Safety issues and mechanisms of lithium-ion battery cell upon mechanical abusive loading: a review, Energy Storage Mater. 24 (2020) 85–112, doi: 10.1016/J.ENSM.2019.06.036.

[24] W. Chen, J. Liang, Z. Yang, G. Li, A review of lithium-ion battery for electric vehicle applications and beyond, Energy Procedia 158 (2019) 4363–4368, doi: 10.1016/J.EGY-PRO.2019.01.783.

[25] X. Han, L. Lu, Y. Zheng, X. Feng, Z. Li, J. Li, et al. A review on the key issues of the lithium ion battery degradation among the whole life cycle, eTransportation 1 (2019) 100005, doi: 10.1016/J.ETRAN.2019.100005.

[26] A. Tomaszewska, Z. Chu, X. Feng, S. O'Kane, X. Liu, J. Chen, et al. Lithium-ion battery fast charging: a review, eTransportation 1 (2019) 100011, doi: 10.1016/J.ETRAN.2019.100011.

[27] D. Zhou, D. Shanmukaraj, A. Tkacheva, M. Armand, G. Wang, Polymer electrolytes for lithium-based batteries: advances and prospects, Chem 5 (2019) 2326–2352, doi: 10.1016/J. CHEMPR.2019.05.009.

[28] G. Yang, Y. Song, Q. Wang, L. Zhang, L. Deng, Review of ionic liquids containing, polymer/inorganic hybrid electrolytes for lithium metal batteries, Mater. Des. 190 (2020) 108563, doi: 10.1016/J.MATDES.2020.108563.

[29] W. Yao, Q. Zhang, F. Qi, J. Zhang, K. Liu, J. Li, et al. Epoxy containing solid polymer electrolyte for lithium ion battery, Electrochim. Acta 318 (2019) 302–313, doi: 10.1016/J. ELECTACTA.2019.06.069.

[30] T.V.S.L. Satyavani, A. Srinivas Kumar, P.S.V. Subba Rao, Methods of synthesis and performance improvement of lithium iron phosphate for high rate Li-ion batteries: a review, Eng. Sci. Technol. Int. J. 19 (2016) 178–188, doi: 10.1016/J.JESTCH.2015.06.002.

[31] C. Liu, J. Lin, H. Cao, Y. Zhang, Z. Sun, Recycling of spent lithium-ion batteries in view of lithium recovery: a critical review, J. Cleaner Prod. 228 (2019) 801–813, doi: 10.1016/J. JCLEPRO.2019.04.304.

[32] N. Bouchhima, M. Gossen, S. Schulte, K.P. Birke, Lifetime of self-reconfigurable batteries compared with conventional batteries, J. Energy Storage 15 (2018) 400–407, doi: 10.1016/J. EST.2017.11.014.

[33] C. Campestrini, P. Keil, S.F. Schuster, A. Jossen, Ageing of lithium-ion battery modules with dissipative balancing compared with single-cell ageing, J. Energy Storage 6 (2016) 142–152, doi: 10.1016/J.EST.2016.03.004.

[34] D. Yuan, J. Zhao, W. Manalastas, S. Kumar, M. Srinivasan, Emerging rechargeable aqueous aluminum ion battery: status, challenges, and outlooks, Nano Mater. Sci. 2 (2019) 248–263, doi: 10.1016/J.NANOMS.2019.11.001.

[35] H. Li, L. Ma, C. Han, Z. Wang, Z. Liu, Z. Tang, et al. Advanced rechargeable zinc-based batteries: recent progress and future perspectives, Nano Energy 62 (2019) 550–587, doi: 10.1016/J.NANOEN.2019.05.059.

[36] L. Zhang, Z. Wang, X. Hu, F. Sun, D.G. Dorrell, A comparative study of equivalent circuit models of ultracapacitors for electric vehicles, J. Power Sources 274 (2015) 899–906, doi: 10.1016/J.JPOWSOUR.2014.10.170.

[37] B. Diarra, A.M. Zungeru, S. Ravi, J. Chuma, B. Basutli, I. Zibani, Design of a photovoltaic system with ultracapacitor energy buffer, Procedia Manuf. 33 (2019) 216–223, doi: 10.1016/J. PROMFG.2019.04.026.

[38] R.B. Choudhary, S. Ansari, B. Purty, Robust electrochemical performance of polypyrrole (PPy) and polyindole (PIn) based hybrid electrode materials for supercapacitor application: a review, J. Energy Storage 29 (2020) 101302, doi: 10.1016/J.EST.2020.101302.

[39] A. Afif, S.M. Rahman, A. Tasfiah Azad, J. Zaini, M.A. Islan, A.K. Azad, Advanced materials and technologies for hybrid supercapacitors for energy storage – a review, J. Energy Storage 25 (2019) 100852, doi: 10.1016/J.EST.2019.100852.

[40] M.S. Masaki, L. Zhang, X. Xia, A hierarchical predictive control for supercapacitor-retrofitted grid-connected hybrid renewable systems, Appl. Energy 242 (2019) 393–402, doi: 10.1016/J. APENERGY.2019.03.049.

[41] D.N. Luta, A.K. Raji, Optimal sizing of hybrid fuel cell-supercapacitor storage system for off-grid renewable applications, Energy 166 (2019) 530–540, doi: 10.1016/J.ENERGY.2018.10.070.

[42] T. Ma, H. Yang, L. Lu, Development of hybrid battery-supercapacitor energy storage for remote area renewable energy systems, Appl. Energy 153 (2015) 56–62, doi: 10.1016/J.APENERGY.2014.12.008.

[43] X. Li, Z. Wang, L. Zhang, Co-estimation of capacity and state-of-charge for lithium-ion batteries in electric vehicles, Energy 174 (2019) 33–44, doi: 10.1016/J.ENERGY.2019.02.147.

[44] M.D. Bhatt, J.Y. Lee, High capacity conversion anodes in Li-ion batteries: a review, Int. J. Hydrogen Energy 44 (2019) 10852–10905, doi: 10.1016/J.IJHYDENE.2019.02.015.

[45] M. Danko, J. Adamec, M. Taraba, P. Drgona, Overview of batteries state of charge estimation methods, Transp. Res. Procedia 40 (2019) 186–192, doi: 10.1016/J.TRPRO.2019.07.029.

[46] P. Krivík, Changes of temperature during pulse charging of lead acid battery cell in a flooded state, J. Energy Storage 14 (2017) 364–371, doi: 10.1016/J.EST.2017.03.018.

[47] T. Nazghelichi, F. Torabi, V. Esfahanian, Prediction of temperature behavior of a lead-acid battery by means of Lewis number, Electrochim. Acta 275 (2018) 192–199, doi: 10.1016/J.ELECTACTA.2018.04.092.

[48] D. Wang, F. Yang, Y. Zhao, K.-L. Tsui, Battery remaining useful life prediction at different discharge rates, Microelectron. Reliab. 78 (2017) 212–219, doi: 10.1016/J.MICROREL.2017.09.009.

[49] P. Wu, G. Shao, C. Guo, Y. Lu, X. Dong, Y. Zhong, et al. Long cycle life, low self-discharge carbon anode for Li-ion batteries with pores and dual-doping, J. Alloys Compd. 802 (2019) 620–627, doi: 10.1016/J.JALLCOM.2019.06.233.

[50] S. Panchal, I. Dincer, M. Agelin-Chaab, R. Fraser, M. Fowler, Experimental and simulated temperature variations in a LiFePO4-20 Ah battery during discharge process, Appl. Energy 180 (2016) 504–515, doi: 10.1016/J.APENERGY.2016.08.008.

[51] E. Ebner, M. Gelbke, E. Zena, M. Wieger, A. Börger, Temperature-dependent formation of vertical concentration gradients in lead-acid-batteries under pSoC operation – Part 2: Sulfate analysis, Electrochim. Acta 262 (2018) 144–152, doi: 10.1016/J.ELECTACTA.2017.12.046.

[52] R. Xiong, H. Chen, C. Wang, F. Sun, Towards a smarter hybrid energy storage system based on battery and ultracapacitor – a critical review on topology and energy management, J. Cleaner Prod. 202 (2018) 1228–1240, doi: 10.1016/J.JCLEPRO.2018.08.134.

[53] C. Wang, H. He, Y. Zhang, H. Mu, A comparative study on the applicability of ultracapacitor models for electric vehicles under different temperatures, Appl. Energy 196 (2017) 268–278, doi: 10.1016/J.APENERGY.2017.03.060.

Chapter 6

Solar Thermal Systems and Thermal Storage

1 Solar thermal systems

Solar energy has been used by humanity since the first years of history. Solar energy was first used for heating and later used for cooking. Thanks to the developing technologies today, solar energy is used in many different ways. Solar energy is directly converted into electrical energy with solar photovoltaic (PV) panels. Another method is to obtain high temperature with concentrated solar energy systems. It is used indirectly with a high heat transfer liquid. This method of use is for electrical energy generation and heating the residential. Concentrated solar energy systems are often used in large-scale applications. Concentrated solar energy systems do not require high technology like solar PV panels [1–3]. It consists of reflectors and collectors with its general structure. The transport of heat is carried out with molten materials (molten salt) that can hold high energy density. The purpose of this is to increase system performance and efficiency. Electric energy generation topology with concentrated solar energy system is given in Fig. 6.1. In these systems the solar irradiation is concentrated and collected in a region. Then, this high-temperature fluid is transferred to the turbine system. Turbine is mechanically rotated with the high-temperature liquid converted into steam phase. Thus solar energy is converted into electrical energy indirectly. These systems are dependent on solar energy and solar energy systems cannot generate power during shadow and night hours. High thermal energy storage systems or auxiliary heaters are used to ensure the continuity of energy in these systems. One of the design parameters in these systems is determining the correct temperature. The correct temperature affects the turbine efficiency and performance that will operate in the solar system. There are many application methods in solar energy systems [4]. The high temperature is obtained in the first method, while low temperature is obtained in the second method. These methods are parabolic trough collectors, linear Fresnel, parabolic dish collector, and central tower power system [5,6]. Concentrated solar energy systems are explained in detail in Chapter 1, Solar System Characteristics, Advantages, and Disadvantages.

The solar irradiation is collected at a single point with reflectors in concentrated solar energy systems. Later, as in a traditional power plant, electrical

Solar Hybrid Systems. http://dx.doi.org/10.1016/B978-0-323-88499-0.00006-9

127

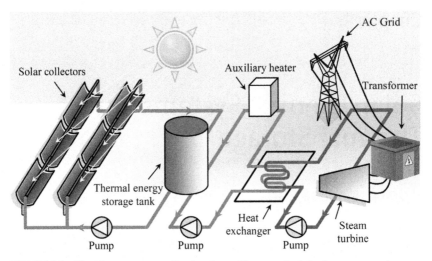

FIGURE 6.1 Electric power generation topology with concentrated solar energy system.

energy is produced by generators using high steam. The purpose of concentrated solar energy systems is to capture a lot of solar irradiation. In these solar energy systems, more space is usually needed. Unproductive areas should be preferred during the selection of these areas. The most widely used method in terms of system efficiency and performance is parabolic trough solar energy systems. In addition, parabolic trough solar energy systems are long-lasting and have low maintenance requirements. Parabolic trough solar energy systems are used with energy storage methods. In these systems, molten salts are used as heat transfer liquid. The heat storage capacity of molten salts is much higher than normal liquids. Thus long-term high-temperature energy storage is possible [7,8]. A large amount of energy is stored from the batteries with this energy storage method. Auxiliary heaters can be used to maintain the temperature of the stored heat energy. The energy source of these auxiliary heaters is renewable energy sources or thermal energy sources. Electrical energy needs can be met at any time with thermal energy storage solar systems. Since the amount of stored heat energy is in high volumes, both long-term and high-power demands are met with thermal energy storage systems [9].

2 Thermal energy storage

Due to the increase in energy need, humankind has caused the search for alternative energy sources. The use of energy resources that do not harm the environment and take their energy from nature is becoming widespread. Energy sources such as sun, wind, ocean, and biogas are among the most preferred renewable energy sources. Energy storage systems are seen as the most suitable solution due to the fluctuation of the power output of renewable energy sources.

The output of solar PV panels and wind energy are intermittent and unstable. Many energy storage methods are used to prevent this power fluctuation. Thermal energy storage method is used especially in solar energy systems. Thermal energy storage systems have a complementary feature for renewable energy sources like other energy storage systems. During periods of intense solar energy, solar energy is charged in the form of thermal. Then, when the system load demands power, the stored thermal energy is used with the help of steam turbines and discharged. The solar energy system, which has a thermal energy storage system in low solar irradiation times, has many benefits. High-capacity thermal energy storage systems are installed with low investment costs. In addition, the efficiency of thermal energy storage systems is high [10]. Thermal energy storage systems can meet high power demands.

Due to all these advantages, thermal energy storage systems become an indispensable part of renewable energy sources. Thermal energy storage systems can be evaluated for use in heating and cooling applications. Thermal energy storage systems are generally used in residential and industrial application areas. Accessibility and reliability are one of the important advantages of thermal energy storage systems. Thermal energy storage systems do not leave any negative environmental impact during the charging and discharging processes. Thanks to the high-power density of thermal energy storage systems, its use in the industrial area is increasing. Research continues on heat transfer materials in order to increase efficiency in thermal energy storage systems [11,12].

Studies are carried out to increase the amount of energy stored per unit volume or per unit mass in thermal energy storage systems [13]. The purpose of these studies is to increase the power density of the stored energy. Phase-change materials are one of the most important factors in thermal energy storage systems. Thus thermal energy storage system performance is increased and thermal energy storage volume is reduced. High steam pressure and corrosive heat carriers affect system performance and reliability in thermal energy storage systems.

Heat can be stored and used in two different ways in thermal energy storage systems. The first of these is that the hot water coming from the collectors is cooled by absorption systems and stored in cold form. The other can be directly stored in hot form. Heating and cooling processes in residential and industrial application areas are generally carried out using natural gas or electricity. Thermal energy storage systems are used to meet these needs. Solar energy is used in the most basic form for heating hot water in houses. This hot water requirement can be met by absorbing tubes and a thermal energy storage tank. The stored hot water is distributed to the house with the help of pipes [14].

There are two different types as thermal energy storage mechanism. One of them is carried out using solid or liquid materials. In the other thermal energy storage mechanism, the phase change process is used. The collector life, system cycle number, storage volume, and heat leaks should be considered in storage design. To improve the thermal energy storage system performance, attention should be paid to the thermal capacitance effect of the storage tank.

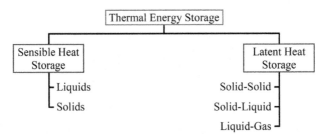

FIGURE 6.2   Thermal energy storage types.

The heat transfer liquid in the collectors must have a very high thermal capability. Although this heat transfer liquid varies according to the application type, generally, water, oil, water–salt, and water–glucose mixtures are used. The reason why water is used in thermal energy storage technologies is that it is low cost, environment friendly, and has a high thermal energy storage capacity. In addition, water is easy to transport with pumps and conventional pipes. It does not require a special installation for the thermal loop to circulate in the system. Another factor affecting thermal energy system efficiency is the location of the storage tank. The storage tank should be located where the system thermal losses are lowest. Placing the storage tank away from the solar system will cause thermal losses during liquid transfer. The tank should be located in a place that is not affected by environmental conditions [15]. The tank must be insulated with a well-insulated outer protection.

Thermal energy storage systems are designed to be economical and with high power density. During system design, the optimum storage unit can be produced by using fluid dynamic methods and using simulation programs. Thermal energy storage types are given in Fig. 6.2. The amount of energy stored in thermal energy storage systems is defined as capacity. This storage capacity varies depending on the storage process, system sizes, and thermal storage process. The ratio of the temperature reaching the user in the thermal energy storage system and the amount of solar energy coming to the collectors gives the total efficiency of the system. Losses in thermal energy storage systems generally occur as heat losses [16]. It is desirable to store the thermal temperature in the storage system as soon as possible. The high energy storage density in the charge and discharge processes is a desired property in thermal energy storage systems. Thermal energy storage systems are carried out by two methods, thermal and chemical. It can be grouped into sensible heat and latent heat.

2.1   Sensible heat energy storage

Sensible heat energy storage method is based on the principle of heating or cooling a liquid or solid material. The material to be used as heat energy storage is low cost. This method is often used for heating a residential or industrial application area. The charging and discharging processes of the sensitive heat energy storage system use the change in the heat capacity of material. In this method, no matter how well the storage system is isolated, there are always some losses. If the duration of the stored energy is long, the thermal insulation of the system should be done very well. The heat loss is directly proportional to the surface area and thermal heat storage material. Thermal energy storage systems are used using air, liquid, and underground storage in this method [17].

2.1.1   Air thermal energy storage system

Air–thermal energy storage, one of the sensible thermal energy storage methods, is widely used. Solid materials are used as hot air collectors. These solid materials are often inexpensive items that are easy to access such as rocks, gravel, and stones. The hot air is used in order to increase the internal temperature of large buildings. Air–thermal energy storage systems are one of the cost-effective methods. The heated air coming from the heater can be used directly. Using gravel material, an air–thermal energy storage system can be installed with an insulated case and air inlet/outlet assembly. Many different designs are used in this method. The case walls can usually be products such as concrete or wood. The air inlet in the case can be horizontal or vertical. High efficiencies are obtained in both designs. The gravel dimensions, air flow, and case geometry used in this method are determined at the design stage. Phase-change materials and water are used in air–thermal energy storage systems. Phase-change materials and water have high thermal energy storage capacities. Air flow can be achieved with the structure designed using this thermal energy storage capacity. These designs increase the universality of use as they add functionality. Systems using water and phase-change materials are very compact. In addition, systems with water and phase-change materials are one-third lower than systems produced with gravel [18].

2.1.2   Liquid thermal energy storage system

The most commonly used material in liquid energy storage is water. Hot water tanks are used as thermal energy storage. Hot water tanks are cost-effective and their performance is high. In this technology, studies are carried out on tank insulations in order to increase the thermal insulation efficiency [19]. Hot water tanks in liquid thermal energy storage systems are of two types, pressure and unpressurized. These systems include internal or external heat exchangers. Hot water tanks are made of galvanized, copper, and concrete materials. High insulation is an important parameter for all tank types. In pressure storage systems with heat exchangers, the exchanger is realized with a coil bundle. More than one hot water tank is used to increase the heat storage capacity. Thus the reliability of the system increases and the system continues to work in the case of fault. An additional pump is required for liquid circulation in external heat

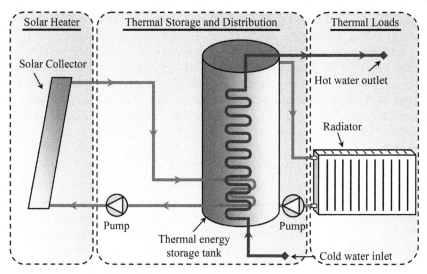

FIGURE 6.3 A solar collector structure with thermal energy storage.

exchanger systems. The liquid thermal energy storage systems are generally designed unpressurised up to 30 m^3. If the volume increases, it will be more appropriate to switch to a pressure design in terms of costs. The mains' water and pressure system fluid are independent and isolated from each other in pressure heat storage systems. The heat transfer process is carried out by using heat exchangers in pressure systems. Fig. 6.3 shows a solar collector structure with thermal energy storage. In this system the liquid is heated by solar collectors and heat transfer is carried out with a heat exchanger. In these designs, hot water is usually taken from the top of the tank. The heated liquid is at the top, the cold liquid is at the bottom. Thus the cold liquid is heated by solar collectors. In this method the heat performance of the tank increases and hot water is always available. As a result, it increases the efficiency of the liquid thermal energy storage system.

In liquid thermal energy storage systems, the stored energy capacity is calculated by the next equation, which is valid for water or any other liquid:

$$Q_{storage} = m_{liq}c_p\Delta t_s \tag{6.1}$$

where $Q_{storage}$ is the heat capacity stored in t_s unit of time, m_{liq} is the mass of water or liquid, and c_p is the specific heat of water or liquid.

High heat capacity, easy installation, wide usage area, low cost, and renewable energy are among the advantages of these systems. These methods allow the installation of heat storage systems even in small scales. Thermal heat storage systems with solar collectors generally have seasonal usage characteristics. It is not possible to obtain hot water with these systems in winter [20].

2.1.3   Underground thermal energy storage system

Underground thermal energy storage systems are the most widely used methods to provide hot and cold storage. In this thermal energy storage system, rocks, soil, sand, and clay are used as storage areas. Pipes are placed to change the temperature of the thermal energy storage area. Heat transfer fluid is pumped into these drill pipes and the desired temperature is achieved. There is no thermal insulation in underground thermal energy storage systems. In these systems, only the floor is insulated. In this method the storage material such as soil, rock, or sand performs heat transfer. These thermal energy storage systems are used in large-volume applications. Since the application volume is large, heat losses are reduced. These thermal energy storage methods are mostly used outside of residential areas. The ground in the application area should be suitable for the placement of drill pipes. These systems are also used for cooling large residential and industrial application areas during the summer months. In this use, cold energy storage is carried out by injecting cold fluid into the drill pipes. Another technique used in underground storage systems is cave or pit storage. Hot or cold water is pumped into these large underground caves or pits. Pumped water is discharged for later use. In this technique, heat losses occur at the point of contact with the water on the surface of the underground cave or pit. A natural underground cave or pit is generally preferred to use this technique [21,22].

2.2   Latent heat energy storage system

Latent heat energy storage system or phase change arises from the isothermal process in the storage material. In this storage technology, material density increases and volume decreases during energy storage. During the heat phase change process, heat exchange occurs directly inside the material. The latent heat energy storage method is the most effective way of thermally storing heat. High energy storage density is an advantage in this method. In this thermal energy storage method, the storage material can be solid–liquid, liquid–gas, and vice versa. The heat absorption or release occurs in this phase change process. The simplest example of a phase change is water [23,24]. Water is solid at low temperature, liquid at medium temperature, and gas at high temperature. Fig. 6.4 shows the solar collector system with the latent heat energy storage. The part where latent heat is used is in the steam turbine part in this system. The water passes from liquid phase to gas phase.

In the latent heat energy storage system, the melting point of the phase-change material should be in the desired temperature range. In this method, there must be a suitable heat exchange surface for the phase change process. The phase-change material must be compatible with the tank in which the heat will be stored. Materials and equipment used in the thermal energy storage system should not contain fire hazards [25].

Phase-change materials are classified as organic, inorganic, and eutectic. Organic phase-change materials can melt and solidify multiple times. A little

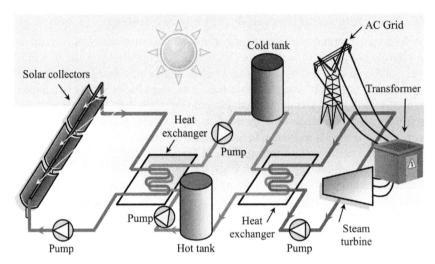

FIGURE 6.4 Solar collector systems with the latent heat energy storage.

crystallization occurs at high heat transitions. As an organic phase-change material, paraffin waxes can be mentioned. Paraffin waxes release a large amount of heat during the crystallization process. Paraffin material is safe, inexpensive, and noncorrosive and can be used in a wide temperature range. Examples of nonparaffin organic materials include esters, fatty acids, and alcohols. These organic materials have high fusion temperatures and low thermal conductivity. Inorganic phase-change materials are generally used in high-temperature solar collector applications. One of the challenges of these inorganic materials is that they require constant maintenance. Inorganic materials freeze at low temperatures and are difficult to use at high temperatures. Salt hydrate can be given as an example of inorganic phase-change materials. Salt hydrates react with water at its melting point to disintegrate hydrate crystals. Salt hydrates have a high heat fusion per unit volume. Salt hydrates have low corrosive properties and are suitable for plastic tanks. Organic phase-change materials are flammable and have low thermal conductivity. On the other hand, inorganic phase-change materials are low-cost, abundant, nonflammable, and have high heat storage capacity. Eutectic phase-change materials are obtained by combining materials with both melting and freezing points. Eutectic materials have high thermal conductivity and their material density is quite high. Eutectic materials have low specific heat capacity [26].

Fig. 6.5 shows the temperature change graph in latent heat and sensible heat methods. When heat is added to a solid material, the temperature of the material increases until a phase change occurs. The molecular bonds break during phase change in solid material. Solid material passes from solid to liquid. If the amount of heat applied to the material increases after the phase change takes place, the amount of heat stored by the material also increases.

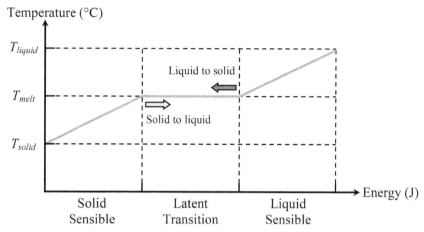

FIGURE 6.5   **Temperature change graph in latent heat and sensible heat methods.**

2.3   Thermal energy storage systems economy, cost

Thermal energy storage systems are very cost-effective compared to other storage technologies. In terms of environmental impact, it is a clean energy storage technology. Thermal energy storage systems are a suitable storage method for large buildings. Thermal energy storage systems are generally used in small-scale applications for hot water and heating. It is also used in the field of electrical energy generation in large-scale applications. Thus electricity generation is realized with the stored heat energy during periods of intense power demand [27]. Thanks to thermal energy storage systems, short-term high load demands can be responded quickly. Storage is made at high temperatures in thermal energy storage systems. While electricity is produced with high temperature, residential heating can be performed with the heat at the turbine outlet. Thus every process of thermal transformation is utilized. Thermal energy storage systems have low initial investment and maintenance costs. Thermal energy storage systems should be specially designed according to the application area. Compressor, pump, storage tank, and distribution lines are installed according to the application area requirement. Optimum thermal energy storage feasibility for the application site is achieved with a rational design [28].

The peak loads in a region can be met with thermal energy storage systems. The thermal energy storage unit operator determines the peak load times and activates the system, thus reducing the load on the grid. In addition, the cost of electricity at high tariff rates is reduced by using a thermal energy storage system. On the contrary to this process, heat is added to the thermal energy storage system in the period when electricity prices are low. There is no fuel cost in thermal energy storage systems with solar collector since the energy source is solar [29].

Thermal energy storage systems are most commonly used to heat or cool a particular area. It is preferred for the water heating in residential or industrial application areas. Thermal energy storage is widely used in agricultural application, especially in greenhouses. It is also used in water pumping systems in the agriculture. Other areas of use are small power plants.

References

[1] I. Baniasad Askari, M. Ameri, A techno-economic review of multi effect desalination systems integrated with different solar thermal sources, Appl. Therm. Eng. 185 (2020) 116323, doi: 10.1016/j.applthermaleng.2020.116323.

[2] W. Lipiński, E. Abbasi-Shavazi, J. Chen, J. Coventry, M. Hangi, S. Iyer, et al. Progress in heat transfer research for high-temperature solar thermal applications, Appl. Therm. Eng. 184 (2021) 116137, doi: 10.1016/j.applthermaleng.2020.116137.

[3] Y. Zheng, R.C. Gonzalez, M.C. Hatzell, K.B. Hatzell, Concentrating solar thermal desalination: performance limitation analysis and possible pathways for improvement, Appl. Therm. Eng. 184 (2021) 116292, doi: 10.1016/j.applthermaleng.2020.116292.

[4] L. Shi, X. Wang, Y. Hu, Y. He, Y. Yan, Solar-thermal conversion and steam generation: a review, Appl. Therm. Eng. 179 (2020) 115691, doi: 10.1016/j.applthermaleng.2020.115691.

[5] S. Khajepour, M. Ameri, Techno-economic analysis of a hybrid solar thermal-PV power plant, Sustain. Energy Technol. Assess. 42 (2020) 100857, doi: 10.1016/j.seta.2020.100857.

[6] M.J. Montes, J.I. Linares, R. Barbero, A. Rovira, Proposal of a new design of source heat exchanger for the technical feasibility of solar thermal plants coupled to supercritical power cycles, Sol. Energy 211 (2020) 1027–1041, doi: 10.1016/j.solener.2020.10.042.

[7] D. Tschopp, Z. Tian, M. Berberich, J. Fan, B. Perers, S. Furbo, Large-scale solar thermal systems in leading countries: a review and comparative study of Denmark, China, Germany and Austria, Appl. Energy 270 (2020) 114997, doi: 10.1016/j.apenergy.2020.114997.

[8] C. Wang, W. Li, Z. Li, B. Fang, Solar thermal harvesting based on self-doped nanocermet: structural merits, design strategies and applications, Renew. Sustain. Energy Rev. 134 (2020) 110277, doi: 10.1016/j.rser.2020.110277.

[9] F.S. Javadi, H.S.C. Metselaar, P. Ganesan, Performance improvement of solar thermal systems integrated with phase change materials (PCM), a review, Sol. Energy 206 (2020) 330–352, doi: 10.1016/j.solener.2020.05.106.

[10] R. Li, Y. Zhang, H. Chen, H. Zhang, Z. Yang, E. Yao, et al. Exploring thermodynamic potential of multiple phase change thermal energy storage for adiabatic compressed air energy storage system, J. Energy Storage 33 (2021) 102054, doi: 10.1016/j.est.2020.102054.

[11] Z. Ma, M.J. Li, K.M. Zhang, F. Yuan, Novel designs of hybrid thermal energy storage system and operation strategies for concentrated solar power plant, Energy 216 (2021) 119281, doi: 10.1016/j.energy.2020.119281.

[12] Z. Li, Y. Lu, R. Huang, J. Chang, X. Yu, R. Jiang, et al. Applications and technological challenges for heat recovery, storage and utilisation with latent thermal energy storage, Appl. Energy 283 (2020) 116277, doi: 10.1016/j.apenergy.2020.116277.

[13] B. Nie, A. Palacios, B. Zou, J. Liu, T. Zhang, Y. Li, Review on phase change materials for cold thermal energy storage applications, Renew. Sustain. Energy Rev. 134 (2020) 110340, doi: 10.1016/j.rser.2020.110340.

[14] G. Mauger, N. Tauveron, Modeling of a cold thermal energy storage for the flexibility of thermal power plants coupled to Brayton cycles, Nucl. Eng. Des. 371 (2021) 110950, doi: 10.1016/j.nucengdes.2020.110950.

[15] M. Barthwal, A. Dhar, S. Powar, The techno-economic and environmental analysis of genetic algorithm (GA) optimized cold thermal energy storage (CTES) for air-conditioning applications, Appl. Energy 283 (2020) 116253, doi: 10.1016/j.apenergy.2020.116253.

[16] B. Koçak, A.I. Fernandez, H. Paksoy, Review on sensible thermal energy storage for industrial solar applications and sustainability aspects, Sol. Energy 209 (2020) 135–169, doi: 10.1016/j.solener.2020.08.081.

[17] H. Chen, Y. Wang, J. Li, B. Cai, F. Zhang, T. Lu, et al. Experimental research on a solar air-source heat pump system with phase change energy storage, Energy Build. 228 (2020) 110451, doi: 10.1016/j.enbuild.2020.110451.

[18] A. Gautam, R.P. Saini, A review on sensible heat based packed bed solar thermal energy storage system for low temperature applications, Sol. Energy 207 (2020) 937–956, doi: 10.1016/j.solener.2020.07.027.

[19] L. Hüttermann, R. Span, P. Maas, V. Scherer, Investigation of a liquid air energy storage (LAES) system with different cryogenic heat storage devices, Energy Procedia 158 (2019) 4410–4415, doi: 10.1016/j.egypro.2019.01.776.

[20] D. Salomone-González, J. González-Ayala, A. Medina, J.M.M. Roco, P.L. Curto-Risso, A. Calvo Hernández, Pumped heat energy storage with liquid media: thermodynamic assessment by a Brayton-like model, Energy Convers. Manage. 226 (2020) 113540, doi: 10.1016/j.enconman.2020.113540.

[21] H. Hasan Ismaeel, R. Yumrutaş, Investigation of a solar assisted heat pump wheat drying system with underground thermal energy storage tank, Sol. Energy 199 (2020) 538–551, doi: 10.1016/j.solener.2020.02.022.

[22] A.N. Mustapha, H. Onyeaka, O. Omoregbe, Y. Ding, Y. Li, Latent heat thermal energy storage: a bibliometric analysis explicating the paradigm from 2000–2019, J. Energy Storage 33 (2021) 102027, doi: 10.1016/j.est.2020.102027.

[23] G. Shen, X. Wang, A. Chan, F. Cao, X. Yin, Investigation on optimal shell-to-tube radius ratio of a vertical shell-and-tube latent heat energy storage system, Sol. Energy 211 (2020) 732–743, doi: 10.1016/j.solener.2020.10.003.

[24] X. Zhou, Y. Xu, X. Zhang, D. Xu, Y. Linghu, H. Guo, et al. Large scale underground seasonal thermal energy storage in China, J. Energy Storage 33 (2021) 102026, doi: 10.1016/j.est.2020.102026.

[25] B. Gürel, Thermal performance evaluation for solidification process of latent heat thermal energy storage in a corrugated plate heat exchanger, Appl. Therm. Eng. 174 (2020) 115312, doi: 10.1016/j.applthermaleng.2020.115312.

[26] R. Karami, B. Kamkari, Experimental investigation of the effect of perforated fins on thermal performance enhancement of vertical shell and tube latent heat energy storage systems, Energy Convers. Manage. 210 (2020) 112679, doi: 10.1016/j.enconman.2020.112679.

[27] Q. Zhou, D. Du, C. Lu, Q. He, W. Liu, A review of thermal energy storage in compressed air energy storage system, Energy 188 (2019) 115993, doi: 10.1016/j.energy.2019.115993.

[28] J. Lv, Z. Wei, J. Zhang, Running and economy performance analysis of ground source heat pump with thermal energy storage devices, Energy Build. 127 (2016) 1108–1116, doi: 10.1016/j.enbuild.2016.06.072.

[29] A. Saxena, P. Verma, G. Srivastava, N. Kishore, Design and thermal performance evaluation of an air heater with low cost thermal energy storage, Appl. Therm. Eng. 167 (2020) 114768, doi: 10.1016/j.applthermaleng.2019.114768.

Chapter 7

Hybrid Energy Storage and Innovative Storage Technologies

1 Hybrid energy storage systems

Due to the nature of renewable energy sources, their output power varies depending on natural conditions. Energy storage systems are the most suitable solution for the intermittent output power characteristic of renewable energy sources. The high power and energy density of energy storage systems supports the use of renewable energy sources. Sources and energy storage must be used together if a reliable renewable energy power system is to be established. Thanks to today's technological developments, energy storage units can supply 1–100 MW power for a long time [1–3].

Energy storage systems mainly consist of four types: electromagnetic, electrochemical, mechanical, and thermal. The classification is made according to material technology and methods used in energy storage units.

The energy storage technology desired to be used in an energy production facility should be determined by considering the performance–cost parameter. Systems established to meet energy needs must meet the long-term energy need as well as sudden load demands [4]. Technologies that respond to power needs are energy storage applications that are used specially to improve power quality. Energy storage applications are used to ensure the constant of the grid frequency, to prevent sudden load changes and voltage fluctuations. In these dynamic power applications, while the response time should be in the millisecond level, the required load power demand may be in the minutes or hours [5–7].

The capacities of energy storage systems can meet the needs in various time intervals such as seconds, minutes, hours, and days/months. The unique structures and characteristics of energy storage systems create these time periods. The usage areas of different energy storage systems, the amount of power they can store, and their durations are compared in Fig. 7.1.

Compressed air and pumped hydroelectric energy storage systems are a storage technology that can meet the load requirement on a day/month basis. These high energy density storage systems are used to support the grid. In addition, superconducting magnetic and hydrogen energy storage systems are also used to support the grid [8]. However, in terms of energy transfer times, the

Solar Hybrid Systems. http://dx.doi.org/10.1016/B978-0-323-88499-0.00007-0

139

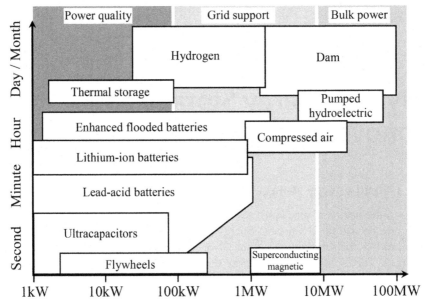

FIGURE 7.1   The comparisons of energy storage technology.

superconducting magnetic energy storage system can provide support for a short time (second), while the hydrogen energy storage system can provide energy on a day/month basis. Fluid batteries are used in long-term and high-energy systems. Nickel–cadmium, lead–acid, and sodium–sulfur battery technologies are used in areas where high power density is required [9]. Lithium-ion batteries are widely used in portable device applications that require high energy density and in applications where energy storage efficiency is important. The ultracapacitor energy storage system is generally preferred in power quality improvement studies. Flywheels and ultracapacitors have higher power density compared to batteries. They can be charged or discharged very quickly at high currents. However, flywheels and ultracapacitors do not transfer energy for a long time as they have low energy density compared to batteries [10,11].

1.1   Hybrid energy storage with battery and ultracapacitor

A hybrid energy storage system is created with the energy storage unit obtained by using two or more energy storage systems together. These energy storage units reveal a complementary structure with their performance characteristics. Battery: it is a more cost-effective energy storage technology compared to other storage types. In addition to being able to provide energy for a long time, batteries can also meet sudden power needs (limited discharge current rates). However, as the batteries are deeply discharged in the case of sudden power demands, their cycle life decreases rapidly [12]. While the energy storage process

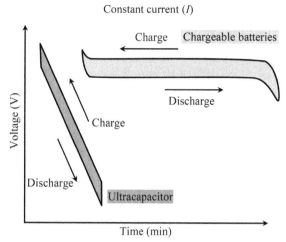

FIGURE 7.2   Charge/discharge of battery and ultracapacitor graph.

takes place electrochemically in a battery, energy storage in the ultracapacitor occurs with a static charging event. Because of these features, ultracapacitors are an energy storage unit with higher power density compared to batteries. Due to these features of the battery and the ultracapacitor, it would be reasonable to combine these two storage technologies for high power and energy needs. The battery and ultracapacitor can be charged/discharged simultaneously thanks to the power electronics converters [13]. The time-dependent voltage change rates of the battery and ultracapacitor storage units are given in Fig. 7.2. While the battery provides energy for a long time, the ultracapacitor can provide energy in a short time.

Hybrid energy storage system consisting of battery and ultracapacitor will be the most appropriate solution to meet the load requirement when instantaneous and long-term power is required on the demand side. Batteries have a power density of 10–1000 W/kg with today's technology, while ultracapacitors have a power density of 10–106 W/kg. When the hybrid energy storage system is used, the ultracapacitor will support it by meeting the demand power that the battery characteristic is insufficient. Thus the cycle life of the batteries that are not operated at high discharge currents will be extended [14,15].

Ultracapacitors or supercapacitors are electric double-layer capacitors. Ultracapacitors are electrochemical storage elements that can allow high charge and discharge currents. High charge/discharge current rates have become a remarkable field of study, especially for electric vehicles, which are becoming widespread today. They are used with batteries in renewable energy sources and distributed energy applications. Ultracapacitors have high costs compared to other battery devices. If the costs of ultracapacitors decrease, their use will become widespread. Ultracapacitors have high power densities and high charge/discharge cycles [16–18]. Electrode types used in ultracapacitors have carbon technology and are different from ordinary capacitors. Thanks to this technology, a larger surface area is provided in the electron flow. Although ultracapacitors store very low energy compared to batteries, their cycle life is quite high. These specifications make ultracapacitors the energy storage unit of the future. Studies are carried out on ultracapacitors within the scope of longer charge/discharge times. High and fast charge/discharge currents will expand the use of ultracapacitors in distributed grid infrastructure in the future. In other possible situations, there will be long-term load demands. In such a case, it will not be sufficient to use only a low energy density ultracapacitor unit. Batteries have 100 times more energy density compared to ultracapacitors due to their structure [19,20]. Therefore it would be a suitable solution to use a high energy density battery unit together with the ultracapacitor unit.

Batteries are used in many different applications. These energy storage technologies have some disadvantages. Batteries have a limited depth of discharge and their cycle life decreases depending on the amount of use. Ultracapacitor energy storage units do not contain this negative specification. At the same time, ultracapacitors have fast response times. In addition, ultracapacitors are more resistant to high temperature and vibration than batteries. Ultracapacitors are widely used in transportation areas such as electric underground trains. Energy transmission lines have been eliminated by integrating the ultracapacitor to the underground train. Trains charge quickly during the passenger drop-off and pickup. It can move forward with the energy stored to the next stop. Ultracapacitors are used in braking times in electric vehicles. When vehicles with large masses are stopped in a short time, high powers arise instantly. This high-power value cannot be stored in batteries in a short time. But ultracapacitors can store this energy with their characteristic structures. Thus they contribute to energy efficiency in the field of transportation [21–24].

Solar photovoltaic (PV) panels and wind energy are renewable energy sources with intermittent and time-dependent variable output characteristics. A hybrid energy storage system with battery and ultracapacitor can ensure the continuity of energy. The hybrid energy storage system helps to increase the cycle life of the batteries and to be used at maximum performance. Especially in off-grid power generation systems, the continuity of energy is provided [25]. The application topology of the hybrid energy storage unit to the solar system is given in Fig. 7.3.

A good system performance is ensured by creating a hybrid energy storage system consisting of battery and ultracapacitor. By using the hybrid energy storage system consisting of battery and ultracapacitor, both the volumetric dimensions of the battery can be reduced and a longer charging state can be achieved. Ultracapacitors have a high-power density compared to batteries. Therefore ultracapacitors can meet the high-power demand in a shorter time. Batteries have a higher energy density than ultracapacitors. These features can meet the long-term energy needs of the batteries. In the hybrid energy storage system, the ultracapacitor meets the sudden power demand, while the battery meets the long-term or continuous power demand [26,27].

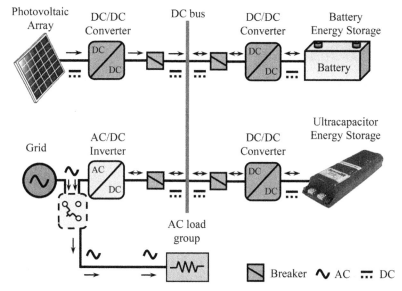

FIGURE 7.3 The topology of hybrid energy storage and solar PV panel system.

1.2 Ultracapacitor EPR calculation experiment

When the ultracapacitor is not used, its energy decreases by itself and the voltage value decreases to zero. This situation is represented in the ultracapacitor by an equivalent parallel resistance (EPR). EPR calculation experiment of Maxwell brand BMOD0006 E160 B02 model ultracapacitor was performed in this chapter. First, the ultracapacitor was charged at a voltage level of 15 V. Then, the voltage value was recorded with certain time periods. These period intervals have been chosen as 10 minutes. As indicated in Fig. 7.4, voltage values were recorded for 1734 minutes.

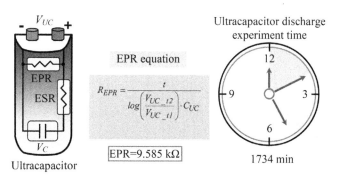

EPR= Equation Parallel Resistance
ESR= Equation Serial Resistance

FIGURE 7.4 The experiment of ultracapacitor EPR.

The voltage value decreasing over time due to the EPR in the ultracapacitor is calculated by the next equation:

$$V_{UC\_t2} = V_{UC\_t1} \cdot e^{-\frac{t}{R_{EPR} \cdot C_{UC}}} \tag{7.1}$$

where V_{UC_t2} is the voltage value of the ultracapacitor read at time t_2, V_{UC_t1} is the voltage value of the ultracapacitor at t_1, t is total experiment time, R_{EPR} is ultracapacitor equivalent parallel resistance, and C_{UC} is the total capacity amount of the ultracapacitor in F. This EPR value found in the ultracapacitor electrical circuit is calculated by the next equation:

$$R_{EPR} = \frac{t}{\log\left(\dfrac{V_{UC\_t2}}{V_{UC\_t1}}\right) \cdot C_{UC}} \tag{7.2}$$

The EPR value of the ultracapacitor with 15 V 58 F capacity used in the experimental studies was calculated as 9.585 kΩ.

This equivalent resistance causes a continuous leakage current to flow through the ultracapacitor and, over time, the ultracapacitor voltage leads to an end. It is seen that the voltage value decreases over time due to the EPR of the ultracapacitor in Fig. 7.5. The voltage value at the beginning decreases rapidly in the first moments. The voltage value decreases to 12.44 V at the end of 1734 minutes. If the EPR of the ultracapacitor is large, the leakage current of the ultracapacitor is also small. This also determines how long the ultracapacitor can hold above the amount of stored capacity. As a result of the EPR experiment, it was observed that the ultracapacitor had a total voltage drop of 2.56 V within 1734 minutes. When a general ratio is established for the ultracapacitor voltage to reach 0 V, it takes approximately 7.05 days for the amount of stored capacity to be finished.

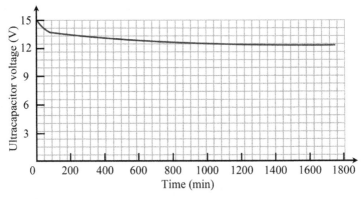

FIGURE 7.5 Ultracapacitor EPR experiment V_{UC_t} change.

2   Innovative energy storage technologies

2.1   Gravity energy storage

The energy storage method can be realized using the force of gravity. A force is exerted by gravity on an object of a certain weight. At first, clock pendulum applications were made using the gravity force. An energy storage and energy production can be realized using gravity. The most important parameters in these applications are mass weight and height [28,29]. The energy to be obtained from the gravity force is calculated by the next equation:

$$E_{gra} = mgh \tag{7.3}$$

where m is the weight of the mass to be used as a storage unit, g is gravity constant (9.80665 m/s^2), and h is height at which the mass will be moved.

Gravity energy storage application can be used especially with renewable energy sources. For this energy storage technology, a mass with a high weight (>3000 t), a channel opened underground, reducer system, pulley system, and generator units are required. Gravity energy storage system is a mechanical storage method. The mass is raised with excess energy produced from renewable energy sources in this energy storage method [30]. The up movement of the mass is carried out by a motor and reducer system. Thus energy will potentially be stored in the system (charged). In the case of energy need, the mass will be released by the gravity force and move down the channel. Meanwhile, electricity will be produced with the generator. Renewable energy–based gravity energy storage system topology is given in Fig. 7.6. Since battery energy

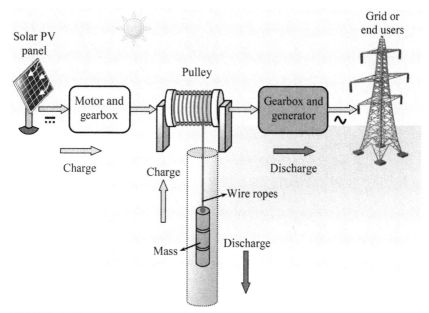

FIGURE 7.6   Renewable energy–based gravity energy storage system topology.

storage units are compared, these systems can store and provide high energy. The high-power charge/discharge rates are not possible in batteries. The cycle life of the batteries decreases and problems such as excessive temperature occur in this situation. There is no problem with cycle life in gravity energy storage system [31]. In addition, sudden high-power values can be provided with gravity energy storage systems that are predicted to have a service life of up to 50 years with periodic maintenance [32]. Gravity energy storage system provides flexibility in terms of use. It can be used at any time of the day. In addition, start-up of the system is very fast. Thus the gravity energy storage system has a fast frequency response. The gravity energy storage system provides a great advantage with a response time of less than 1 second at full load. Opening a high depth channel for gravity energy storage system installation is high in terms of initial setup cost. Since it has a mechanical structure, it needs periodic maintenance.

2.2   Flywheel energy storage

One of the mechanical energy storage methods is flywheel. It was one of the energy storage methods used by establishing a motor and generator system before chemical energy storage technologies. Basically, flywheels have a very simple working logic. Energy is stored rotatable in a rotating mass [33]. When energy is needed, electrical energy is provided by coupling the rotating mass with the generator. The energy stored in the flywheel energy storage system is calculated by the next equation:

$$E_{fly} = \frac{1}{2} \int \rho(x) r^2 \omega^2 \, dx \qquad (7.4)$$

The amount of energy stored in the rotating mass depends on the density of the rotating axis ($\rho(x)$), the diameter of the rotating mass (r), and the square of the angular velocity of the rotating mass shaft. Fig. 7.7 shows the power capacity comparison of the flywheel energy storage.

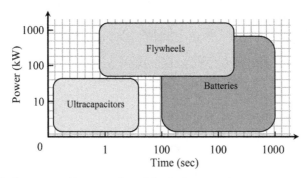

FIGURE 7.7   Power capacity comparison of flywheel energy storage.

The flywheel energy system has a fast response time compared to electrochemical energy storage systems. It is used in grid power cuts with this feature. Thanks to the power electronics and composite material technology, the flywheel energy storage system performances are increasing. In conventional flywheel energy storage systems, a motor is connected to a rotating mass shaft and the motor performs energy storage. Energy is taken with another generator connected to the rotating mass (discharge). With the developments in power electronics, the same process is carried out with a single motor/generator. The system efficiency is increased by providing bidirectional energy flow thanks to the power electronic circuits [34–36]. Flywheel energy storage unit capacities can be increased by increasing the material moment of inertia or by higher rotation speeds. The material moment of inertia can be realized using design and composite material. In some applications the inside of the rotating mass is hollow and the bulk density is concentrated on the outer diameter [37]. Thus the weight of the rotating mass is reduced by design. Flywheel energy storage systems have a moving and rotating structure. Friction is minimized by using a magnetic system in the bearing of the rotating body. In addition, the environment with the rotating object is vacuumed, and the air friction during rotation is reduced. All these improve the performance of the flywheel energy storage system. One of the methods used in flywheel energy storage systems is that a mass with a large radius is increased up to several thousand cycles. Energy storage can be provided by using a conventional motor and power electronics circuits. This method is generally used in large flywheel energy storage systems (heavy mass). In this method the rotation speed of the rotating object is not high [38]. Energy storage is performed by radius and weight parameters in this method. Fig. 7.8 shows the integration of the flywheel energy storage system with the grid. In this method the stored energy is transferred to the grid by a generator, alternative current (AC)/direct current (DC) rectifier circuit, and DC/AC inverter circuit.

Another method used in flywheel energy storage systems is to store energy with high speed. In this method the rotating object is rotated up to 100,000 rpm [39]. The rotating object weight is low in this method. This method is used in small applications in terms of volume and weight. Small applications connected in parallel can be used instead of large flywheel energy storage systems. There

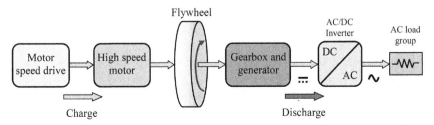

FIGURE 7.8   **Flywheel energy storage system topology.**

are losses due to air friction and bearing in flywheel energy storage systems. These cause energy losses with self-discharge in the flywheel energy storage system. The high speeds have been achieved in the rotating body with the developments in the field of composite materials. Composite material technology has enabled it to work with low losses, especially at high rotational tip speeds [40]. The very high speeds are achieved with permanent magnet brushless DC motors. Power electronics converter circuits enable motors to operate at high speeds. In addition, bidirectional energy flow is provided by a single-power electronic converter.

Flywheel energy storage systems have high power density. They can provide very high powers sudden thanks to these features. Flywheel energy storage systems allow high charge/discharge powers. Flywheel energy storage systems do not cause environmental pollution since they have a mechanical technology. Their efficiency is high during energy storage and energy transfer (>90 %). The performance of flywheel energy storage systems operating in magnetic bearing and vacuum is high. Flywheel energy storage systems have a long working life if periodically maintained (>25 years). The cycle numbers of flywheel energy storage systems are very high (>100,000). In addition, this storage technology is not affected by weather and climatic conditions [41].

One of the most important issues of flywheel energy storage systems is safety. As a result of mechanical failure, the rotating object fails during high rotational speed poses a serious danger. One of the disadvantages of these storage systems is noise. It is generally located underground to eliminate this problem.

2.3   Superconductive magnetic energy storage

Superconducting materials are the state in which the electrical resistance of the material is zero, and the magnetic exchange fields disappear. This case occurs when cooled to temperatures below a characteristic critical temperature. In superconducting magnetic energy storage systems, electrical energy is stored as a magnetic field in the form of DC [42]. Fig. 7.9 shows the superconducting magnetic energy storage system topology.

The electrical energy is obtained by moving the coil group around an electromagnetic field. This electric energy produced is explained by the principle of magnetomotive force. The voltage value induced in a coil group is calculated by Eq. (7.5). The amount of energy stored in a coil group is calculated by Eq. (7.6). According to this equation, the amount of stored energy varies depending on the inductance value and the square of the current. The energy density obtained in a magnetic field in the superconducting magnetic energy storage system is calculated by Eq. (7.7). The high energy density values can only be achieved with high current values in the superconducting magnetic energy storage system.

$$e_{emf} = -N \frac{d\theta}{dt} \qquad (7.5)$$

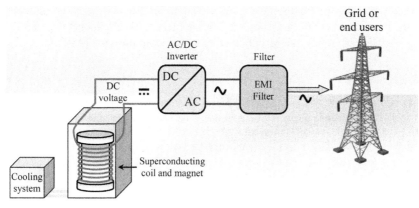

FIGURE 7.9   Superconducting magnetic energy storage system topology.

$$E_{(t)} = \frac{1}{2} L i_{(t)}^2 \qquad\qquad (7.6)$$

$$E_{mag} = \frac{B^2}{2\mu} \qquad\qquad (7.7)$$

In the superconducting magnetic energy storage system, special alloyed (niobium–titanium) conductive wires are used to create a high-density magnetic field. The current carrying capacity of the conductive wire varies depending on the temperature and magnetic field. Closed vessel and low-temperature environment with vacuum provide this working environment. The electrical resistance reaches a very low value at this critical temperature value. In order to maintain a superconductor's environment, the temperature must be cooled to approximately 269 °C [43].

The superconducting magnetic energy storage system is an important technology to ensure proper power quality and increase the penetration of renewable resources. Superconducting magnetic energy storage offers significant advantages, including fast response time from standby to full power.

It has the power capacity of superconducting magnetic energy storage systems (1–10 MW). Superconducting magnetic energy storage systems have a fast discharge structure at nominal capacity. The response time in this energy storage technology is around milliseconds and it is a long-life energy storage system [44]. Despite the low maintenance costs, the installation cost is high. There is no heat loss due to resistance and system efficiency is high in this energy storage technology. Energy is stored magnetically in the form of DC current in superconducting magnetic energy storage systems. It is sufficient to use a DC/AC inverter power circuit to connect to the grid. It has no harmful effect on the environment during the energy storage system operation. The long conductors are needed to storage energy at high powers in this energy storage system. In order for the energy storage system to operate in the superconductor state, it must be continuously cooled. With the recent developments in technology, low temperatures can be obtained by using liquid nitrogen [45].

References

[1] A. Awasthi, A.K. Shukla, M.M.S.R.U. Akram, M. Nadarajah, R. Shah, F. Milano, A review on rapid responsive energy storage technologies for frequency regulation in modern power systems, Renew. Sustain. Energy Rev. 120 (2020) 109626, doi: 10.1016/j.rser.2019.109626.

[2] E. Bullich-Massagué, F.J. Cifuentes-García, I. Glenny-Crende, M. Cheah-Mañé, M. Aragüés-Peñalba, F. Díaz-González, et al. A review of energy storage technologies for large scale photovoltaic power plants, Appl. Energy 274 (2020) 115213, doi: 10.1016/j.apenergy.2020.115213.

[3] Y. Liu, J. liang Du, A multi criteria decision support framework for renewable energy storage technology selection, J. Cleaner Prod. 277 (2020) 122183, doi: 10.1016/j.jclepro.2020.122183.

[4] M.M. Rahman, A.O. Oni, E. Gemechu, A. Kumar, Assessment of energy storage technologies: a review, Energy Convers. Manage. 223 (2020) 113295, doi: 10.1016/j.enconman.2020.113295.

[5] L. Sun, J. Qiu, X. Han, X. Yin, Z.Y. Dong, Capacity and energy sharing platform with hybrid energy storage system: an example of hospitality industry, Appl. Energy 280 (2020) 115897, doi: 10.1016/j.apenergy.2020.115897.

[6] Y. Sun, W. Pei, D. Jia, G. Zhang, H. Wang, L. Zhao, et al. Application of integrated energy storage system in wind power fluctuation mitigation, J. Energy Storage 32 (2020) 101835, doi: 10.1016/j.est.2020.101835.

[7] T.D. Hutty, S. Dong, S. Brown, Suitability of energy storage with reversible solid oxide cells for microgrid applications, Energy Convers. Manage. 226 (2020) 113499, doi: 10.1016/j.enconman.2020.113499.

[8] W. Cao, Y. Qiu, P. Peng, F. Jiang, A full-scale electrical-thermal-fluidic coupling model for li-ion battery energy storage systems, Appl. Therm. Eng. 185 (2020) 116360, doi: 10.1016/j.applthermaleng.2020.116360.

[9] H. Chen, H. Wang, R. Li, Y. Zhang, X. He, Thermo-dynamic and economic analysis of s a novel near- isothermal pumped hydro compressed air energy storage system, J. Energy Storage 30 (2020) 101487, doi: 10.1016/j.est.2020.101487.

[10] D. Erdemir, I. Dincer, Assessment of renewable energy-driven and flywheel integrated fast-charging station for electric buses: a case study, J. Energy Storage 30 (2020) 101576, doi: 10.1016/j.est.2020.101576.

[11] A. Demircalı, P. Sergeant, S. Koroglu, S. Kesler, E. Öztürk, M. Tumbek, Influence of the temperature on energy management in battery-ultracapacitor electric vehicles, J. Cleaner Prod. 176 (2018) 716–725, doi: 10.1016/j.jclepro.2017.12.066.

[12] Z. Fu, L. Zhu, F. Tao, P. Si, L. Sun, Optimization based energy management strategy for fuel cell/battery/ultracapacitor hybrid vehicle considering fuel economy and fuel cell lifespan, Int. J. Hydrogen Energy 45 (2020) 8875–8886, doi: 10.1016/j.ijhydene.2020.01.017.

[13] A. Aktas,, Y. Kırçiçek, A novel optimal energy management strategy for offshore wind/marine current/battery/ultracapacitor hybrid renewable energy system, Energy 199 (2020) 117425, doi: 10.1016/j.energy.2020.117425.

[14] M.K. Azeem, H. Armghan, Z. e. Huma, I. Ahmad, M. Hassan, Multistage adaptive nonlinear control of battery-ultracapacitor based plugin hybrid electric vehicles, J. Energy Storage 32 (2020) 101813, doi: 10.1016/j.est.2020.101813.

[15] H. Peng, J. Wang, W. Shen, D. Shi, Y. Huang, Compound control for energy management of the hybrid ultracapacitor-battery electric drive systems, Energy 175 (2019) 309–319, doi: 10.1016/j.energy.2019.03.088.

[16] P. Naresh, N. Sai Vinay Kishore, V. Seshadri Sravan Kumar, Mathematical modeling and stability analysis of an ultracapacitor based energy storage system considering non-idealities, J. Energy Storage 33 (2020) 102112, doi: 10.1016/j.est.2020.102112.

[17] Y. Wang, Z. Sun, X. Li, X. Yang, Z. Chen, A comparative study of power allocation strategies used in fuel cell and ultracapacitor hybrid systems, Energy 189 (2019) 116142, doi: 10.1016/j.energy.2019.116142.

[18] J. Tian, R. Xiong, W. Shen, J. Wang, Frequency and time domain modelling and online state of charge monitoring for ultracapacitors, Energy 176 (2019) 874–887, doi: 10.1016/j.energy.2019.04.034.

[19] M.R. Kumar, S. Ghosh, S. Das, Charge-discharge energy efficiency analysis of ultracapacitor with fractional-order dynamics using hybrid optimization and its experimental validation, AEU – Int. J. Electron. Commun. 78 (2017) 274–280, doi: 10.1016/j.aeue.2017.05.011.

[20] B. Diarra, A.M. Zungeru, S. Ravi, J. Chuma, B. Basutli, I. Zibani, Design of a photovoltaic system with ultracapacitor energy buffer, Procedia Manuf. 33 (2019) 216–223, doi: 10.1016/j.promfg.2019.04.026.

[21] R. Xiong, Y. Duan, J. Cao, Q. Yu, Battery and ultracapacitor in-the-loop approach to validate a real-time power management method for an all-climate electric vehicle, Appl. Energy 217 (2018) 153–165, doi: 10.1016/j.apenergy.2018.02.128.

[22] J. Yang, X. Xu, Y. Peng, J. Zhang, P. Song, Modeling and optimal energy management strategy for a catenary-battery-ultracapacitor based hybrid tramway, Energy 183 (2019) 1123–1135, doi: 10.1016/j.energy.2019.07.010.

[23] C. Liu, Y. Wang, Z. Chen, Q. Ling, A variable capacitance based modeling and power capability predicting method for ultracapacitor, J. Power Sources 374 (2018) 121–133, doi: 10.1016/j.jpowsour.2017.11.033.

[24] C. Liu, Y. Wang, L. Wang, Z. Chen, Load-adaptive real-time energy management strategy for battery/ultracapacitor hybrid energy storage system using dynamic programming optimization, J. Power Sources 438 (2019) 227024, doi: 10.1016/j.jpowsour.2019.227024.

[25] S.V. Rajani, V.J. Pandya, V.A. Shah, Experimental validation of the ultracapacitor parameters using the method of averaging for photovoltaic applications, J. Energy Storage 5 (2016) 120–126, doi: 10.1016/j.est.2015.12.002.

[26] S.V. Rajani, V.J. Pandya, Experimental verification of the rate of charge improvement using photovoltaic MPPT hardware for the battery and ultracapacitor storage devices, Sol. Energy 139 (2016) 142–148, doi: 10.1016/j.solener.2016.09.037.

[27] T.D. Atmaja, Amin, Energy storage system using battery and ultracapacitor on mobile charging station for electric vehicle, Energy Procedia 68 (2015) 429–437, doi: 10.1016/j.egypro.2015.03.274.

[28] A. Berrada, K. Loudiyi, I. Zorkani, System design and economic performance of gravity energy storage, J. Cleaner Prod. 156 (2017) 317–326, doi: 10.1016/j.jclepro.2017.04.043.

[29] A. Berrada, K. Loudiyi, I. Zorkani, Dynamic modeling and design considerations for gravity energy storage, J. Cleaner Prod. 159 (2017) 336–345, doi: 10.1016/j.jclepro.2017.05.054.

[30] H. Hou, T. Xu, X. Wu, H. Wang, A. Tang, Y. Chen, Optimal capacity configuration of the wind-photovoltaic-storage hybrid power system based on gravity energy storage system, Appl. Energy 271 (2020) 115052, doi: 10.1016/j.apenergy.2020.115052.

[31] C.D. Botha, M.J. Kamper, Capability study of dry gravity energy storage, J. Energy Storage 23 (2019) 159–174, doi: 10.1016/j.est.2019.03.015.

[32] T. Morstyn, M. Chilcott, M.D. McCulloch, Gravity energy storage with suspended weights for abandoned mine shafts, Appl. Energy 239 (2019) 201–206, doi: 10.1016/j.apenergy.2019.01.226.

[33] M. Shadnam Zarbil, A. Vahedi, H. Azizi Moghaddam, M. Saeidi, Design and implementation of flywheel energy storage system control with the ability to withstand measurement error, J. Energy Storage 33 (2020) 102047, doi: 10.1016/j.est.2020.102047.

[34] V. Kale, M. Secanell, A comparative study between optimal metal and composite rotors for flywheel energy storage systems, Energy Rep. 4 (2018) 576–585, doi: 10.1016/j.egyr.2018.09.003.

[35] B. Xiang, W. Wong, Power compensation mechanism for AMB system in magnetically suspended flywheel energy storage system, Measurement 173 (2020) 108646, doi: 10.1016/j.measurement.2020.108646.

[36] L. Shen, Q. Cheng, Y. Cheng, L. Wei, Y. Wang, Hierarchical control of DC micro-grid for photovoltaic EV charging station based on flywheel and battery energy storage system, Electr. Power Syst. Res. 179 (2020) 106079, doi: 10.1016/j.epsr.2019.106079.

[37] M. Mansour, M.N. Mansouri, S. Bendoukha, M.F. Mimouni, A grid-connected variable-speed wind generator driving a fuzzy-controlled PMSG and associated to a flywheel energy storage system, Electr Power Syst. Res. 180 (2020) 106137, doi: 10.1016/j.epsr.2019.106137.

[38] J. Kondoh, T. Funamoto, T. Nakanishi, R. Arai, Energy characteristics of a fixed-speed flywheel energy storage system with direct grid-connection, Energy 165 (2018) 701–708, doi: 10.1016/j.energy.2018.09.197.

[39] J. Šonský, V. Tesař, Design of a stabilised flywheel unit for efficient energy storage, J. Energy Storage 24 (2019) 100765, doi: 10.1016/j.est.2019.100765.

[40] A. Soomro, M.E. Amiryar, K.R. Pullen, D. Nankoo, Comparison of performance and controlling schemes of synchronous and induction machines used in flywheel energy storage systems, Energy Procedia 151 (2018) 100–110, doi: 10.1016/j.egypro.2018.09.034.

[41] S. Wicki, E.G. Hansen, Clean energy storage technology in the making: an innovation systems perspective on flywheel energy storage, J. Cleaner Prod. 162 (2017) 1118–1134, doi: 10.1016/j.jclepro.2017.05.132.

[42] A. Kumar, J.V. Muruga Lal Jeyan, A. Agarwal, Electromagnetic analysis on 2.5 MJ high temperature superconducting magnetic energy storage (SMES) Coil to be used in uninterruptible power applications, Mater. Today Proc. 21 (2020) 1755–1762, doi: 10.1016/j.matpr.2020.01.228.

[43] H.S. Salama, I. Vokony, Comparison of different electric vehicle integration approaches in presence of photovoltaic and superconducting magnetic energy storage systems, J. Cleaner Prod. 260 (2020) 121099, doi: 10.1016/j.jclepro.2020.121099.

[44] G. Vyas, R.S. Dondapati, AC losses in the development of superconducting magnetic energy storage devices, J. Energy Storage 27 (2020) 101073, doi: 10.1016/j.est.2019.101073.

[45] G. Vyas, R.S. Dondapati, Investigation on the structural behavior of superconducting magnetic energy storage (SMES) devices, J. Energy Storage 28 (2020) 101212, doi: 10.1016/j.est.2020.101212.

Chapter 8

Solar Hybrid Systems for Smart Grids

1 Smart grids and solar hybrid systems

When electrical energy was introduced in a usable form in the late 19th century, it was in the form of regional power plants that only served consumers. Although power consumption and production were in certain regions, long-distance transmission was not in question. As the demand requirement increased, so did the distribution grid line gradually. The grid systems are primarily isolated from each other and connected to each other from a single point. Fig. 8.1 shows the basic operating structure of the conventional electricity grid. Electricity is produced in large power plants that transfer the electricity to the transmission line with the help of step-up transformers to carry it over long distances. The electrical energy in the transmission lines is reduced to the voltage level in the distribution line with the step-down transformers and transmits it to the end users. While the basic operating structure of the grid has generally remained the same for many years, the application structures used for its planning and operation have changed over time. Taking into consideration the amount of energy demanded in the conventional electricity grid, the control of electrical energy is ensured by the power plants coming into operation.

There are large power plants that meet the basic demand power and peak power plants that activated at peak demand times. Especially the operating costs of peak power plants are quite high and this affects the total operating cost.

1.1 Smart grid structures

Smart grid can be defined as the fact that all units or nodes (such as production units, transformer centers, distribution centers, and consumption units) in the electric power system are active, adaptive (able to adapt to changing conditions), flexible, and have a strong structure that any production or consumption unit can easily be connected [1–3]. In addition, smart grid can display the operation of all interconnected components, from central and distributed production facilities to high voltage lines to industrial users, energy storage applications, end users, and electric vehicles. Existing transmission and distribution networks are not designed in the smart grid logic. The requirements of smart grids are quite different, and, therefore, a detailed study is required for conventional

Solar Hybrid Systems. http://dx.doi.org/10.1016/B978-0-323-88499-0.00008-2

153

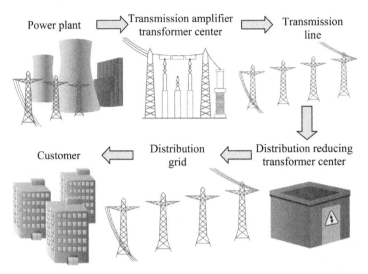

FIGURE 8.1 Conventional electricity grid transmission structure.

grids. It includes many steps for developing and expanding the conventional grid, maintenance activities, integration of distributed production and storage, and the development of a comprehensive two-way communication system. Fig. 8.2 is given to show the smart grid electricity transmission structure.

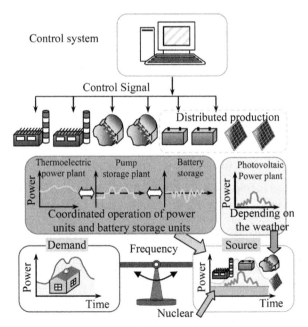

FIGURE 8.2 Smart grid electricity transmission structure.

With a smart grid structure, electrical energy should be presented to all consumers in a sufficient, continuous, quality, low-cost, and environment-conscious method. Electricity transmission and distribution services should be provided in a quality and should be sufficient to meet consumer satisfaction. Necessary measures should be taken considering the activities of climate change and environmental impacts in all areas of the sector. Minimizing losses and increasing efficiency in electrical energy, production, transmission, distribution, and use are an important issue for smart grid structure. In order to strengthen the smart grid structure, energy resource diversity should be increased and renewable energy sources should be used at the maximum level [4–6].

A smart grid is selective and it can also enable or disable the energy sources needed. Different power generation units can communicate with each other within the smart grid structure. By controlling all the powers in the system, power flow can be directed as necessary. A smart grid analyzes present situations, creates estimates and measures, and provides ease of operation/maintenance. In addition, its smart grid structure provides superior power quality, reliability, and efficiency. In the case of smart grid, many arrangements are needed in the process from energy production to its use within the framework of the objectives of ensuring energy supply security, reducing the risks arising from external dependence, increasing the effectiveness of combating climate change, and protecting the environment. With the smart grid structure, technical losses are minimized in electricity production, transmission, and the prevention of illegal use in distribution. In addition, end-user satisfaction can be increased by making regulations regarding demand-side management [7,8].

The smart grid is a concept that forms the future of electrical system and includes technology components from many different disciplines. These components are high-level energy management, transmission system, monitoring and control system, digital control technology, smart distribution system, information technology–based energy services, distributed energy generation systems, and microgrid structures. Table 8.1 presents the features of the smart grid structure and the conventional electricity grid [9–13].

There are some factors in the transition process of the electricity grid to the modernization phase. Political and legal factors especially with government regulations are of great importance in the development of smart grids. The structure to be revealed within the scope of the smart grid will reveal new job opportunities and business models. In terms of consumer satisfaction, service quality, and sustainability, the issue of ensuring energy reliability and security at all times is an important factor that requires smart grids. With the support of adding renewable and distributed production to the grid, significant breakthroughs are needed in terms of environmental sustainability within the scope of the smart grid.

TABLE 8.1 Comparison of conventional and smart grid structures.

Conventional grid structure	Smart grid structure
Consumers do not have enough options to reduce energy consumption and costs	Detailed price information is available; many programs, prices, and payment terms are selected
Central production is used, and there is very limited distributed production and storage	Distributed production resources are also included in central production
There is a limited wholesale market for the integration of new technological developments	The development of new electricity markets, taking into account the technological developments and different production options, is provided
It focuses more on interruptions than on power quality	Power quality is a priority for smart grids
There is limited artificial intelligence and automatic control grid structure	In grid applications, efficiency increases with artificial intelligence in energy management
It focuses on the protection of assets after failure	It takes precautions to prevent malfunction before the malfunction minimizes the effects on malfunctions and ensuring operational continuity
It is very weak against natural disasters and other dangers	In situations such as natural disasters and attacks, the system quickly recovers and continues to operate

1.2   Microgrid structures

One of the most important subcomponents that make up the smart grid is the microgrid structure. Different definitions can be introduced to the concept of microgrid. Microgrid with smart energy management unit generates its own energy in local energy production units and ensures that the loads are fed in a balanced and controlled manner. In addition, the energy systems of commercial/residential buildings that can be the grid connected and independent from grid can also be defined as microgrid. Moreover, a regional distribution system that can control the production and consumption units with smart systems and provide a balanced operation when the connection to the main electricity grid is lost can be defined as a microgrid. An example of a microgrid structure is given in Fig. 8.3.

The microgrid consists of on-site power generation (distributed generation), local loads, and electrical breaker. Smart microgrid can be defined as the electricity grid that makes electricity generation, distribution, and adjustment of the electricity flow given to local electrical consumers in a smarter way. The advantages of the smart microgrid can be expressed as increased reliability, increased

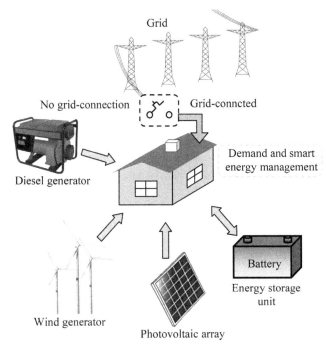

FIGURE 8.3   **Microgrid electricity transmission structure.**

efficiency, increased sustainability, and gain [14–17]. Thus the integration of distributed energy sources into the grid becomes easier, reducing energy costs/interruptions and improving power quality.

The most important problems in the microgrid structure are the synchronization of distributed energy generation sources, starting currents drawn by loads, and three-phase imbalance problems. In a smart grid structure the number of resources in the system can be increased by synchronizing energy storage systems with distributed energy generation resources. In the smart grid, fluctuations in electricity production and consumption can easily be met with the help of the energy storage unit connected to the system. With the distributed generation structure in the system, the loads on the demand side are not affected by voltage fluctuation and drop, three-phase imbalance, and voltage harmonics. During a long energy interruption, if the power consumption is more than production, the microgrid structure ensures the continuity of the energy by disabling some loads to continue feeding important loads. When a load in the system is temporarily increased excessively, the unit feeding the load can meet the electrical power via an additional power line.

Since the microgrid structure consists of too many units, a classification is needed. Therefore microgrid structures are classified as residential type, distribution plant, and commercial, industrial, and remote settlement microgrids [18].

Electrical energy production and consumption units can be controlled in residential micro grids thanks to smart applications. Smart home systems with this type of energy management unit are in the development process. The local production units in the microgrid, energy storage units, and loads are connected to the local distribution grid with a controllable interface in smart homes. The residential microgrid structure evaluates the factors such as energy usage time, prices, or tariff information. Factors such as the duration of use, energy prices or tariff information are evaluated in the residential micro grid structure. The amount of energy produced in the local generation unit, the amount of energy in the energy storage unit or the amount of energy that can be drawn from the grid are calculated. Among these units, choices are made to provide the most economical and reliable operation.

The distribution plant may also be considered as a partial part of the microgrid systems for the residential area or for an industrial zone. Possible problems in the distribution network can be prevented by the proper operation of local distributed generation systems (wind turbines, biogas plants, energy storage applications, etc.) against loads in such microgrids. Quality energy can be provided to consumers by operating certain parts of the distribution grid as microgrid, operating these parts in island mode in the case of maintenance or failure [19–22].

The microgrid structure can provide energy reliability especially for consumers with critical and sensitive loads in commercial and industrial microgrids. Especially in large consumption units such as universities, hospitals, schools, and shopping centers, microgrid applications can be used to fully control all energy consumption units in their systems. In this way, energy savings as well as economic benefits can be achieved with various applications in a fully controlled system. Microgrids may be suitable for remote location areas with poor or no connection to the grids. In such places, it should be ensured that they utilize local energy sources as much as possible and be loaded into the mains as little as possible. For such residential areas, microgrid application can be very suitable for the integration of renewable local distributed generation facilities and the entire system under control [23].

1.3 Dynamic control in microgrids

The most important feature of a microgrid is that it can be separated from the main grid very quickly, when necessary. Another feature is that it can be put into operation in island mode and, if necessary, can be switched from the island mode back to the grid. In this case the microgrid should be able to provide the energy needs of the loads dynamically without the need for information from the loads. This situation requires using energy storage units in the system. In terms of control method, different control strategies should be applied in the case of grid-connected state and no grid-connected situations. In the grid-connected state the grid provides reference values for voltage and frequency. However, when it is switched from the grid to the independent state, the microgrid must establish the voltage and frequency reference values as well as it must ensure the load balance and production balance. In this way the need for two different operating modes and a different control strategy for each different operating mode are the main challenges of switching modes. The main concern is the transition from island operating mode to grid-connected mode, because the reference values used in island mode differ from those used in grid-connected mode [24–26].

An superior control unit is required to decide the operating modes and to produce reference values for the specified operating mode. The signals are received from the connection point of the system to the grid and the networked mode or island mode, and thus, the operating modes are decided. The microgrid operates as a power generator in grid-connected operation. There are peak clipping, valley filling, and energy efficiency for grid-connected mode. These statuses are valid only for mains-connected operation mode. Load shifting is essential in island mode operation independent of the grid. Sample graphics for dynamic load control applications are given in Fig. 8.4 [27–30].

Peak clipping is the reduction of peak loads in the system. It is one of the classic ways of load control management. Peak clipping mode is user-preferential and user-defined values. The user defines a demand load value that must be fed continuously from the grid and in cases where an excess load exceeds this load value, these excess loads are not drawn from the grid, but from renewable sources or stored energy.

Valley filling is the second classic form of load control management. Valley filling involves creating loads for nonpeak time periods. This is especially used to increase the load factor (rate of base load to peak load) over a period of time.

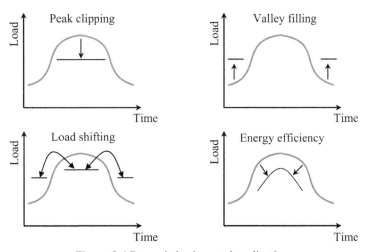

Figure 8.4 Dynamic load control applications.

FIGURE 8.4   **Dynamic load control applications.**

Load shifting is the final form of load control management. This approach involves shifting loads from peak to nonpeak hours. In this way the production side arrangements required to meet the peak demand can be prevented.

Energy efficiency can be evaluated within the scope of load control management. It is a changing situation that usually occurs according to the technological or usage features of the device used. The unit spent for a service without reducing the comfort conditions of the consumer is the reduction of energy and thus the downward shift of the load pattern is carried out. Consumers and suppliers can act independently to change the way they load. However, the demand-side interaction is aimed at a bilateral gain between suppliers and consumers [31,32].

1.4 Distributed production grid structures

Distributed generation grid structures have become a part of smart and microgrids. The developments in the electric power system in recent years encourage energy production units and energy storage units connected to the power system from the distribution level. Taken together, these applications are defined as distributed generation units, although new technologies that can be applied as distributed generation units are being marketed and applied commercially. The various studies are continuing to integrate these systems into the grid, to use them more efficiently, and to reduce costs. Examples of distributed generation units are micro turbines, fuel cells, photovoltaic (PV) systems, wind turbines, diesel generators, and gas turbine systems. With these technological developments, the electric power system is undergoing a complete change. The goal of this change is that the highest utilization of clean and renewable energy potential is to provide consumers with more reliable and quality energy [33–35].

1.5 Grid-connected distributed solar systems

Distributed generation systems with grid connection are seen as indispensable components of increasing importance within the system as an energy source. However, the expansion of PV systems is a major challenge for grid designers, operators, and engineers working in this field. This difficulty can be summarized as the negative power quality effects of PV systems due to their widespread use in the grid that they are connected with due to continuously variable and intermittent power output. In addition, the disadvantage is that the PV system output power and demand power peaks do not match and can lead to inefficient use, which is a problem to be solved. To solve these problems, grid-connected energy storage applications located close to PV systems are very successful [36].

The reason why the output power is unstable in the PV power generation system is natural conditions such as cloud transitions, air temperature, and seasonal changes. The large and sudden drops in PV system output power can be seen with such effects. These major power changes are important problems for grid operators. Partial cloud transitions, especially visible in summer, cause a sharp shadow transition and can cause a power fluctuation of approximately 25 % in PV panels, which changes over a few seconds. These changes are very difficult to predict. Such sudden drops in the PV system power output must be filled with a supporting system. It is also known that PV systems can only produce power from sunrise to sunlight and reach peak values of power generation, especially at noon hours when the sun reaches a right angle. The daily load requirement of a house and the power graph produced by the PV system are given as examples in Fig. 8.5. When compared with the load demand curve, it is seen that demand values are low in this range and PV power generation is not reachable in the evening hours when demand values reach peak values. In this case, there is a need to provide drivability in order to match the PV system power generation and the demand graph. In balancing such daily power changes, it can be seen that PV solar systems are supported by wind energy systems [37–40].

Another remarkable risk arising from the PV systems, especially in the distribution systems, which dominate the house loads, is that the generation values of the small power PV systems scattered in the distribution grid exceed the demand values in that region and have a serious impact on the voltage level in the region. Another problem that occurs in this case is called reverse power flow. In this case, redundant PV power generation in the distribution region is transferred to the grid and current flows from the secondary of the distribution transformer to the primary. Conventional distribution transformers are designed as a unidirectional current transition, so they are not designed for a bidirectional power flow. With the decrease in PV system installation costs, this problem is expected to gain importance in the coming years.

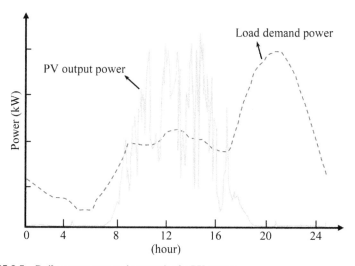

FIGURE 8.5   Daily power generation graph of a PV system.

1.6 Energy storage application in distributed solar systems

Energy storage system is very important for a smart microgrid structure. The energy storage system used to ensure the proper functioning of the loads in the grid can also improve the power quality. Depending on the climatic conditions, the output power fluctuation of solar energy and wind energy, output power of which varies, can be coped with energy storage applications. The concept of smart microgrid based on renewable energy is a good way to reduce the use of resources and fossil fuel consumption. The use of energy storage applications becomes mandatory by using microgrids to provide quality power to loads. There are two operating modes: grid-connected and isolated/island modes in microgrids. The main responsibility of the energy storage unit during the isolated/island mode operation is to ensure the energy balance [41]. The purpose of the grid-connected mode is to prevent intermittent and load fluctuation of renewable energy sources from affecting the grid. In a microgrid that uses renewable power with uniform energy storage, not all of these functions can be performed effectively. Renewable energy sources such as PV require the use of high energy density storage. At the same time, sudden changes in the load require high power density storage.

In the PV system integration the energy storage application can be connected directly to the grid as a whole with the PV system as well as to the central or distributed independent energy storage units in the regions where the PV system installations are concentrated. In all of these options, grid risks originating from the PV system can be prevented.

First of all, it is possible to obtain a smoother character of the PV system power output graphic with the energy storage. Power over certain limits can be stored and discharged for power values below certain limits. Thus a smoother power curve can be reached within a certain range at the output of the PV system and storage unit. The negative effects of the sudden fluctuating power structure on the grid can be suppressed. Fig. 1.7 is given the positive effect of the system on the grid that is given by using the PV energy production system and the energy storage system together [42].

Energy is stored when demand values are low by using PV energy production system and energy storage unit together. Stored energy is offered for consumption in the range where there is no PV generation and the load demand reaches peak values. The most economical operation is provided on the consumer side, while the loading rate is reduced on the grid side with such an application. If the energy storage system has power electronics or power converter structure, the energy storage unit can also be operated with voltage and reactive power regulation by controlling power quality.

PV energy systems have become an indispensable power source in the power systems of the future with their environment-friendly structures and simple, maintenance-free, and cost-effective operation possibilities. However, as mentioned earlier, the increase in the prevalence of these systems in electrical grids poses serious risks. The necessity and efficiency of energy storage applications

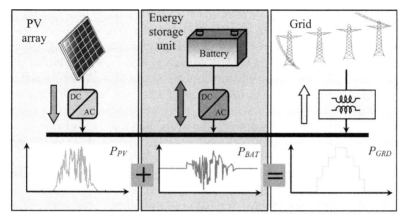

FIGURE 8.6 Effect of PV and energy storage system on the grid.

are remarkable in terms of solution applications. In the grid structure, different energy storage applications should be available to meet the needs in the most appropriate way. One of the most important key elements of the smart grids of the future is seen as energy storage applications. Costs are seen as the biggest obstacle to the spread of energy storage applications. If energy storage applications are integrated into small powerful residential PV system applications, access to these benefits will be achieved in a much more economical way, because, in such applications, costs such as wiring, communication, control system, power electronics devices, and labor are common for PV system and energy storage application [43]. Fig. 8.6 is depicted effect of PV and energy storage system on the grid.

Energy storage systems must be controlled with energy management systems to ensure uninterrupted supplying of critical loads. Successful management of the microgrid is possible with suitable control and operation of storage systems. Electronically interconnected micro sources are advantageous in terms of maintenance of the grid without interruption. Energy storage systems support this situation and support micro sources in low-voltage transients, motor start-ups, and other short-term overload conditions, especially in island mode operation. Energy storage systems should be able to respond immediately to changes in load demand. Therefore they should be managed by local control units. Some energy storage sources, such as ultracapacitors, are discharged in a short time, although they have high power density. Flywheels can provide energy for a long time, even if they have low power density [44–46].

2   Inverter structures in smart grids

In grid-connected systems, there are usually two or more converter stages. It may be necessary to increase or decrease this voltage level depending on the available PV array voltage. In the first stage, it usually includes a special direct current (DC)/DC power converter to perform voltage level conversion and maximum power point tracking (MPPT) algorithm. The structure of the single-stage and dual-stage converter topology is given in Fig. 8.7. The disadvantages of the dual-stage system are less efficiency, larger size, and higher cost. For this reason a single-stage grid-connected structure is used that is common today because of its small size, low cost, high efficiency, and high reliability. Only DC/alternative current (AC) power converter is available in this system topology. Single-stage PV systems can be easily used for compact module and plug-and-play applications. DC/DC converter stage is not used in a single-stage structure. The MPPT takes place at the inverter power stage in single-stage grid-connected PV systems. Factors such as overall efficiency, reliability, and control complexity are important issues in grid-connected PV systems. There are cases of efficiency and reliability issues as the number of stages increases in the double-stage system. Therefore it is desirable to reduce the number of stages contained in such systems. There are two alternatives to achieve this: either to use a step-up transformer to the power converter output or to use a high-voltage PV array. The first option causes an increase in system dimensions (power density). The second option can increase the possibility of leakage current by reducing safety and parasitic capacitance. This situation leads to dangerous spots during the partial shading of the PV array. An ideal solution is to use an intermediate power converter, with only a single-stage low-voltage PV array [47]. In this single-stage low-voltage PV array, the voltage is increased, converted into a quality AC waveform, and at the same time, maximum power is taken from the PV array.

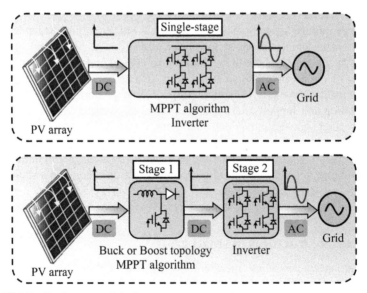

FIGURE 8.7 Single-stage and dual-stage converter topology.

Single-stage inverters must both transfer sinusoidal power to the grid and perform the MPPT algorithm. Single-stage grid-connected PV systems perform both cycles in single power conversion simultaneously. Thus the system topology is simplified. The MPPT method should be run at low speed to maintain system stability. MPPT algorithm calculates the maximum power point of PV panels with voltage and current sensors. In order to prevent voltage sags events, monitoring is performed with a minimum step length to change the reference value of the power. It is quite difficult due to the nonlinear I–V characteristic of PV array. The MPPT algorithm sets the reference current or voltage value that the PV panel should work with. One of the methods commonly used in grid-connected single-stage PV systems is perturb and observe technique. This technique determines the reference current and then provides a synchronous current to the grid. The reference voltage is used to calculate the grid reference current.

In other control methods, it gives accurate results when operating in the negative slope region of the PV array or in the voltage source region of the P–V characteristic. Based on these methods, the controller is designed in such a way that the reference voltage and reference mains current operate in reverse proportion in this region. The PV array continues this proportion as if it entered the current source region. When sudden weather change occurs, the system remains unstable and the control fails as the working point moves to the positive slope region. In addition, only the step size change can be made to prevent grid current disturbance [48–50].

2.1 Isolation and leakage currents in solar inverter structures

Isolation and safety are an important issue in PV systems without transformers. Small powerful PV single-phase systems are usually installed with a power capacity of 5–6 kW. While there is a DC source at the inverter input, there is AC power at the inverter output. Large powerful DC capacitors in single-phase systems reduce the reliability and long working life of the system. On the other hand, large capacitors are not needed in three-phase systems. The reliability and long-term operation of the system can be provided at a small cost. As for security, galvanic isolation transformers are used in PV systems. High-frequency transformer is used in DC/DC converter structures and low-frequency transformer is used in AC output. Adding a galvanic isolated transformer increases the cost and size of the system and reduces overall efficiency. A high-efficiency inverter can be made without including an isolation transformer. The transformerless structures are generally advantageous, when preferred. However, on the side of the solar panel parasitic capacitance, grounding safety issues occur. System grounding is done by connecting a capacitor between the PV panel terminals and the ground. A model with parasitic capacitance capacitors connected to a PV system is given in Fig. 8.8. Leakage and fault currents are grounded to ensure the reliability of the grid-connected PV system. Parasitic capacitance capacitor value is chosen depending on many factors such as PV panel and PV

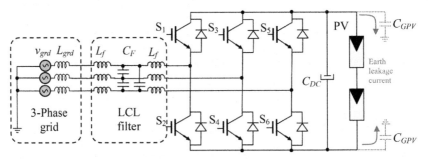

FIGURE 8.8 Three-phase full-bridge inverter topology.

model frame, cell surfaces, and distance between cells, weather conditions, and humidity/dust.

The three-phase full-bridge inverter topology given in Fig. 8.8 is the simplest and most widely used structure for grid-connected PV systems. A galvanically isolated system was installed between the PV array and the AC grid. If no current flows into the parasitic capacitance capacitors, the grid current is in a smooth sinusoidal curve. Under the same conditions, if there is no galvanic isolation between the PV array and the AC grid, leakage current flows to the ground. The same grid current draws a larger oscillating sinusoidal curve than the previous current curve [51]. Also in this topology structure, it contains high-frequency components at common-mode voltage. Due to the high switching frequency between the DC buses, the leakage current value is also very high. These values go beyond the grid-connected PV system standards.

The three-phase full-bridge voltage-sourced inverter structure with split connection is given in Fig. 8.9. Unlike other topology structures, capacitors and PV array are connected in a similar way to the input side. In this structure, PV array and capacitors are divided into two equal parts and their midpoints are connected to the neutral point of the grid. This topology is equivalent to three independent single-phase half-bridge inverters. In this structure, current control

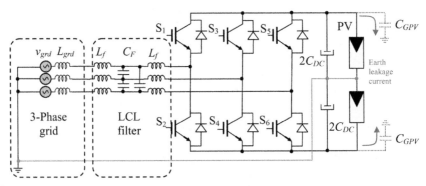

FIGURE 8.9 Three-phase full-bridge inverter topology with split capacitor.

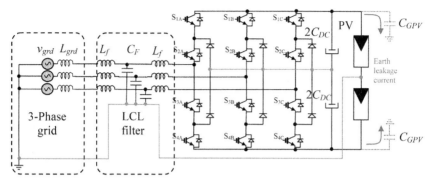

FIGURE 8.10   Three-phase neutral-point-clamped inverter topology.

is used to control the switches. Two strategies are used for pulse width modulation (PWM). The first is to use a single triangular carrier signal for all three phases. The second is switching using three triangular signals with a 120° difference. The purpose of this method is to eliminate switching harmonics in the grid neutral current. By using triangular signals for PWM, the common-mode voltage for the three phases can be reduced.

The DC bus voltage fluctuations in this inverter structure are much smaller, as they keep the neutral at zero potential. In this topology, few fluctuations occur in the ground and DC bus voltage and leakage current is greatly reduced. These values are within the grid-connected PV system standards and have a reliable structure.

The three-phase inverter structure with three-phase neutral point clamped is given in Fig. 8.10. This structure is interesting for PV and other renewable applications, compared to six-switch full-bridge inverters with some important advantages [52]. These advantages are as follows:

- Since the switches are connected in series and the voltage is split, the voltage stress on the switches decreases.
- Since the output phase voltage has two levels, the voltage at the output contains lower harmonics.
- Output filter size decreases because of low *dv/dt* ratio.
- Since low switching voltage is used, switching losses decrease and efficiency increases.

There are three separate current controllers for each phase and the output current is always synchronized with its own phase voltage. In this topology structure, it is more convenient to use a transformerless PV inverter as there is no voltage fluctuation and earth leakage between the PV panel terminals. Since these values are among the standards, it becomes a suitable solution for PV systems.

When examined in terms of power converters, six switching elements and six reverse diodes are used in the three-phase neutral-point-clamped inverter structure. In this case, only half of the voltage value of the switching elements is needed compared to the other two topologies. Since the midpoint of the capacitors is connected to the neutral line, an extra function is required to control the voltage imbalance. If the load is directly connected to the capacitors, it will cause voltage unbalance between the upper and the lower capacitors as the load current will be drawn from the capacitors.

Earth leakage is one of the important issues for transformerless PV systems. Leakage current is very high due to the three-phase full-bridge inverter structure having high common-mode voltage. On the contrary, there is almost no voltage fluctuation in the three-phase neutral-point-clamped inverter and three-phase full-bridge inverter with split capacitor structure. Since there is a small voltage fluctuation, the earth leakage current to be formed is more advantageous as it complies with the standards. If three-phase full-bridge inverter topology is used, lower fluctuation occurs in the grid current with galvanic isolation. In fact, in the structure with transformer, the triple harmonics of the grid current are canceled. The grid current is subjected to subharmonics within limit values.

The fluctuation in the grid current is in the lowest three-phase full-bridge inverter with split capacitor, then in the three-phase neutral-point-clamped inverter, and the highest in the three-phase full-bridge inverter topology. Different inductance-capacitance-inductance (LCL) filters must be selected for each topology to comply with the IEEE 929 standard. When the three-phase neutral-point-clamped inverter structure is used, the filter size is 20 % smaller than the three-phase full-bridge inverter structure [53].

With a high efficiency of up to 98 %, three-phase neutral-point-clamped inverter structure becomes the most effective topology for the transformerless PV inverter. Table 8.2 presents the comparison of these topology structures, switching number, voltage stress on switches, number of diodes, voltage unbalance control, common-mode voltage, earth leakage current, transformer requirement, grid unbalance effect, neutral leakage current effect, and LCL filter size [54].

2.2   Grid-connected three-phase inverter structures in smart grids

The transfer of energy obtained from the PV panels to the grid is provided by distributed power generation systems. At the output of the solar panels, the DC voltage must be converted to AC voltage in order to energize the grid. This conversion is carried out with the inverter power unit. Three-phase inverters are used in high-power applications. While energy is being transferred to the grid, it is requested that the energy produced by the inverter be of high quality in order not to create a disruptive effect on the grid. Three-phase currents produced at the inverter output should be close to sinusoidal for high energy quality. The inverter should also work synchronously with the grid. Therefore the current control of the inverter must be done properly. The inverter must have high power density, high efficiency, low cost, and simple circuit structure [55].

TABLE 8.2 Comparison of inverter topology structures.

Inverter topology	Three-phase full bridge	With split capacitor	Neutral point clamped
Switching number	6	6	12
Voltage stress on switches	V_{DC}	V_{DC}	$V_{DC}/2$
Diode number	6	6	18
Voltage unbalance control	No	Yes	Yes
Common-mode voltage	$\pm V_{DC}$	$\pm 1\% V_{DC}$	$\pm 1\% V_{DC}$
Earth leakage current	High	Low	Low
Transformer requirement	Required	No required	No required
Grid unbalance effect	Low	No	No
Neutral leakage current effect	No neutral line	High	High
LCL filter size (per unit)	1	0.87	0.80

The grid-connected inverter must synchronize with the grid and produce currents in the same phase in order to realize the power transmission. The harmonic contents of these currents should be low. According to the IEEE 519 harmonic standard, the currents transferred to the grid must have a total harmonic content less than 5 %. According to the EN 50160 standard, voltage harmonics should be maximum 8 %, voltage imbalance maximum 3 %, voltage amplitude maximum ±10 %, and frequency value change should be maximum ±1 % [56].

2.3   Three-phase voltage supply inverters

Three-phase inverters are used to generate AC output voltages from the voltage generated from the PV panel. A standard three-phase voltage-fed inverter structure is given in Fig. 8.11. The PV voltage is generated from the AC voltage using the S_1–S_6 switching elements. Each switching element in the circuit remains in transmission for 180°. After one of the switching elements in the high-side arm enters the cut, the corresponding switching element in the low-side arm is transmitted after the dead time value. Thus two switching elements in the same phase are prevented from being transmitted simultaneously [57].

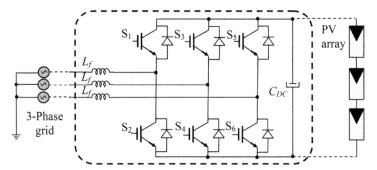

FIGURE 8.11 Circuit diagram of the three-phase voltage-fed inverter.

2.4 Modulation methods in three-phase voltage fed inverters

PWM is a method in which voltage pulses are produced by properly switching semiconductor power elements for the desired output frequency and voltage. A typical modulator generates an average voltage value equal to the reference voltage within the PWM period. Considering that the PWM period is too short, the reference voltage gives the basic component of the switched pulse sample. Among the numerous methods used to generate the output voltage, the main are sinusoidal PWM, hysteresis band PWM, and space vector PWM methods. The sinusoidal PWM and hysteresis band PWM methods can be performed using analog methods, while other methods require the use of microprocessors or digital signal processors [58].

2.4.1 Sinusoidal PWM method

In the sinusoidal PWM method, PWM signals are generated as a result of a comparison of a high-frequency triangular carrier wave with a sinusoidal modulation signal. This control method is frequently used in industrial applications. The frequency of the modulation signal determines the frequency of the output voltage and the peak value modulation index. The modulation index changes the effective value of the output voltage. As the modulation index increases, the inverter output voltage also increases and the index value gets the largest value of 1. The carrier wave at a high frequency, according to the modulation signal, determines the switching frequency of the inverter. The output harmonics produced occur in the form of switching frequency and its multiples.

To obtain the modulation signal, first, the current measured from the reference current is subtracted and current error is obtained. This current error is given to the proportional integral controller and the reference voltage is obtained at the controller output. This is done separately for each phase. Three-phase switching signals are generated by comparing the obtained reference voltages with the carrier wave. In Fig. 8.12 the inverter output voltage produced for one

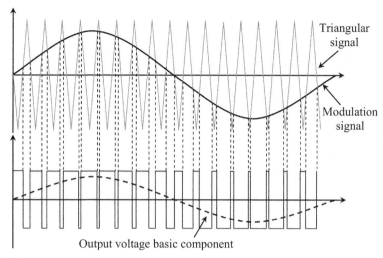

FIGURE 8.12 Sinusoidal pulse width modulation.

phase is given as a result of the comparison between the carrier wave and the modulation signal [58].

2.4.2 Hysteresis band PWM method

Hysteresis current control method is a preferred control method for ease of application. Besides its fast dynamic response and natural current protection features, it is not affected by changes in system parameters. However, in constant band current control, the switching frequency of the inverter is variable depending on the voltage difference between the inverter output and the grid. Since the difference between the inverter output voltage and the grid voltage is high in the regions close to the zero crossing of the grid, the switching frequency increases and decreases toward the peak values of the grid voltage. This variable switching frequency complicates the input filter design. In order to control the current in the defined band, the input sampling should also be done quickly. The block diagram of the hysteresis current control method is given in Fig. 8.13. In this control method, currents for each phase are obtained by

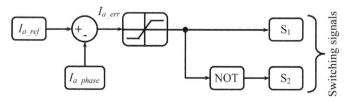

FIGURE 8.13 Hysteresis current control method block diagram.

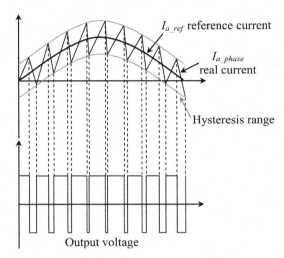

FIGURE 8.14   **Hysteresis current control PWM generation principle.**

comparing the phase currents measured at the inverter output with the reference phase currents. This current error is applied to the hysteresis controller and its signals are obtained.

The current of each phase in the inverter is tried to be kept within the band defined by the current error as given in Fig. 8.14 by the hysteresis controller. The inverter output currents move in the band as the current error remains in the band. Since the current measured when the current error reaches the upper limit of the band, the upper limit switch of the related phase is transmitted to increase the current, since the current reaches the lower limit of the band. With the increasing current the error starts to decrease and continues to increase until the upper band is reached. When the current reaches the upper band, the current error reaches the lower band and the upper switch is inserted into the sector and a signal is sent to the lower switch. Thus the inverter output current starts to decrease. This process repeats during each switching period and the output current is produced to move within the specified band [59].

2.4.3   Space vector PWM method

Space vector PWM method has found a wide application in recent years with features such as no need for reference, sinusoidal input, lower harmonic content, and wider linear working area. According to the sinusoidal PWM method, DC input voltage is used 15 % more efficiently. This method is popular for its use in AC drives with a voltage-fed inverter structure. Since this control structure will allow a control system with a lower current to be used, the transmission losses in the voltage fed inverter can be reduced. In addition, the harmonic content and switching losses of the inverter output voltage and currents are reduced [60].

2.5   Three-phase, four-arm, four-wire inverter structures

Distribution network must have a three-phase converter topology for low-voltage four-wire distributed network units. The distributed generation unit can also be used as an active power filter to improve the mains power quality. When an effective active power filter is desired, the converter structure must be a system that allows the neutral current to flow. There are four-wire topology, four-arm topology, and split-link capacitor converter topologies that allow this structure. The benchmarks can be classified as ease of control, circuit complexity, midpoint control, DC bus voltage utilization, and electromagnetic compatibility performance. The split-link capacitor converter topology offers the best option compared to the four-arm structure. For the same circuit complexity and the use of DC bus voltage control, split-link capacitor topology results in easy current control and less electromagnetic compatibility problems.

The choice of converter topology has a significant effect on its connection to the four-wire distribution grid. The neutral line can be achieved by adding a fourth arm to the converter (four-arm converter) or by connecting the midpoint of the split-link capacitors to the neutral line.

Classic three-phase, three-wire inverter topology is given in Fig. 8.15. Since the low-voltage distribution grid is a typical three-phase four-wire configuration, a converter transformer is used to connect to the grid. Since this transformer is heavy and expensive, it is not desired in many applications. In some cases the transformer can be used as part of the filter impedance. Another disadvantage is that the absence of a neutral line restricts the use of active power filters, since various filter methods require neutral current.

Instead of using Δ/Y converter transformer, neutral line can be provided with three-phase four-wire converter topology. The neutral line can be made by adding a fourth arm to the inverter structure or by adding two capacitors to the DC bus. Neutral wire will allow the use of active power filtering techniques. The four-wire inverters are more preferred than three-wire ones.

The four-arm converter topology is given in Fig. 8.16. Two extra S_7 and S_8 switches are used to control the neutral line. In order to weaken the high-

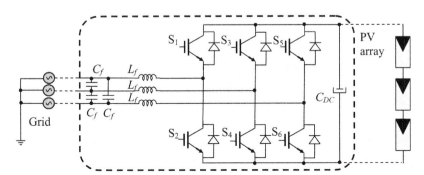

FIGURE 8.15   Three-phase, three-wire inverter structure.

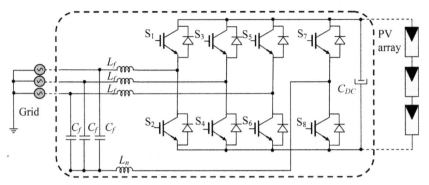

FIGURE 8.16   Three-phase, four-arm, four-wire inverter structure.

frequency currents due to the switching of the fourth arm, Ln choke coil must be connected to its neutral line. It is sufficient to connect a C_{DC} capacitor to prevent voltage fluctuations in the DC bus.

There is a split-link capacitor, a four-wire inverter structure, in Fig. 8.17. This structure has fewer switching elements due to its simple topology structure compared to the four-arm inverter. The DC bus line is divided by 2 C_{DC} capacitors. Since there is no switching at this point, there is no need to connect a choke coil to the neutral line. The split-link capacitor converter can be considered as a three-phase half bridge inverter, which can be controlled independently. The voltage unbalance of the DC bus capacitors is a known issue in the split-link capacitor converter topology. High voltage fluctuations due to neutral line currents are another problem. These voltage fluctuations should be reduced with capacitors.

A neutral line is preferred for programming distributed units with a secondary filter function. Δ/Y transformer is needed to obtain the neutral line in the three-wire converter structure. However, this situation becomes disadvantageous as it will increase the system size. Since distributed generation systems

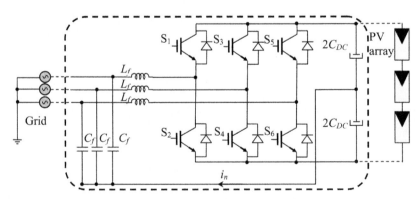

FIGURE 8.17   Split-link capacitor, four-wire inverter structure.

are widely used in the grid, these inverters can provide energy quality improvement when programmed with a secondary active parallel filter function. A neutral line is needed to use all this filtering potential. Therefore four-wire inverters offer an important advantage [61].

References

[1] M. Asaad, F. Ahmad, M.S. Alam, M. Sarfraz, Smart grid and Indian experience: a review, Resour. Policy (2019) 101499., doi: 10.1016/J.RESOURPOL.2019.101499. In press.

[2] G. Dileep, A survey on smart grid technologies and applications, Renew. Energy 146 (2020) 2589–2625, doi: 10.1016/J.RENENE.2019.08.092.

[3] A.A.G. Agung, R. Handayani, Blockchain for smart grid, J. King Saud Univ. – Comput. Inf. Sci. (2020)doi: 10.1016/J.JKSUCI.2020.01.002. In press.

[4] S. Dorahaki, M. Rashidinejad, A. Abdollahi, M. Mollahassani-pour, A novel two-stage structure for coordination of energy efficiency and demand response in the smart grid environment, Int. J. Electr. Power Energy Syst. 97 (2018) 353–362, doi: 10.1016/J.IJEPES.2017.11.026.

[5] M. Shrestha, C. Johansen, J. Noll, D. Roverso, A methodology for security classification applied to smart grid infrastructures, Int. J. Crit. Infrastruct. Prot. 28 (2020) 100342, doi: 10.1016/J.IJCIP.2020.100342.

[6] M. Abujubbeh, F. Al-Turjman, M. Fahrioglu, Software-defined wireless sensor networks in smart grids: an overview, Sustain. Cities Soc. 51 (2019) 101754, doi: 10.1016/J.SCS.2019.101754.

[7] D.B. Avancini, J.J.P.C. Rodrigues, S.G.B. Martins, R.A.L. Rabêlo, J. Al-Muhtadi, P. Solic, Energy meters evolution in smart grids: a review, J. Cleaner Prod. 217 (2019) 702–715, doi: 10.1016/J.JCLEPRO.2019.01.229.

[8] M. Ourahou, W. Ayrir, B.E.L. Hassouni, A. Haddi, Review on smart grid control and reliability in presence of renewable energies: challenges and prospects, Math. Comput. Simul. 167 (2020) 19–31, doi: 10.1016/J.MATCOM.2018.11.009.

[9] P. Dato, T. Durmaz, A. Pommeret, Smart grids and renewable electricity generation by households, Energy Econ. 86 (2020) 104511, doi: 10.1016/J.ENECO.2019.104511.

[10] G.G. Dranka, P. Ferreira, Towards a smart grid power system in Brazil: challenges and opportunities, Energy Policy 136 (2020) 111033, doi: 10.1016/J.ENPOL.2019.111033.

[11] H. Meschede, E.A. Esparcia, P. Holzapfel, P. Bertheau, R.C. Ang, A.C. Blanco, et al. On the transferability of smart energy systems on off-grid islands using cluster analysis – a case study for the Philippine archipelago, Appl. Energy 251 (2019) 113290, doi: 10.1016/J.APENERGY.2019.05.093.

[12] I. González, A.J. Calderón, Integration of open source hardware Arduino platform in automation systems applied to smart grids/micro-grids, Sustain. Energy Technol. Assess. 36 (2019) 100557, doi: 10.1016/J.SETA.2019.100557.

[13] A.D. Georgakarakos, E.A. Hathway, Battery storage systems in smart grid optimised buildings, Energy Procedia 151 (2018) 23–30, doi: 10.1016/J.EGYPRO.2018.09.022.

[14] D. Murakami, Y. Yamagata, Micro grids clustering for electricity sharing: an approach considering micro urban structure, Energy Procedia 142 (2017) 2748–2753, doi: 10.1016/J.EGYPRO.2017.12.220.

[15] S. Mohseni, A.C. Brent, D. Burmester, A comparison of metaheuristics for the optimal capacity planning of an isolated, battery-less, hydrogen-based micro-grid, Appl. Energy 259 (2020) 114224, doi: 10.1016/J.APENERGY.2019.114224.

[16] M.J. Salehpour, S.M.M. Tafreshi, Contract-based utilization of plug-in electric vehicle batteries for day-ahead optimal operation of a smart micro-grid, J. Energy Storage 27 (2020) 101157, doi: 10.1016/J.EST.2019.101157.

[17] S.H.C. Cherukuri, B. Saravanan, G. Arunkumar, Experimental evaluation of the performance of virtual storage units in hybrid micro grids, Int. J. Electr. Power Energy Syst. 114 (2020) 105379, doi: 10.1016/J.IJEPES.2019.105379.

[18] F.M. Shakeel, O.P. Malik, On-line self-tuning adaptive control of an inverter in a grid-tied micro-grid, Electr. Power Syst. Res. 178 (2020) 106045, doi: 10.1016/J.EPSR.2019.106045.

[19] M.U. Hassan, M. Humayun, R. Ullah, B. Liu, Z. Fang, Control strategy of hybrid energy storage system in diesel generator based isolated AC micro-grids, J. Electr. Syst. Inf. Technol. 5 (2018) 964–976, doi: 10.1016/J.JESIT.2016.12.002.

[20] J. Zhang, L. Huang, J. Shu, H. Wang, J. Ding, Energy management of PV-diesel-battery hybrid power system for island stand-alone micro-grid, Energy Procedia 105 (2017) 2201–2206, doi: 10.1016/J.EGYPRO.2017.03.622.

[21] A. Baldinelli, L. Barelli, G. Bidini, G. Discepoli, Economics of innovative high capacity-to-power energy storage technologies pointing at 100% renewable micro-grids, J. Energy Storage 28 (2020) 101198, doi: 10.1016/J.EST.2020.101198.

[22] W. Liu, N. Li, Z. Jiang, Z. Chen, S. Wang, J. Han, et al. Smart micro-grid system with wind/PV/battery, Energy Procedia 152 (2018) 1212–1217, doi: 10.1016/J.EGYPRO.2018.09.171.

[23] S. Weckmann, A. Sauer, DC micro grid for energy efficient and flexible production, Procedia Manuf. 39 (2019) 655–664, doi: 10.1016/J.PROMFG.2020.01.440.

[24] N. Wu, H. Wang, Deep learning adaptive dynamic programming for real time energy management and control strategy of micro-grid, J. Cleaner Prod. 204 (2018) 1169–1177, doi: 10.1016/J.JCLEPRO.2018.09.052.

[25] Y. Liu, K. Zuo, X. (Amy) Liu, J. Liu, J.M. Kennedy, Dynamic pricing for decentralized energy trading in micro-grids, Appl. Energy 228 (2018) 689–699, doi: 10.1016/J.APENERGY.2018.06.124.

[26] X. Luo, P. Peng, Y. Shao, J. Li, G. Yu, A coordinated power control strategy for urban micro-grid, Energy Procedia 158 (2019) 6626–6631, doi: 10.1016/J.EGYPRO.2019.01.043.

[27] D. Kumar, H.D. Mathur, S. Bhanot, R.C. Bansal, Modeling and frequency control of community micro-grids under stochastic solar and wind sources, Eng. Sci. Technol. Int. J. 23 (2020) 1084–1099, doi: 10.1016/J.JESTCH.2020.02.005.

[28] S. Numminen, S. Yoon, J. Urpelainen, P. Lund, An evaluation of dynamic electricity pricing for solar micro-grids in rural India, Energy Strategy Rev. 21 (2018) 130–136, doi: 10.1016/J.ESR.2018.05.007.

[29] C. Hua, Y. Wang, S. Wu, Stability analysis of micro-grid frequency control system with two additive time-varying delay, J. Franklin Inst. 357 (2020) 4949–4963, doi: 10.1016/J.JFRANKLIN.2019.08.013.

[30] L. Raju, A.A. Morais, R. Rathnakumar, P. Soundaryaa, L.D. Thavam, Micro-grid grid outage management using multi agent systems, Energy Procedia 117 (2017) 112–119, doi: 10.1016/J.EGYPRO.2017.05.113.

[31] S.M. Moghaddas-Tafreshi, S. Mohseni, M.E. Karami, S. Kelly, Optimal energy management of a grid-connected multiple energy carrier micro-grid, Appl. Therm. Eng. 152 (2019) 796–806, doi: 10.1016/J.APPLTHERMALENG.2019.02.113.

[32] L. Barelli, G. Bidini, F. Bonucci, A. Ottaviano, Residential micro-grid load management through artificial neural networks, J. Energy Storage 17 (2018) 287–298, doi: 10.1016/J.EST.2018.03.011.

[33] N. Dkhili, J. Eynard, S. Thil, S. Grieu, A survey of modelling and smart management tools for power grids with prolific distributed generation, Sustain. Energy Grids Networks 21 (2020) 100284, doi: 10.1016/J.SEGAN.2019.100284.

[34] Q. Wang, W. Yao, J. Fang, X. Ai, J. Wen, X. Yang, et al. Dynamic modeling and small signal stability analysis of distributed photovoltaic grid-connected system with large scale of panel level DC optimizers, Appl. Energy 259 (2020) 114132, doi: 10.1016/J.APENER-GY.2019.114132.

[35] C. Ma, J. Dasenbrock, J.-C. Töbermann, M. Braun, A novel indicator for evaluation of the impact of distributed generations on the energy losses of low voltage distribution grids, Appl. Energy 242 (2019) 674–683, doi: 10.1016/J.APENERGY.2019.03.090.

[36] C. Ma, J.-H. Menke, J. Dasenbrock, M. Braun, M. Haslbeck, K.-H. Schmid, Evaluation of energy losses in low voltage distribution grids with high penetration of distributed generation, Appl. Energy 256 (2019) 113907, doi: 10.1016/J.APENERGY.2019.113907.

[37] X. Luo, X. Wang, M. Zhang, X. Guan, Distributed detection and isolation of bias injection attack in smart energy grid via interval observer, Appl. Energy 256 (2019) 113703, doi: 10.1016/J.APENERGY.2019.113703.

[38] M. Azab, Multi-objective design approach of passive filters for single-phase distributed energy grid integration systems using particle swarm optimization, Energy Rep. 6 (2020) 157–172, doi: 10.1016/J.EGYR.2019.12.015.

[39] C. Gavriluta, C. Boudinet, F. Kupzog, A. Gomez-Exposito, R. Caire, Cyber-physical framework for emulating distributed control systems in smart grids, Int. J. Electr. Power Energy Syst. 114 (2020) 105375, doi: 10.1016/J.IJEPES.2019.06.033.

[40] J.-H. Menke, N. Bornhorst, M. Braun, Distribution system monitoring for smart power grids with distributed generation using artificial neural networks, Int. J. Electr. Power Energy Syst. 113 (2019) 472–480, doi: 10.1016/J.IJEPES.2019.05.057.

[41] C. Li, H. Zhou, J. Li, Z. Dong, Economic dispatching strategy of distributed energy storage for deferring substation expansion in the distribution network with distributed generation and electric vehicle, J. Cleaner Prod. 253 (2020) 119862, doi: 10.1016/J.JCLEPRO.2019.119862.

[42] S.A. Bozorgavari, J. Aghaei, S. Pirouzi, A. Nikoobakht, H. Farahmand, M. Korpås, Robust planning of distributed battery energy storage systems in flexible smart distribution networks: a comprehensive study, Renew. Sustain. Energy Rev. 123 (2020) 109739, doi: 10.1016/J.RSER.2020.109739.

[43] J. Yuan, C. Cui, Z. Xiao, C. Zhang, W. Gang, Performance analysis of thermal energy storage in distributed energy system under different load profiles, Energy Convers. Manage. 208 (2020) 112596, doi: 10.1016/J.ENCONMAN.2020.112596.

[44] L. Tao, Y. Gao, Real-time pricing for smart grid with distributed energy and storage: a non-cooperative game method considering spatially and temporally coupled constraints, Int. J. Electr. Power Energy Syst. 115 (2020) 105487, doi: 10.1016/J.IJEPES.2019.105487.

[45] C. Sun, K. Yuan, T. Zhao, G. Song, X. Yang, Y. Song, Operational strategy based evaluation method of distributed energy storage system in active distribution networks, Energy Procedia 158 (2019) 1027–1032, doi: 10.1016/J.EGYPRO.2019.01.249.

[46] A. Kumar, N.K. Meena, A.R. Singh, Y. Deng, X. He, R.C. Bansal, et al. Strategic integration of battery energy storage systems with the provision of distributed ancillary services in active distribution systems, Appl. Energy 253 (2019) 113503, doi: 10.1016/J.APENER-GY.2019.113503.

[47] A.M. Howlader, S. Sadoyama, L.R. Roose, Y. Chen, Active power control to mitigate voltage and frequency deviations for the smart grid using smart PV inverters, Appl. Energy 258 (2020) 114000, doi: 10.1016/J.APENERGY.2019.114000.

[48] H. Almasalma, S. Claeys, G. Deconinck, Peer-to-peer-based integrated grid voltage support function for smart photovoltaic inverters, Appl. Energy 239 (2019) 1037–1048, doi: 10.1016/J.APENERGY.2019.01.249.

[49] A.M. Howlader, S. Sadoyama, L.R. Roose, S. Sepasi, Distributed voltage regulation using Volt-Var controls of a smart PV inverter in a smart grid: an experimental study, Renew. Energy 127 (2018) 145–157, doi: 10.1016/J.RENENE.2018.04.058.

[50] L. Yang, J. Liu, C. Wang, A novel parameter design for level grid-connected smart inverters, Cogn. Syst. Res. 52 (2018) 775–784, doi: 10.1016/J.COGSYS.2018.09.016.

[51] E. Kabalci, Review on novel single-phase grid-connected solar inverters: circuits and control methods, Sol. Energy 198 (2020) 247–274, doi: 10.1016/J.SOLENER.2020.01.063.

[52] R. Dogga, M.K. Pathak, Recent trends in solar PV inverter topologies, Sol. Energy 183 (2019) 57–73, doi: 10.1016/J.SOLENER.2019.02.065.

[53] M.H. Mahlooji, H.R. Mohammadi, M. Rahimi, A review on modeling and control of grid-connected photovoltaic inverters with LCL filter, Renew. Sustain. Energy Rev. 81 (2018) 563–578, doi: 10.1016/J.RSER.2017.08.002.

[54] N.N. Nam, M. Choi, Y. Il Lee, Model predictive control of a grid-connected inverter with LCL filter using robust disturbance observer, IFAC-PapersOnLine 52 (2019) 135–140, doi: 10.1016/J.IFACOL.2019.08.168.

[55] C.C. Gomes, A.F. Cupertino, H.A. Pereira, Damping techniques for grid-connected voltage source converters based on LCL filter: an overview, Renew. Sustain. Energy Rev. 81 (2018) 116–135, doi: 10.1016/J.RSER.2017.07.050.

[56] S. Li, X. Fu, M. Ramezani, Y. Sun, H. Won, A novel direct-current vector control technique for single-phase inverter with L, LC and LCL filters, Electr. Power Syst. Res. 125 (2015) 235–244, doi: 10.1016/J.EPSR.2015.04.006.

[57] D.I. Brandao, L.S. de Araújo, T. Caldognetto, J.A. Pomilio, Coordinated control of three- and single-phase inverters coexisting in low-voltage microgrids, Appl. Energy 228 (2018) 2050–2060, doi: 10.1016/J.APENERGY.2018.07.082.

[58] A. Ibrahim, M.Z. Sujod, Variable switching frequency hybrid PWM technique for switching loss reduction in a three-phase two-level voltage source inverter, Measurement 151 (2020) 107192, doi: 10.1016/J.MEASUREMENT.2019.107192.

[59] M.R. Miveh, M.F. Rahmat, A.A. Ghadimi, M.W. Mustafa, Control techniques for three-phase four-leg voltage source inverters in autonomous microgrids: a review, Renew. Sustain. Energy Rev. 54 (2016) 1592–1610, doi: 10.1016/J.RSER.2015.10.079.

[60] J. Hu, M. Marinelli, M. Coppo, A. Zecchino, H.W. Bindner, Coordinated voltage control of a decoupled three-phase on-load tap changer transformer and photovoltaic inverters for managing unbalanced networks, Electr. Power Syst. Res. 131 (2016) 264–274, doi: 10.1016/J.EPSR.2015.10.025.

[61] S. Dowruang, P. Bumrungsri, C. Jeraputra, Improved voltage vector sequences on model predictive control for a grid connected three phase voltage source inverter, Procedia Comput. Sci. 86 (2016) 393–396, doi: 10.1016/J.PROCS.2016.05.041.

Chapter 9

The Role and Importance of Energy Storage Systems in Solar Hybrid Applications

1 Energy storage systems in solar hybrid systems

Today, solar energy is one of the various energy sources, which is the subject of research as an alternative to fossil and nuclear fuels. Solar energy is rapidly becoming widespread due to its advantages such as being an infinite and widespread source, and it can easily be converted into electrical energy. Another important feature of solar energy systems is that it offers a wide range of power from a few watts (W) to megawatts (MW) [1,2].

Photovoltaic (PV) panels produce direct current (DC). For this reason, all kinds of electrically powered loads can be fed directly or with the help of a converter from the PV panel. There are two types of application areas as the topology of the use of PV panel technology. These topologies can be examined under the heading of stand-alone (off-grid) and grid-connected systems. Off-grid systems are road lighting, small powerful autonomous applications, water pumping systems, signaling systems, and domestic applications located in remote areas. There are generally PV panels and battery energy storage system in off-grid systems. Energy from PV panels is used directly or stored in batteries. When the PV panel does not generate energy, the energy stored in the batteries is used [3–5].

Depending on the system structure, PV panels are operated as DC or alternating current with an inverter. Grid-connected systems work synchronously with the grid and transfer energy directly to the grid. Grid-connected systems may be small-powered examples of domestic applications. There may be large power plants to meet the energy needs of a particular region [6,7].

In this chapter, an exemplary solar hybrid system design has been realized. A grid-connected system was simulated by using PV panel array and inverter. PV panels were connected to the DC bus via a converter. PV panels were controlled by the perturb&observe algorithm using maximum power point tracing (MPPT) technique. Thus maximum power transfer is ensured from the PV

Solar Hybrid Systems. http://dx.doi.org/10.1016/B978-0-323-88499-0.00009-4

179

panels to transfer maximum power to the grid and the load. A three-phase four-arm inverter circuit was used as inverter structure. The reason for using the three-phase four-arm inverter structure is to allow feeding of unbalanced loads that may take place in the system. The inverter works synchronously with the grid and transfers the energy generated from the PV panels to the grid when there is no load.

2   Simulation of solar hybrid system

A MATLAB/Simulink block diagram of a simulation solar hybrid system is given in Fig. 9.1. The PV panel array, battery group, and three-phase grid were used in this simulation study. The MPPT technique and hysteresis current controller were used as control methods in solar hybrid system.

In the simulation study, it is aimed at operating the grid-connected PV inverter structure at the maximum power point with MPPT technique. As the MPPT method, the perturb&observe algorithm was used, which is widely used

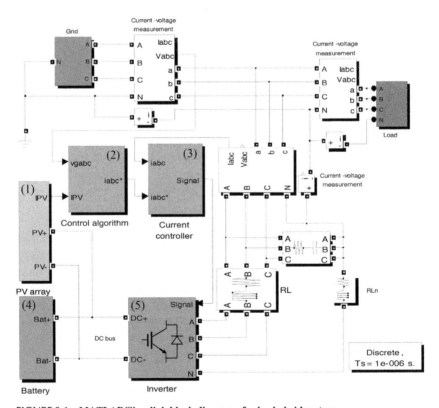

FIGURE 9.1   MATLAB/Simulink block diagram of solar hybrid system.

in terms of applicability and cost, and also has a fast response time against changes in weather conditions. With this algorithm the system gets more efficient operation by taking maximum power from PV panels. The reference signal produced in the perturb&observe algorithm is processed in the current controller unit, and suitable operating signals are produced for the single-stage inverter.

For simulation of real weather conditions, working conditions under dynamic condition were examined. Thus the dynamic behavior of the perturb&observe algorithm on the system was observed. In the simulation study the weather condition was changed in 0.3 second of the system. The response time of the MPPT algorithm and the speed of capturing the maximum power point were determined.

PV panel inverter structure transfers energy synchronously depending on the grid. The inverter structure feeds the load group at the same time and transfers energy to the grid. The solar hybrid system can feed balanced loads with its neutral line structure as well as respond to single-phase and unbalanced loads. Under the same conditions the current and voltage of the inverter are examined by replacing the balanced load group connected to the system with the unbalanced load group.

The six serial, two orders of parallel PV panel arms were created and 12 PV panels were simulated. In simulation, current, voltage, and power values were obtained by simulating PV panel array. System parameters used in simulation are given in Table 9.1.

TABLE 9.1 Solar hybrid system simulation parameters.

Parameters		Value
PV array	Open circuit voltage (V_{oc})	500 V
	Short circuit current (I_{ph})	2.5 A
	MPPT voltage (V_m)	410 V
	MPPT current (I_m)	2 A
Grid	Voltage (V_{gabc})	110 V_{rms}/phase-neutral
	Frequency (f)	50 Hz
	Impedance (R_g, L_g)	5 mΩ, 2 mH
Battery (lead–acid)	Voltage	48 V
	Current capacity	25 Ah
	Power capacity	1200 Wh
Load	Ohmic load	Single-/three-phase

2.1 PV array and perturb&observe control algorithm

In Fig. 9.1, block (1) contains PV panels and the perturb&observe control algorithm with MPPT method. A MATLAB/Simulink block diagram of PV panels and MPPT algorithm is given in Fig. 9.2. The perturb&observe method observes the PV panel voltage or current that fluctuates depending on the weather conditions and changes the working points. It works by comparing the PV panel output power with the previous power value. PV panel operating voltage changes and if the power increases, the control system changes PV panel operates by pointing in this direction. Otherwise, the working point moves in the opposite direction. In the next cycle the algorithm continues in the same way.

The PV panel output power is constantly monitored in the deflection method. It is decided to decrease or increase the reference by establishing a relation between the movement of the control variable and the movement of the power [8,9]. The program codes of the perturb&observe algorithm are given in Table 9.2.

The input and output parameters of the system must be known in order to create the PV panel array in the simulation. Using Eq. (9.1) obtained from the electrical model of the PV panel, the PV panel array has simulated. In the PV panel model the input parameters are PV input current and radiation, and the output parameters are current and power. Switching signals are produced by processing the PV panel output current in the control unit. The PV panel power value is continuously operated at the maximum power point with the MPPT

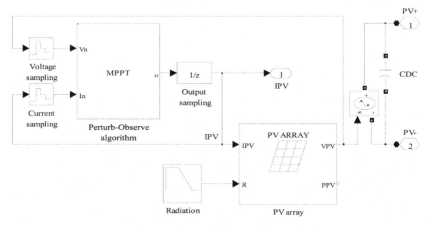

FIGURE 9.2 PV panel and perturb&observe algorithm block diagram.

TABLE 9.2 Program code of the MPPT algorithm.

Program code of the perturb&observe algorithm (MATLAB):

```
function o = MPPT_PO(vn,in)
persistent io;
persistent po;
persistent vo;
persistent iro;
persistent dsign;
if isempty(io)
io = 0;
po = 0;
vo = 0;
iro = 0.6;
dsign = 1;
end
delta = 0.02;
pn = vn*in;
if pn > = po
dsign = dsign;
else
dsign = 0-dsign;
end
irn = iro + dsign*delta;
vo = vn;
io = in;
po = pn;
iro = irn;
o = irn;
end
```

algorithm according to the entered radiation value [10–12]. The block diagram of the PV panel is given in Fig. 9.3.

$$I_{PV} = N_p I_{ph} - I_o \left[e^{\left(\frac{q\left(\frac{V_d}{R_s} + \frac{IR_s}{R_p} \right)}{N_s KFT_{PV}} \right)} - 1 \right] - \frac{N_p V_d / N_s}{R_p} \tag{9.1}$$

2.2   Battery model and bidirectional DC/DC converter model

In the battery simulation the lead–acid battery model in the MATLAB/Simulink library was used. This model simulates the battery's voltage, capacity, and the battery state of charge (SOC) parameters. The battery-equivalent circuit used in the simulation is given in Fig. 9.4.

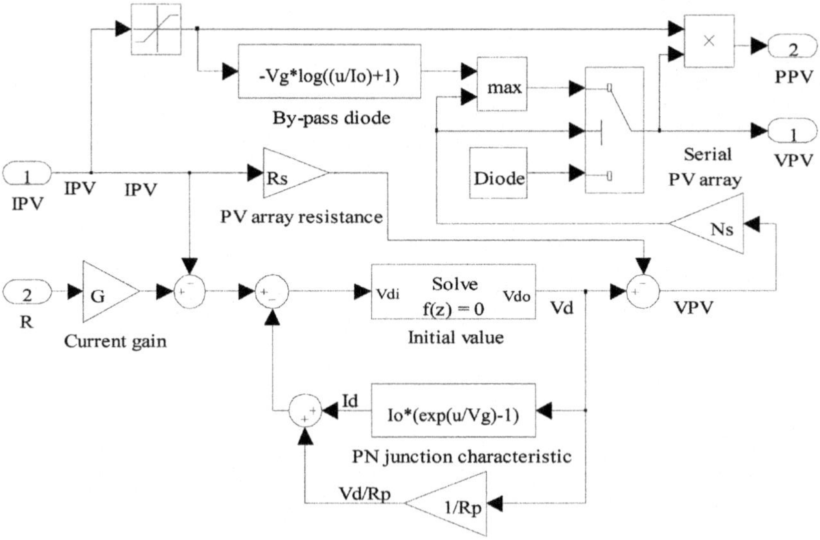

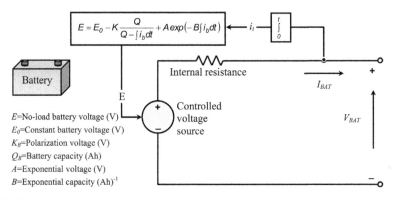

FIGURE 9.3 PV panel block diagram.

$$E = E_0 - K\frac{Q}{Q - \int i_b dt} + A\,exp\left(-B\int i_b dt\right)$$

FIGURE 9.4 Battery equivalent circuits.

E=No-load battery voltage (V)
E_0=Constant battery voltage (V)
K_B=Polarization voltage (V)
Q_B=Battery capacity (Ah)
A=Exponential voltage (V)
B=Exponential capacity $(Ah)^{-1}$

A bidirectional DC/DC converter structure is used in the battery group, which is given in Fig. 9.5A. DC bus voltage is determined as 400 V in solar hybrid system. The voltage level of the battery group is 48 V, which is lower than the DC bus. It is not suitable to use only a buck or boost DC/DC converter in the system. For this purpose a bidirectional DC/DC converter is used, which can charge and discharge the battery. Increasing the voltage level of the battery group to 400 V and transferring energy to the DC bus is undertaken by the bidirectional DC/DC converter unit. In this case the S_{B5} switch is passive as in Fig. 9.5B and the pulse with modulation (PWM) signal is applied to the S_{B6} switch. Likewise, it provides the battery group to be charged with the excess

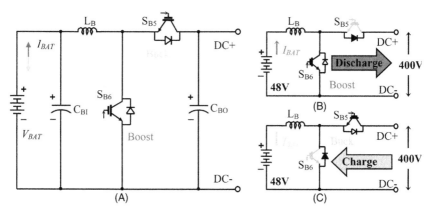

FIGURE 9.5 Bidirectional DC/DC converter structure.

energy produced by the solar system. In charging mode the circuit reduces from 400 to –48 V. In this case, as in Fig. 9.5C, S_{B6} switch is inactive and PWM signal is applied to S_{B5} switch. In the bidirectional DC/DC converter structure, a coil, two semiconductor switching elements and capacitors are used.

2.3   Hysteresis band current controller and phase-locked loop (PLL)

The hysteresis band current controller compares the i_{abc*} reference current calculated at the control unit with the i_{abc} current at the inverter output. As a result of comparison, switching signals of the inverter are produced. Depending on the current fault, the inverter current is tried to be kept in the band defined by the hysteresis controller [13–15]. A block diagram where inverter switching signals are generated using a hysteresis band current controller is given in Fig. 9.6.

One of the important issues in grid-connected PV power systems is grid synchronization. The angle of the grid voltage vector is determined by the synchronization algorithm used. Depending on this angle value, the switching elements of the inverter are triggered and MPPT control is performed in the system. Depending on the angle of the grid voltage vector, three-phase variables are reduced to the axis set rotating at synchronous speed.

Phase-locked loop (PLL) is used to give a pure sinusoidal current reference and to enable the inverter to operate at the unit power factor. PLL provides a synchronization of mains voltage and inverter output current. It is possible to determine the placement time and damping coefficients directly with the proportional integral (PI) control parameters of the PLL structure. PLL structure is also used to obtain amplitude and grid voltage frequency values and to monitor grid voltage. The angle produced is crucial for generating the reference i_{abc*} signal in the dq0/abc transformation (park transformation) [16–19]. In Fig. 9.7, there is a park transformation and PLL control unit that processes the PV current value coming from the MPPT so that the three-phase inverter can transfer energy to the grid synchronously.

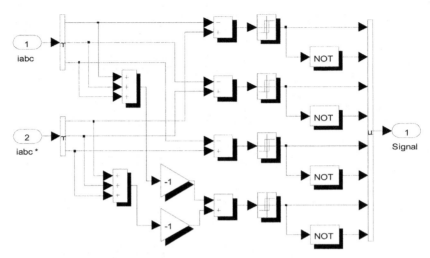

FIGURE 9.6 Hysteresis band current controller block diagram.

Park transformation is used to reduce three-phase sizes to the axis set (dq0) rotating at synchronous speed. The same conversion is used to obtain three-phase sizes with the DC component. Using the I_{PV} value generated in the MPPT method in the control algorithm block, i_{abc*} reference signals are generated for the inverter to the current controller. With this transformation the rotating vector in the fixed axis tool takes a constant value. The d-axis component is equal to the amplitude of the three-phase magnitudes, while the q-axis component is zero in the balanced system. The dq-axis set rotates at the same speed as the variables in the three-phase system, and the angle produced in PLL is used for conversion. The resultant vector moves at a constant speed relative to the fixed axis set, while the dq remains constant relative to the axis set. Therefore the sizes in the dq-axis tool are DC. When it is desired to reduce the sizes in the three-phase system directly to the axis set rotating at synchronous speed, the park transformation given in Eq. (9.2) is used [20]. The d-axis is I_{PV}, which is the PV array panel current, and the q-axis is 0.

$$\begin{bmatrix} I_{PVd} \\ I_{PVq} \end{bmatrix} = \frac{2}{3} \begin{bmatrix} \cos\theta & \cos\left(\theta - \frac{2\pi}{3}\right) & \cos\left(\theta - \frac{4\pi}{3}\right) \\ -\sin\theta & -\sin\left(\theta - \frac{2\pi}{3}\right) & -\sin\left(\theta - \frac{4\pi}{3}\right) \end{bmatrix} \begin{bmatrix} i_a * \\ i_b * \\ i_c * \end{bmatrix} \tag{9.2}$$

3 Simulation results of solar hybrid system

The grid-connected PV system was operated according to two different situations, and PV current, voltage, power, grid voltage, current, inverter current, and load currents are examined. In the first case, it was examined that there is

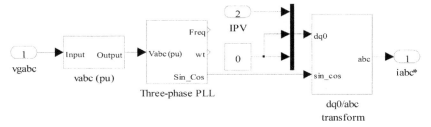

FIGURE 9.7 **PLL and park transform (dq0/abc) block diagram.**

no battery energy storage unit in the system. In the case of no battery group the load power is supplied from the grid due to the PV panel power fluctuation. Since the PV output characteristic is instability, the power demand of the load from the grid is also fluctuation.

In the second case where battery energy storage unit is added to the PV solar system, the power fluctuation is covered by the energy storage unit. Battery energy storage unit discharges depend on the load power during power surge transitions. Thus there is no fluctuating power demand from the grid. Through solar hybrid systems, there is a more stable energy demand from the grid. Power stresses on the grid are compensated by energy storage units.

Simulation results of control structures were obtained using MATLAB/ Simulink Power System Toolbox software. Depending on these two different situations, the perturb&observe algorithm is shown to operate the response time and PV panels at the maximum power point. The PV inverter structure is operated as balanced and unbalanced load groups under the same radiation conditions. In the balanced state, there is a resistance of 75 Ω in each phase; in the case of unbalanced load, there is a resistance of 90 Ω in the C phase, and in A and B phases, it is 75 Ω. A power flow diagram of solar system (not including battery) and balanced load status is given in Fig. 9.8.

The simulation works for a total of 0.5 second. The radiation is 1000 W/m^2 for 0.2 second, decreases to 500 W/m^2 for 0.4 second, and is operated at a radiation value of 500 W/m^2 in the range of 0.4–0.5 second. This radiation change can occur depending on the atmospheric conditions. The purpose is to see the

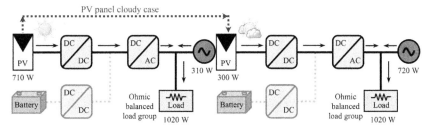

FIGURE 9.8 **Solar system (not include battery) and balanced load status.**

fast response time of the MPPT algorithm and to examine how the system can react in the case of possible rapid changes. In the first case, there is only the PV system, the grid, and the load group. Comprehensive simulation results for this situation are given in Figs. 9.9–9.11 and Table 9.3.

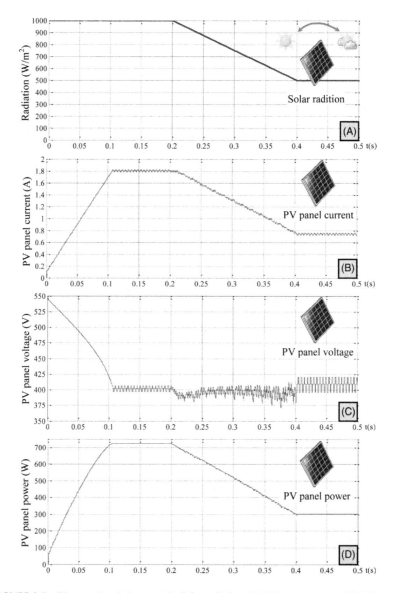

FIGURE 9.9 **The graphs of change.** (A) Solar radiation, (B) PV panel current, (C) PV panel voltage, (D) PV panel power.

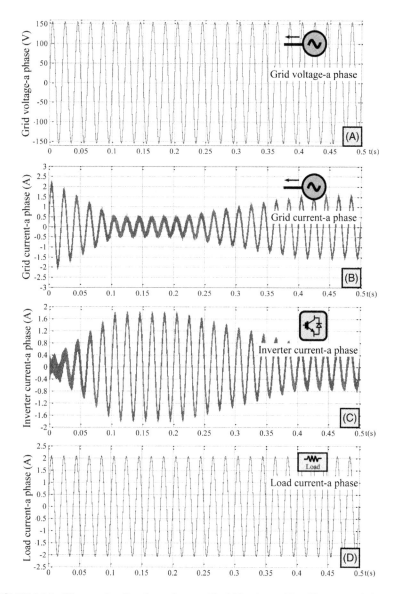

FIGURE 9.10 The graphs of a-phase change. (A) Grid voltage, (B) grid current, (C) inverter current, (D) load current.

Since the neutral load current is zero in a balanced load status, the grid and inverter neutral current is also zero. The neutral current generated when there is an unbalanced load status is provided by the inverter depending on the PV panel power. A neutral current close to zero passes through the grid.

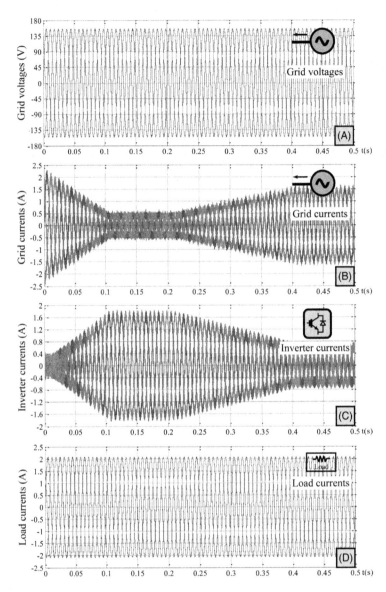

FIGURE 9.11   The graphs of three-phase change. (A) Grid voltages, (B) grid currents, (C) inverter currents, (D) load currents.

In this case the solar radiation value is simulated according to its characteristic, which includes a variable weather. As soon as the system is first working, the MPPT algorithm finds the maximum power point in a short time in 0.1 second. This time is the time until the PV panel current starts from zero at first, until the current value at the maximum power point reaches the curve on the graph.

TABLE 9.3 Summary of the measured values in the PV solar system.

Parameters		0.1 second			0.2 second			0.3 second			0.4 second			0.5 second		
Radiation (W/m²)		1000			1000			752.5			500			500		
PV current (A)		1.69			1.80			1.31			0.74			0.74		
PV voltage (V)		420.3			402			400.3			409.7			409.7		
PV power (W)		714.5			723.6			528.3			303.2			303.2		
Balanced load (rms)	Grid current (A)	0.28			0.18			0.53			0.91			0.91		
	Inverter current (A)	1.18			1.28			0.93			0.55			0.55		
	Load current (A)	1.46			1.46			1.46			1.46			1.46		
Unbalanced load (rms)	Phases	A	B	C	A	B	C	A	B	C	A	B	C	A	B	C
	Grid current (A)	0.28	0.28	0.14	0.18	0.18	0.07	0.53	0.53	0.43	0.91	0.91	0.79	0.91	0.91	0.79
	Inverter current (A)	1.18	1.18	1.08	1.28	1.28	1.15	0.93	0.93	0.79	0.55	0.55	0.43	0.55	0.55	0.43
	Load current (A)	1.46	1.46	1.22	1.46	1.46	1.22	1.46	1.46	1.22	1.46	1.46	1.22	1.46	1.46	1.22

This period varies depending on the radiation value entered. When the balanced ohmic load (75 Ω, three-phase) is connected to the system, the load continuously draws a total of 1.46 A. Due to the radiation change, the inverter transfers current to the grid for 1.18 A for 0.1 second. Inverter provides some of the total current that the load must draw. The remaining current is drawn from the 0.28 A grid and the PV system's contribution to the grid is shown.

If the unbalanced ohmic load (A, B phase 75 Ω, C phase 90 Ω) is connected to the system, the load in C phase is 1.22 A. Loads in A and C phase draw a current of 1.46 A. Since the PV panel is operated under the same radiation conditions, it transfers the same current value to the grid for A, B phases. C phase varies depending on the load. For 0.1 second the inverter transfers 1.18 A current to the grid. The current value of 1.22 A of the load in C phase is drawn from the grid as 0.14 A. 1.18 A of the loads that draw a total of 1.46 A in A, and B phase is drawn from the inverter and 0.28 A from the grid. These values are given in detail in Table 9.3. Depending on the unbalanced load, the inverter is transferring different currents to the grid, and when the load is unbalanced, the unbalance is met by the inverter depending on the PV panel power. Since the neutral current is zero in the balanced load case, the inverter and the grid neutral current are also zero. In the case of unbalanced load, the difference current in the neutral line is provided by the inverter. Grid neutral current is close to zero. Inverter phase currents are given in detail in Figs. 9.12 and 9.13.

The solar system supports single and unbalanced loads thanks to its three-phase four-arm inverter structure. An unbalanced load is connected to the system under the first solar radiation conditions. In the load group, there is an ohmic load of 280 W in A phase and 380 W in B and C phases. Thanks to the

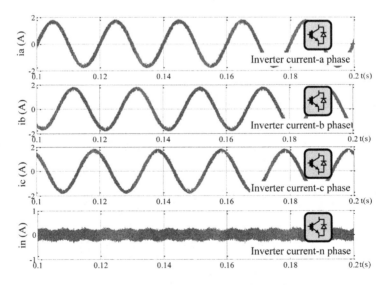

FIGURE 9.12   Currents of inverter A phase, B phase, C phase, and neutral.

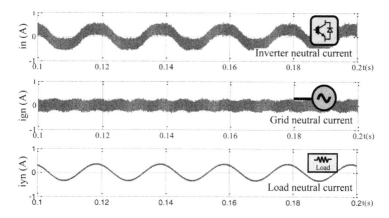

FIGURE 9.13   Currents of inverter, grid, and load neutral.

inverter structure, balanced load is demanded from the grid continuously. The unbalance in the load group is met with the inverter. Solar system (not including battery) and unbalanced load status power flow is given in Fig. 9.14. First, the PV panel produces 710 W. The load group demands an unbalanced load of 960 W in total. In this case, there is only 250 W power flow from the grid to the load group. When it is cloudy, the PV panel generates 300 W. The difference power requested by the load is 660 W provided from the grid. Although PV panels have produced low power, the unbalance load demand is provided by the inverter thanks to its inverter structure. The graphs of three-phase change under unbalanced load, grid currents, inverter currents, and load currents, respectively, are given in detail in Fig. 9.15.

The battery group was added to the solar system to which the PV panels are connected to the grid, and the results were examined. Lead–acid battery group with 48 V, 25 Ah, and 1200 Wh power capacity was added to the PV panels. Thus a solar hybrid structure containing an energy storage system was created. As in the previous tests, solar radiation changes are at the same value. In these test conditions the contribution of the battery group to the PV panels is clearly

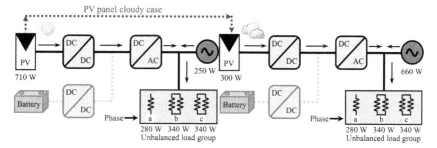

FIGURE 9.14   Solar system (not including battery) and unbalanced load status.

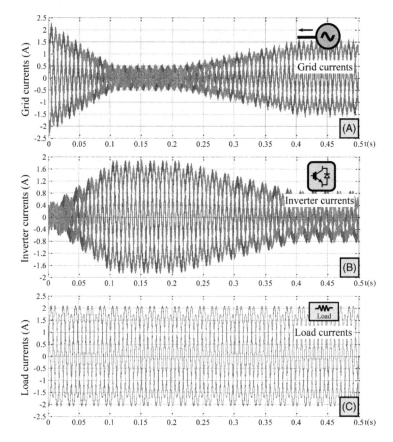

FIGURE 9.15 **The graphs of three-phase change under unbalanced load.** (A) Grid currents, (B) inverter currents, (C) load currents.

shown together with the graphic results. The power flow diagram of the solar hybrid system and the balanced load case containing the battery energy storage system is given in Fig. 9.17.

In this case, PV panel power generates power such as 710 W and then 300 W. The battery group is connected to the DC bus with a bidirectional DC/DC converter. DC bus voltage and battery group output voltage is 400 V. Thanks to the battery group, the voltage and the power fluctuations of the solar system in the DC bus are prevented. Depending on the power output of the PV panels, the battery group discharges at different current values for the continuous providing of the load. The battery group discharges with 0.9 A in the case of high solar radiation. Power demand of the load group is provided by solar hybrid system. In this case, there is no energy demand from the grid. When the solar radiation value changes, PV panels generate 300 W. Thanks to the solar hybrid system structure, the discharge current of the battery group changes in real time and meets

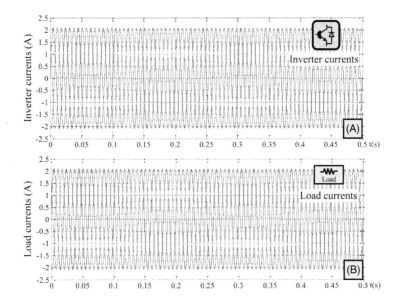

FIGURE 9.16 **The graphs of three-phase change.** (A) Inverter currents, (B) load currents.

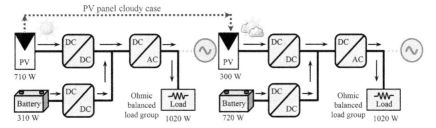

FIGURE 9.17 **Solar hybrid system (include battery) and balanced load status.**

the load demand power with 1.8 A. The battery SOC is 77 % and the discharge decreases over time depending on the amount of current. Voltage, current, and SOC values of the battery group are given in Fig. 9.18, respectively.

The inverter and the load of three-phase currents are given in Fig. 9.16. In this case, since there is no energy demand from the grid, the grid phase currents are 0. All power demanded by the load is provided by the PV panel and the battery group.

One of the most important advantages of solar hybrid systems is a complementary factor for PV panels whose output characteristics change depending on the weather conditions. With the energy storage unit in the system the ripple effects are prevented in the DC bus. Thanks to the battery group, quality energy

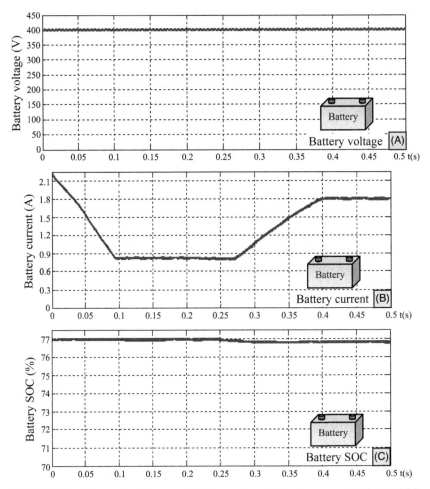

FIGURE 9.18 The graphs of change. (A) Battery voltage, (B) battery current, (C) battery SOC.

is provided and the continuity of energy is maintained. Solar hybrid systems occupy the grid as little as possible regarding energy use. In contrast, solar hybrid systems support the grid structure.

References

[1] A. Awasthi, A.K. Shukla, S.R. M.M., C. Dondariya, K.N. Shukla, D. Porwal, et al. Review on sun tracking technology in solar PV system, Energy Rep. 6 (2020) 392–405, doi: 10.1016/J. EGYR.2020.02.004.

[2] P.-C. Hsu, B.-J. Huang, P.-H. Wu, W.-H. Wu, M.-J. Lee, J.-F. Yeh, et al. Long-term energy generation efficiency of solar PV system for self-consumption, Energy Procedia 141 (2017) 91–95, doi: 10.1016/J.EGYPRO.2017.11.018.

[3] W.M.P.U. Wijeratne, R.J. Yang, E. Too, R. Wakefield, Design and development of distributed solar PV systems: do the current tools work?, Sustain. Cities Soc. 45 (2019) 553–578, doi: 10.1016/J.SCS.2018.11.035.

[4] A.S. Aziz, M.F.N. Tajuddin, M.R. Adzman, M.F. Mohammed, M.A.M. Ramli, Feasibility analysis of grid-connected and islanded operation of a solar PV microgrid system: a case study of Iraq, Energy 191 (2020) 116591, doi: 10.1016/J.ENERGY.2019.116591.

[5] K.N. Nwaigwe, P. Mutabilwa, E. Dintwa, An overview of solar power (PV systems) integration into electricity grids, Mater. Sci. Energy Technol. 2 (2019) 629–633, doi: 10.1016/J.MSET.2019.07.002.

[6] T. Jamal, C. Carter, T. Schmidt, G.M. Shafiullah, M. Calais, T. Urmee, An energy flow simulation tool for incorporating short-term PV forecasting in a diesel-PV-battery off-grid power supply system, Appl. Energy 254 (2019) 113718, doi: 10.1016/J.APENERGY.2019.113718.

[7] B.D. Mert, F. Ekinci, T. Demirdelen, Effect of partial shading conditions on off-grid solar PV/ Hydrogen production in high solar energy index regions, Int. J. Hydrogen Energy 44 (2019) 27713–27725, doi: 10.1016/J.IJHYDENE.2019.09.011.

[8] D. Verma, S. Nema, A.M. Shandilya, S.K. Dash, Maximum power point tracking (MPPT) techniques: recapitulation in solar photovoltaic systems, Renew. Sustain. Energy Rev. 54 (2016) 1018–1034, doi: 10.1016/J.RSER.2015.10.068.

[9] A. Manmohan, A. Prasad, R. Dharavath, S.P. Karthikeyan, I.J. Raglend, Up and down conversion of photons with modified perturb&observe MPPT technique for efficient solar energy generation, Energy Procedia 117 (2017) 786–793, doi: 10.1016/J.EGYPRO.2017.05.195.

[10] E. Moshksar, T. Ghanbari, A model-based algorithm for maximum power point tracking of PV systems using exact analytical solution of single-diode equivalent model, Sol. Energy 162 (2018) 117–131, doi: 10.1016/J.SOLENER.2017.12.054.

[11] M. Al-Addous, Z. Dalala, F. Alawneh, C.B. Class, Modeling and quantifying dust accumulation impact on PV module performance, Sol. Energy 194 (2019) 86–102, doi: 10.1016/J.SOLENER.2019.09.086.

[12] A.M. Humada, S.Y. Darweesh, K.G. Mohammed, M. Kamil, S.F. Mohammed, N.K. Kasim, et al. Modeling of PV system and parameter extraction based on experimental data: review and investigation, Sol. Energy 199 (2020) 742–760, doi: 10.1016/J.SOLENER.2020.02.068.

[13] H. Komurcugil, Double-band hysteresis current-controlled single-phase shunt active filter for switching frequency mitigation, Int. J. Electr. Power Energy Syst. 69 (2015) 131–140, doi: 10.1016/J.IJEPES.2015.01.010.

[14] T. Arun Srinivas, G. Themozhi, S. Nagarajan, Current mode controlled fuzzy logic based inter leaved cuk converter SVM inverter fed induction motor drive system, Microprocess. Microsyst. 74 (2020) 103002, doi: 10.1016/J.MICPRO.2020.103002.

[15] A. Chatterjee, K.B. Mohanty, Current control strategies for single phase grid integrated inverters for photovoltaic applications—a review, Renew. Sustain. Energy Rev. 92 (2018) 554–569, doi: 10.1016/J.RSER.2018.04.115.

[16] R. Sharma, S. Suhag, Virtual impedance based phase locked loop for control of parallel inverters connected to islanded microgrid, Comput. Electr. Eng. 73 (2019) 58–70, doi: 10.1016/J.COMPELECENG.2018.11.005.

[17] A. Kherbachi, A. Chouder, A. Bendib, K. Kara, S. Barkat, Enhanced structure of second-order generalized integrator frequency-locked loop suitable for DC-offset rejection in single-phase systems, Electr. Power Syst. Res. 170 (2019) 348–357, doi: 10.1016/J.EPSR.2019.01.029.

[18] W. Subsingha, Design and analysis three phase three level diode-clamped grid connected inverter, Energy Procedia 89 (2016) 130–136, doi: 10.1016/J.EGYPRO.2016.05.019.

[19] Y. Wang, X. Chen, Y. Wang, C. Gong, Analysis of frequency characteristics of phase-locked loops and effects on stability of three-phase grid-connected inverter, Int. J. Electr. Power Energy Syst. 113 (2019) 652–663, doi: 10.1016/J.IJEPES.2019.06.016.

[20] R. Escudero, J. Noel, J. Elizondo, J. Kirtley, Microgrid fault detection based on wavelet transformation and Park's vector approach, Electr. Power Syst. Res. 152 (2017) 401–410, doi: 10.1016/J.EPSR.2017.07.028.

Chapter 10

Distributed Solar Hybrid Generation Systems

1 Distributed solar hybrid systems

Electricity distribution means delivering quality to the home or workplace of the end user, in line with the grid standards. The on-site generation and distribution can be made with the distributed generation. Instead of distributed generation, it is also referred to as on-site generation, embedded generation, or decentralized generation. There are many small energy sources in the distributed energy system structure [1–3]. The distributed generation source species is given in Fig. 10.1. The heat and alternative fuels are produced in addition to electricity with these sources. The lost energy rate in transmission lines is reduced due to on-site generation and on-site consumption in distributed energy systems. Thus the number and the size of transmission lines required for the electricity needs of the same amount of receiver are reduced. Electricity energy saving and load management are carried out with distributed energy systems. In addition, with a distributed energy system structure, the need for the construction of transformer centers in the conventional network infrastructure is eliminated. The distributed grid structure can supply to the demands of peak loads [4].

It enables the establishment of a low-capacity power generation unit and energy storage systems in the distributed generation system. Distributed generation systems help one to solve energy quality problems by supporting the existing electricity grid with this hybrid structure. Power generation sources in the distributed generation system require low maintenance. In addition, these power generation units are environment friendly and have high efficiency rates. Experts, engineers, and complex systems are required to reduce environmental pollution in existing power generation plants. Since renewable energy sources are generally used as power generation units in the distributed generation system, these problems are eliminated [5–8]. The operation units that enable the system to work in distributed generation are natural processes such as sunlight, wind, and geothermal. All these natural processes reduce the size and cost of the power plant. The future power grid will not include high-capacity fossil fuel–fired power plants or high-voltage transmission lines. Each user in this new electricity energy supply infrastructure will have renewable energy technologies such as solar, wind, and biomass [9].

Solar Hybrid Systems. http://dx.doi.org/10.1016/B978-0-323-88499-0.00010-0

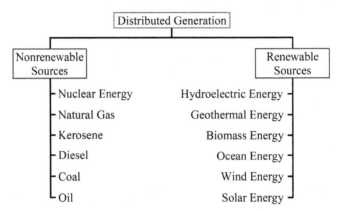

FIGURE 10.1   Distributed generation sources.

Solar and wind energy will be the primary source in distributed generation systems. In cases where the sun and wind are not enough, energy storage units will be complementary power. In addition, biomass energy will be used as both complementary power and heat energy source. The alternating power generated by solar and wind energy needs to be balanced. This power surge can be provided by chemical or thermal energy storage solutions. Future energy storage units can come in many sizes. While low-capacity energy storage units will be sufficient for a family or an apartment, high-capacity applications will be able to meet the needs of a region. Small-scale energy storage technologies are low in cost [10]. Distributed generation systems are an alternative to the traditional electricity grid system. In addition, small-scale power generation technologies enable the improvement of electrical energy quality for the existing grid infrastructure [11]. The cost of renewable energy technologies is falling and government support programs are increasing. All these developments enable the increase and development of distributed energy generation systems.

Distributed power generation units are in most cases environment-friendly technologies compared to traditional electricity generation technologies. It has become possible to reach the electricity generation structure with small-scale and plug-and-play features with low costs. In addition, these technologies can provide high power reliability. Conventional electrical power systems generally produce far from the user. However, in the distributed power generation unit, the electricity generation source is located very close to the consumer (such as building roof and home garden). This increases the reliability of access to energy. Distributed power generation sources can provide power between a few kW of low power capacity and several MW of high power capacity [12,13]. Thanks to the distributed power generation system capacities, it is possible to produce electrical energy close to the power values produced by thermal power plants with high capacity.

1.1 Distribution generation model

Generally, as a fuel source in high-capacity thermal-based electricity power generation plants, fossil-based fuels such as coal, oil, natural gas, or nuclear energy are used. High-voltage transmission lines are used to transmit the electrical energy generated by these power plants to end users. These power plants are often far from end users. These distances are tens to hundreds of kilometers away. This requires the transport of electrical energy over a long distance. These central power generation plants have many disadvantages [14]. Large-scale power plants have greenhouse gas emissions and generate nuclear waste. Power loss occurs in the transmission of electrical energy to long distances, which reduces the system efficiency. It provides negative environmental effects near the power plant area. Most of these negative effects can be avoided by the distributed power generation model [15]. By placing the electrical energy source close to the end user, the problems in the transmission lines are overcome. The distributed power generation system can be easily implemented, for example, with an array of solar photovoltaic (PV) panels. The solar PV panel system that has become widespread in many parts of the world can be grid connected as well as off-grid [16,17].

Thanks to the developing technological innovations, the use of renewable energy sources in the world is becoming popular. Increasing use of renewable energy sources helps to increase the percentage of electricity generation in the distributed power generation system. As a result, distributed power generation systems will grow significantly in the near future. The problem of the connection between existing grid networks and distributed power generation systems is an issue that needs to be solved. In this case, establishing technical requirements regarding the connection issue and the implementation of regulations are among the biggest difficulties [18]. Another problem in transferring the excess energy generated by distributed power generation systems to the existing grid is the issue of tariff pricing. The solution to this issue will be provided mostly by regulations and agreements made with public institutions. Distributed power generation systems today can generate satisfactory energy and can be connected synchronously with the grid frequency [19]. However, existing wholesale electricity markets are not designed to accommodate such resources and pay according to the amount of energy generated [20].

There are many challenges associated with the increased development of distributed power generation systems resources. The high number of renewable energy sources affects the system infrastructure. Connection of these sources to the existing grid is another problem and technical regulations are required [21]. A long-term guarantee payment system should be established for distributed power generation connected to the grid. Distributed power generation systems should be facilitated to join the wholesale electricity markets. However, the commercialization roles and limits of distributed power generation systems in the future distribution network should be determined. A fair market mechanism is required for pricing the electricity generated by these sources [22,23].

Globally, solar and wind source capacities in renewable energy resources are increasing very seriously. In this general trend, wind energy sources are located in windy regions away from the city center. On the other hand, rooftop solar PV panels are directly connected to the consumption grid of the end users. These sources may vary in the energy generation process. The dependence of these sources on location and nature creates a partial problem of unpredictability. Distributed power generation systems must transfer active power injection to the common connection point. These sources located close to the electricity grid can cause voltage increase in the grid [24].

1.2 Distribution generation grid

Distributed power generation systems are often installed by private owners in their homes. Depending on the installed power system capacity, it is usually connected to the distribution grid at low-voltage or rarely at medium-voltage level. Currently, technical regulations regarding the connection of distributed power generation systems to the existing distribution grid are minimal. It also does not seem possible to allow each distributed power generation system to be randomly connected to the grid. For this reason, there should be a technical arrangement to connect the distributed power generation systems to the distribution grid and to each other in a safe manner without any negative effects [25–27].

Distributed power generation system owners need to achieve a kind of steady income for their long-term economic sustainability. Since most distributed power generation systems will be located in a regional distribution system, the decision on the distribution system should be the responsibility of the regulatory authorities. This distributed power generation system revenue should be applied within the scope of tariffs determined by regulations. Distributed power generation system owners should be protected using long-term energy metering and planning such as feed-in tariffs. Solar PV panels on the roof of a house are owned by individual customers and are located on the customer's premises [28]. The remaining energy production from these sources can be sold back to the power company. The operator of the new generation distribution system must accurately measure the amount of energy transferred to the grid and sell it with a fair marketing. A user with a distributed power generation system should have equal conditions with all other sources in the grid [29].

Someone with a distributed power generation system gets benefited from the renewable electricity generation tariff set by local governments in order to join the electricity grid. In this case the distributed power generation system that will connect to the grid should be installed at the consumption place or it should only produce its own electricity (off-grid). The electricity generated in the distributed power generation system can cause a great challenge to the running operation of the distribution grid, because the traditional grid structure was designed according to the one-way power flow from the power production point to the users connected to the medium- and low-voltage grids. If distributed power generation systems are connected to the grid, low-capacity load demands are met by these sources [30,31]. However, when the energy produced by distributed power generation systems is excessive, the energy must be transferred in the opposite direction. In this case, there will be a challenge in energy flow between the grid and the distributed power generation system. Power generation plants in the existing grid structure are generally owned by third-party organizations. The distribution system operator has limited or no control over these exchanges. This causes security and protection problems in the grid operations. Another problem that arises in terms of grid infrastructure planning is that the location of distributed power generation systems is not planned centrally by the distribution system operator. Initial grid planning and dimensioning are done according to the highest demand levels [32]. The impact of distributed power generation systems on the grid infrastructure has not been properly taken into account.

The number of distributed power generation systems that will be connected to the existing grid system poses challenges for the future distribution system market. The distributed power generation system must have certain characteristics to properly accommodate its increased penetration. This situation raises the issue of communication between distributed power generation systems and the distribution system operator. But the high number of distributed power generation system units is another concern about control. Regional control can be carried out in this regard. Generally, the connection of any power plant to the grid is managed by grid code operation [33]. The grid operation code includes technical information such as when and at what power amount the plant will be activated. The grid operation code differs from country to country due to national power security, requirements, and features. Another purpose of this grid operation code is to reach electrical energy at the voltage and frequency values specified in the standards whenever the end users wish. In order to ensure safety and stability in distributed power generation systems, it is to keep the voltage and the frequency within the specified standard values [34]. There is a communication between the grid operation code and high-capacity power generation plants, and the power flow is under control.

Recently, there has been a significant increase in their quantity, with many users supporting the distributed power generation system. The increase in the percentage share of distributed power generation system resources in the present grid structure is high. Due to the impact on the stability and reliability of distributed power generation systems, the grid operation code needs to be developed or updated. Because of its nature, some distributed power generation system output powers are intermittent. Therefore a large distributed power generation penetration into the grid can affect the grid reliability. In addition, if the number of distributed power generation systems is high, there should be some regulations to make its connection cost-effective. As a result, a distributed power generation connection directive should be established for a stable and reliable grid infrastructure. There are several standards for enabling the communication and the power system on the distributed power generation connection with the new smart grid infrastructure. Some of the new standards on this grid connection are IEC 62477, IEC 62109, and IEEE 1547. These standards cover important requirements for connecting a distributed power generation system to the grid, and also equipment requirements to be used in system [35–38].

Connecting a distributed power generation system to the grid has both positive and negative effects. In general, it is predicted that the distributed power generation system will have a negative impact on the distribution network. These concerns include voltage fluctuation, frequency change, and harmonics. A distributed power generation system also makes important contributions such as reduction of power losses, active/reactive power control, security of energy supply, invoice saving, and energy balancing to the present distribution grid [39].

The operating costs are reduced and savings are achieved with the distributed power generation system concept. Loss and leakage problems can be eliminated, thanks to on-site generation and on-site consumption. In addition, thanks to the system dimensions and renewable energy technologies, ease of operation and maintenance is provided. Efficient use of renewable energy resources is ensured. Greenhouse gas emissions are reduced, thanks to the use of renewable energy sources. Real-time demand balancing can be achieved by using the energy storage unit in the distributed power generation system. Thanks to the energy storage feature, energy cuts are prevented by installing smart and adaptive systems for the end user. Thus the end user reaches and uses electrical energy with a quality feature. A stable, reliable, and safe energy system is provided with the distributed power generation system [40–42]. Fig. 10.2 shows a distributed generation grid structure with an alternative current (AC) busbar distribution grid.

The rate of integration with the network is increasing in distributed generation systems. These power generation units are generally smaller than conventional power plants and are more geographically dispersed. Although a large rate of electrical energy is still produced by conventional power plants, distributed generation systems have a rapidly growing share of energy generation. In addition, it encourages energy policies in governments to use renewable energy-based distributed generation systems. Some types of renewable energy source–based distributed generation such as solar PV panels and fuel cells generate direct current (DC) power. Energy storage technologies used in these systems also work based on DC voltage. Therefore two conversion stages are required to connect these sources to AC grids. Accordingly, the integration of most distributed generation systems into DC infrastructure is simpler and more economical than the integration of these resources into the AC grid. Most renewable energy sources can be integrated into the AC grid after they are collected through a DC bus. Using DC bus distribution systems contributes to loss/cost reduction. It can also increase the reliability of the grid as some power conversion stages are eliminated [43–45]. Fig. 10.3 shows a distributed generation grid structure

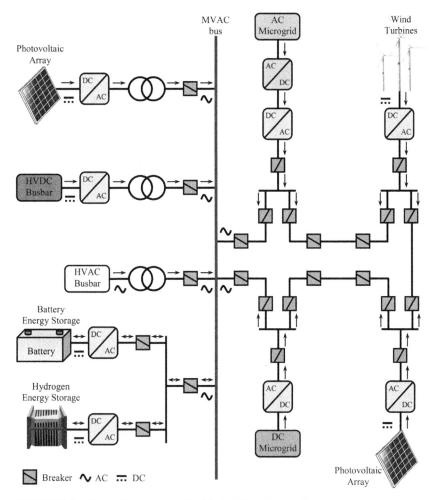

FIGURE 10.2 Alternative current grid with distributed generation sources.

with a DC bus distribution grid. In this DC bus–distributed power generation system, by using DC/AC inverters on the load side, the transfer of energy is ensured with a single power converter. Thanks to these converters, both DC and AC buses are controlled.

1.3 The role of energy storage in distributed solar hybrid systems

The challenge to be encountered in transmission grids is the integration of renewable energy sources. This issue of integration has been carefully handled and researched around the world. For solar and wind-based renewable energy sources, which have a high prevalence value in the grid, power generation does

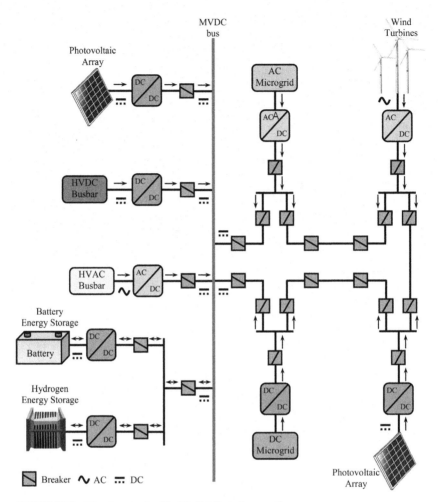

FIGURE 10.3 Direct current grid with distributed generation sources.

not exactly match the consumption values in the daily time schedule. Most of the time, regions with high wind and solar potential are in areas far from residential centers. Transmission grids are not designed for such situations. Therefore this difference between generation and load may lead to the violation of technical limits regarding the integration of renewable energy sources, especially the thermal load capacity of the transmission system [46].

Necessary measures can be taken to overcome the difficulties that transmission systems will face in the future. First of all, a transmission system infrastructure strengthening and expansion investment that can provide safe operation should be provided. Transmission system investments are long and difficult processes in terms of social (land expropriation needs, social acceptance problems, etc.)

and technical (physical conditions of the lands where the lines will pass, etc.), and they are high-cost investments.

A storage system installed with a suitable positioning in the grid can regulate deviations in the thermal load and voltage level of the transmission grid by applying charge–discharge processes according to need, for example, near a renewable energy source. Thus additional investment costs for the transmission system can be avoided [47,48].

The contribution of energy storage systems to transmission grid planning and operation depends largely on the positioning of energy storage systems within the transmission grid. Energy storage systems located close to the generation units can be considered as a good method in terms of operating the transmission grid.

Renewable energy sources are mostly connected to the electricity grid from the distribution level. While solar PV panel energy systems, especially for domestic applications, are connected to the low-voltage grid, large-scale applications are connected to medium-voltage grid systems. High power capacity wind energy systems can generally be connected to the transmission or distribution grid at medium voltage and high voltage levels [49].

Grid-connected large-scale renewable energy resource installations are usually carried out in remote areas for the convenience of finding suitable areas. As a result, the production values in the region where renewable energy resource systems are located may exceed the consumption values. Therefore renewable energy source power generation is first transferred to the distribution system and then to the transmission system to meet the demand in more remote regions [50]. In such a case the power flow direction changes in the electricity grid. Throughout its historical development, electrical grids have not been designed for bidirectional power flow, which leads to exceeding the accepted limits in thermal load and voltage values.

Expansion investments to be made in the distribution grid will require additional investment in the direct transmission grid. Distribution grid strengthening investments can be achieved by establishing new lines parallel to the existing ones. Another method is to reduce the thermal loads of transformers by adding new distribution transformer facilities using the existing distribution grid. With this type of measure, deviations in the voltage level that may occur in subdistribution systems can be prevented as well as thermal load values can be lowered [51].

Reducing thermal load values in the distribution system, it can be realized with active power control for generation units, load groups, and small-scale energy storage systems connected to the distribution grid. Small-scale energy storage systems connected to the distribution grid have an easily controllable feature. In addition, energy storage systems and reactive power control and voltage control are performed. They can prevent grid reinforcement needs with the benefits such as active power control capability they offer in the distribution grid operation [52,53]. As a result, they can contribute to a much more flexible and reliable grid operation. The benefits that energy storage systems can provide for distribution grids will depend on their positioning and operational characteristics within the grid.

2 Transmission lines in distributed generation systems

The transfer of electrical energy from power generation plants to city centers is provided by transformers and high voltage transmission lines. Interconnected transmission lines are called high voltage transmission grids. In order to reduce the energy lost in long distance transmission, electrical energy is transmitted with high voltage. Understanding the characteristics of the conductor performing the electrical energy transport process will help us understand the limiting factors for the power transmission capacity of the transmission line. Using sensors and communication technology during power transmission will help in energy optimization [54]. Conductor is one of the main determinants of a transmission line design. For optimum operating efficiency of the transmission system, the most suitable conductor type and size should be selected.

The limiting factors for power transmission are voltage drop, voltage stability, and thermal rating. For short transmission line less than about 80 km, power transmission is limited by thermal rating. Thermal rating is limited by the maximum operating temperature of the conductor. This maximum operating temperature is limited by the maximum line sag limit and the maximum allowable operating temperature of the conductive material [55,56].

2.1 Classification of transmission lines

The classification of transmission lines depends on the voltage and the length of the conductor. The transmission line is the means to transfer power from the generation station to the load center. Transmission line performance depends on three parameters, R, L, and C, that are evenly distributed over the entire length of the line. In a transmission line, resistance and inductance create series impedance. Capacitance is found between line and grounding conductors. To know transmission line performance, it is necessary to know how capacitance is taken into account [57]. Fig. 10.4 shows the general topology structure of AC transmission lines. Transmission lines are classified as follows:

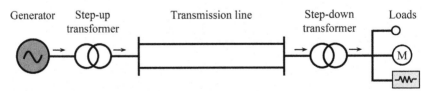

FIGURE 10.4   General topology of AC transmission line.

Two-Port Circuit

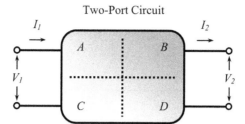

FIGURE 10.5   **Two-port circuit with *ABCD* parameters.**

- AC transmission line
  - short transmission line,
  - medium transmission line (π and T model), and
  - long transmission line.
- DC transmission line

2.1.1   Transmission line models

Although an energy transmission line differs depending on its classes, it is expressed with a two-port circuit representation on its basis. The two-port circuit model presented for an overhead transmission line is given in Fig. 10.5. The relationship between line head and end of line voltage and current in a two-port circuit is explained by the following equations:

$$V_1 = AV_2 + BI_2 \tag{10.1}$$

$$I_1 = CV_2 + DI_2 \tag{10.2}$$

These equations are given in the next equation in a matrix form:

$$\begin{bmatrix} V_1 \\ I_1 \end{bmatrix} = \begin{bmatrix} A & B \\ C & D \end{bmatrix} \begin{bmatrix} V_2 \\ I_2 \end{bmatrix} \tag{10.3}$$

where the constants A, B, C, and D are called general circuit constants of transmission lines. These are generally complex numbers. If the two-port circuit is symmetrical when viewed from either end of the line, A and D are equal to each other. Unit of B is Ω and unit of C is S ($1/\Omega$). If this symmetry is provided, it is $AD - BC = 1$ and the system is balanced.

2.1.2   Short transmission line

When the length of a power transmission line is less than 80 km and the line voltage is less than 20 kV, it is generally considered to be a short transmission

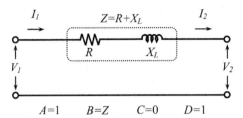

FIGURE 10.6 Short transmission line model.

line. Because of the smaller length and lower voltage of the line, the capacitance effects of the line are extremely small and can be neglected. Therefore when examining short transmission line performance, only the resistance and inductance of the line are taken into account [58]. Fig. 10.6 shows the equivalent circuit model of the short transmission line.

2.1.3 Medium transmission line

When the length of a power transmission line is about 80–240 km and the line voltage is high that is 20–100 kV, it is generally considered as a medium transmission line. Due to the sufficient length and voltage of the line, capacitance effects are also taken into account when examining medium transmission line performance. Although the capacitance is distributed evenly over the entire length of the line, a solution is obtained by taking into account the line capacitance collected in one or more places. Therefore there are two models in which the capacity effect is given. The first is the π model and half of the capacitance is considered to be concentrated at both ends of the line [59]. Fig. 10.7 shows the π equivalent circuit model of the medium transmission line.

The second is the T model and the capacitance is considered to be concentrated in the center of the line. The equivalent circuit model of T of the medium transmission line is given in Fig. 10.8.

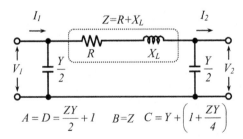

FIGURE 10.7 Medium transmission line π model.

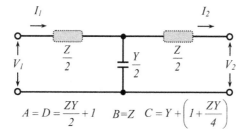

$$A = D = \frac{ZY}{2} + 1 \quad B = Z \quad C = Y + \left(1 + \frac{ZY}{4}\right)$$

FIGURE 10.8   Medium transmission line T model.

2.1.4   Long transmission line

When the length of a power transmission line is more than 240 km and the line voltage is more than 100 kV, it is generally considered as a long transmission line. To study the performance of such a transmission line, it is assumed that the transmission line constants are evenly distributed over the entire length of the line [60]. Fig. 10.9 shows the equivalent circuit model of the long transmission line.

2.1.5   Direct current transmission line

The power carrying capacity of AC power transmission lines depends on thermal limits, required reactive power, and stability problems. The capacity of DC energy transmission lines depends only on thermal limits. In fact, due to the absence of reactive current component, current magnitude and cable losses are reduced when DC power transmission systems are used. Fig. 10.10 shows the general topology structure of DC transmission lines.

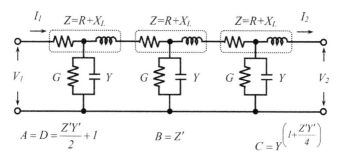

$$A = D = \frac{Z'Y'}{2} + 1 \qquad B = Z' \qquad C = Y\left(1 + \frac{Z'Y'}{4}\right)$$

FIGURE 10.9   Long transmission line model.

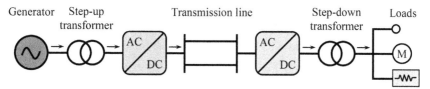

FIGURE 10.10   General topology of DC transmission line.

A certain load group can be transported by DC energy transmission lines with a lower current compared to AC power transmission lines. The earth is used as return conductive in DC power transmission line. Thus the load in the DC power transmission line can work with a cable with a smaller cross section. It reduces the cost of equipment used in low cross section cable distribution grid. The resistance of AC power transmission line cables varies due to skin effect. With DC power transmission line, more use is made of the insulation surface of a conductor. In addition, the corona losses of DC energy transmission lines are less. Therefore the resistance of a cable in a DC power transmission line is lower than the resistance of the AC power transmission line. While there is effective resistance in the AC power transmission line, there is an ohmic resistance in the DC power transmission line. The energy losses in the AC power transmission line are higher. As a result, the use of DC power transmission lines can reduce the total losses in the grid's network [61–63]. A load group can be met with a lower current with the DC power transmission line. By converting a conventional AC power transmission line into a DC power transmission line, it will be possible to increase the total power transmitted with the same transmission line. While three cables are used for a circuit in the AC power transmission line, this number of conductors is one in the DC power transmission line. Thus while the power consumption of a region increases, new transmission line costs will be eliminated. In addition to all these positive results, it is difficult to cut the power because the current flow in the DC energy transmission line is continuous. On the AC power transmission line, the cutting of energy becomes easier when the zero point is reached depending on the time. It is very difficult to raise and lower the voltage in the DC power transmission line. Power electronics elements used in DC power transmission line can cause the formation of harmonics [64].

3 Inverter structures in solar hybrid systems

Power electronic converters are very important units in distributed generation systems to keep the flow of energy between sources at an optimum level. Many converter topologies are used to transfer the DC voltage received from solar PV panels to the energy storage unit and to provide AC energy flow to the grid. For example, in order to transfer energy from a solar PV panel, it must control many parameters such as operating the panels at the maximum power point and keeping them in a certain voltage range [65]. At the same time, power electronic converters are equipped with many requirements such as wide operating range, ability to work in different seasonal conditions, and reaching the highest possible efficiency. Inverters operating synchronously with the grid in distributed energy conversion systems can be classified into the following three groups:

- centralized inverter topology,
- string inverter topology, and
- multistring inverter topology.

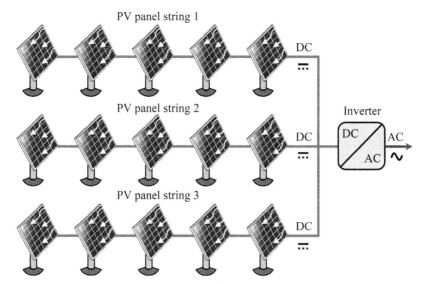

FIGURE 10.11   Centralized inverter topology.

3.1   Centralized inverter topology

Central inverter system topology is used in small facilities and solar PV panels with the same output characteristics. A single inverter structure has economic advantages in terms of reducing initial investment and maintenance costs. However, fault of a single inverter causes power generation to stop in the entire system. Also, this topology creates an unprotected structure in the system against overcurrent. Shadow or fault in solar PV panels will significantly reduce the total power generation. Therefore it is not appropriate to establish this topology structure at very high powers. Central inverter system topology structure is shown in Fig. 10.11. In this topology, solar PV panels are connected in series and parallel to obtain the required current and voltage levels. In this topology a common DC bus is obtained and an inverter is used. Generally, an inverter with a large power rating is used to convert the DC power output of solar PV panels to AC power. In this topology structure, series-connected array diodes are required between solar PV panels. In the central inverter system, the power losses of the inverter are higher than the string inverter or multistring inverter topologies. In this topology, DC bus voltage can be adjusted with series-connected solar PV panels [66].

3.2   String inverter topology

The string converter topology is given in Fig. 10.12. The separate inverter is used for each solar PV panel string. In this topology, if enough solar PV panels in each string are connected in series, the voltage does not need to be stepped

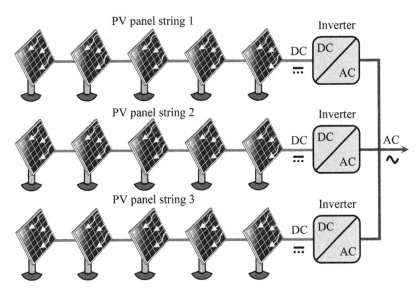

FIGURE 10.12 String inverter topology.

up. In this topology structure the voltage can be provided by the high-frequency DC/DC boost circuit in the DC bus. Separate maximum power point tracking can be applied to each solar PV panel string to increase the overall efficiency of the system. In addition, in this topology, the system performance will not decrease as the shadowing that will occur in a solar PV panel array will not affect other parts. In this topology structure, low power capacity (50–300 W) inverter is used in each solar PV panel arrays. Thus low power capacity inverter structure will reduce costs. In a plant with a medium power range, each solar PV panel can be connected directly to its own inverter. In this topology the blocking diode that prevents the energy from flowing against the flow direction is usually located in the inverter unit. This topology can also provide protection against atmospheric overcurrent and overvoltage in the DC bus. Another advantage of this topology is that solar PV panel arrays with different properties can be used in different modules. Thus the efficiency and reliability of the entire facility can be increased [67].

3.3 Multistring inverter topology

In the multistring inverter topology, different solar PV panel strings are first connected to the DC bus together with their own DC/DC converter structure. Then the common DC bus is connected to a single high-capacity inverter. Fig. 10.13 shows the multistring inverter topology. In this topology structure, capacity can be increased by adding a series of solar PV panels to the system. Thanks to this feature, this topology has a highly efficient flexible design. In this topology, each solar PV panel module has an integrated power electronics

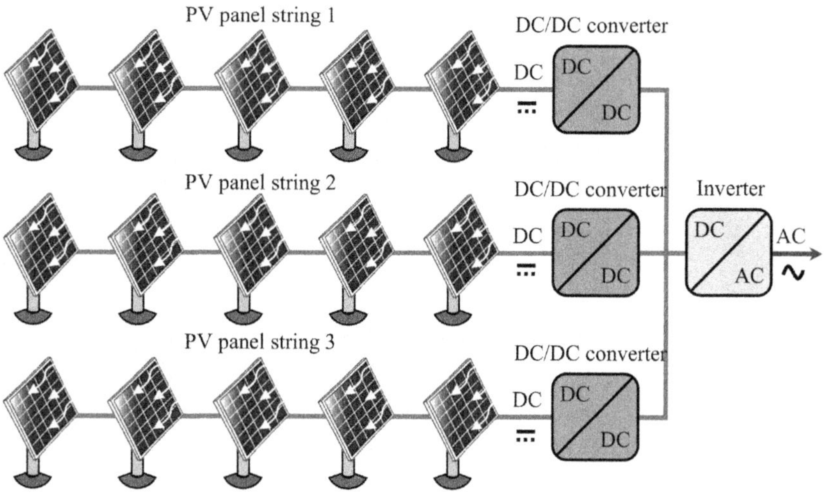

FIGURE 10.13   Multistring inverter topology.

converter. The power loss of the system is lower as the incompatibility between the solar PV panel modules decreases. However, the fixed losses in the inverter may be the same as the string inverter topology. In this topology, each solar PV panel module supports optimum operation, which generally ensures optimum performance in the system. This is because each solar PV panel string has its own DC/DC converter and maximum power levels can be achieved for each panel string individually [68–70].

In this topology, in large-scale facilities, the solar PV panel area is generally divided into more subdivisions. There are fewer inverters in this topology compared to the string inverter topology. Thus investment and maintenance costs are reduced. The shadowing situations are eliminated with the separate arrays structure. Also, the fault of an inverter in this topology does not affect the power generation loss of the entire facility. Necessary operation and maintenance can be done without taking all solar PV panels out of service. Protection against overcurrent can be realized by means of a thermomagnetic circuit breaker or a fuse. Protection against reverse current in this topology is achieved through blocking diodes.

3.4   On-grid solar generation systems

In distributed generation systems, especially renewable energy sources operating and based on DC voltage must be converted to AC voltage. Electrical devices used in homes around the world are designed to work with AC voltage. This AC voltage requirement makes the development and use of inverters widespread. Grid-connected solar PV panels operate on the principle that the generated electricity is consumed at the generation site instead of being stored

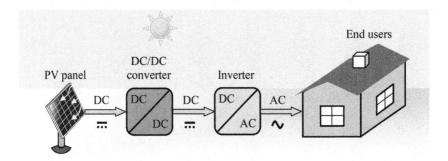

FIGURE 10.14 Application of off-grid connected solar PV panel.

in batteries. During the project design the amount of energy that is desired or required to be produced is determined. DC electrical energy is generated by the contact of solar irradiation coming from the atmosphere on solar PV panels. The energy produced is connected to the central grid with inverters with high cycle power that can be connected to the central grid. Thus the energy generated from solar PV panels is sent directly to the grid system. If the field and solar irradiation conditions are suitable, it is possible to produce electrical energy at the desired power with the electricity generation system connected to the grid. Electricity generation with solar energy is an easy-to-install power generation tool as well as being preferred because of its long-time, no operating cost, and practical and portable nature [71]. Fig. 10.14 gives an example application structure with grid connection.

Grid-connected systems can be installed in high power plant size as well as in smaller power installations for domestic needs. For example, while the electricity requirement of a house can be met in these systems, the excess energy generated is supplied to the electricity grid. In cases where sufficient energy is not produced, energy is taken from the grid. In such a system, there is no need to store energy; it is enough to convert the generated DC electricity to AC electricity and be compatible with the grid [72].

Since energy storage units such as batteries will not be used in grid-connected systems, there is no additional cost for storage. There will be consumption close to grid-connected systems; the losses will be minimal due to low energy conversion units. The produced energy is connected to the grid in a synchronized method; the grid is activated when the energy produced is not sufficient. Thus the end user will reach the energy continuously. While the system is designed, it has the flexibility to design according to the desired amount or area, as there is no obligation to meet the entire load. In addition, if the area is sufficient, the installed power of the system can be increased.

3.5 Off-grid solar generation systems

Off-grid systems provide an energy source for the end user in areas far from the existing electricity grid. It has the same reliability and quality as in the electricity

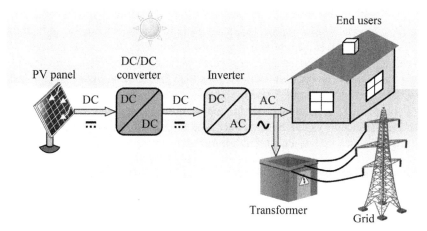

FIGURE 10.15   Application of grid-connected solar PV panel.

grid. Many energy sources can be included in these systems. Electric energy can be produced by renewable energy sources such as solar, wind, or geothermal. An example of the application of off-grid structure is given in Fig. 10.15. Off-grid structures show the future of energy resources in distributed generation systems.

Off-grid structures are an example of an off-grid, renewable energy source–based, and micro-distributed generation system that includes energy storage units. The distributed generation system was also classified as grid-connected and independent. Although the amount of energy produced in buildings independent of the grid is low, it has an intermittent structure due to natural resources. In order to eliminate this intermittent energy feature, there is usually at least one energy storage unit in independent structures. Thanks to this energy storage unit, the continuity and reliability of energy is ensured. In addition, there is a central inverter for end users to run their loads in off-grid systems. As widespread usage areas today, water pumping, radio stations, weather stations, base stations, lightning protection systems, sign systems on highways, sign systems for ports/airports, and advertising sign applications can be shown [73].

3.6   Basic features of inverters in solar hybrid systems

Grid-connected inverters are expected to have fast response time, low power factor, frequency control, low harmonic output, and grid synchronization features. The response time of the inverters must be extremely fast. The use of high-efficiency semiconductor switches enables the inverters to respond in milliseconds. Power factor of inverters is weak due to harmonics. However, with the latest developments in inverter technology, it is possible to keep the power factor low. With many control methods, it is possible to prevent voltage fluctuation by controlling the reactive power flow of the inverters. While this feature helps the grid structure, it helps to produce quality electrical energy [74].

The output waveform of the inverters is synchronous with the grid frequency. Inverter output voltage frequency and grid frequency are transmitted to the end user as 50 Hz. Harmonic output produced by inverters is generally very weak. Old thyristor-based inverters were operating at slow switching speeds and pulse width modulation was not possible. Therefore harmonics produced by old inverters were transferred to the grid. As a result, it caused losses, heating of devices, activation of protection equipment, and low power quality. The inverter controls can be done very well today with the advances in power electronics technology. Pulse width modulated inverters produce high-quality sine waves (voltage/current). Harmonic levels of inverters are much lower than ordinary household devices [75]. If harmonics are present in the grid voltage waveform, harmonic currents can be induced in the inverter. These harmonic currents, especially produced by a voltage-controlled inverter, will help and support the grid. These are useful harmonic currents.

Synchronization of the inverter with the grid is carried out automatically by control methods. It is generally used by detecting zero-crossing points on the voltage waveform. Thanks to the reference waveforms in the inverter, it can be jumped to any point required within a sampling period. This time may take a few seconds if phase locked loops are used. Phase locked loops are used for more stable operation against noise that may occur on the grid side. These control structures work by evaluating several zero-crossing points' information so that the synchronization is stable.

Solar PV panels generate power in direct proportion to the amount of solar irradiation falling on them. Solar PV panels are normally scaled to produce 1000 W/m^2 at 25 °C. If the solar irradiation value is low, the maximum current to be generated will be less than the rated full load current. As a result, solar PV panel systems cannot provide short circuit power capacity to the grid. If there is a battery energy storage unit in the solar PV panel system, it is possible to ensure the continuity of the energy with inverters when there is no energy. Inverters have a specified power capacity value in their catalog values. If there is a power demand exceeding its capacity from the inverter, it can meet for a certain period of time and then it will be disabled with circuit breakers. Inverters must also protect themselves against short circuits. This is because power electronics components can typically fault very quickly until a protection device such as a circuit breaker trips [76,77].

In the case of fault of inverters, inverters can inject DC components into the grid. Many methods are used to prevent these. One of these methods is to use a transformer between DC and AC buses. In DC component injection the signal generating a positive half cycle different from the negative half cycle resulting in a DC component is generated at the inverter output. This DC component may be caused by a power electronic device. If DC component can be measured at the inverter output, then it can be added to the inverter control unit in the feedback path to eliminate the amount of DC component.

In order to prevent chronic faults in an inverter, methods to detect over-voltage, over- and underfrequency conditions should be available. These fail-safe limit values must be set by the manufacturer or programmable by the end user. Voltage operating ranges can be adjusted in a limited band, if necessary. Inverters must include active protection as well as passive protection methods. If more than one inverter works in parallel, the frequency reference value must be shared. The reason for this is to prevent the situation in which islanding may occur. Inverters must have an active island avoidance method accepted when there are grid faults such as frequency offset and impedance measurement. This function will also protect the inverter output against passive protection tolerances. Thus it helps the inverter to separate from the active grid system. The maximum combined operating time of both passive and active protections should not exceed two seconds. If the frequency shifting method is used in the inverter control structure, the shift direction should be down. The inverter must remain separate from the grid until reconnection procedures take place. Inverters performing low load transfer generate a dangerous high voltage for electronic equipment. These voltage values must be within the voltage limit values specified in IEEE P929 standard. Inverters should also have the frequency characteristics included in the standard. The inverter must have adequate protection against short circuit, other faults and overheating of the inverter components [78,79].

The selection of the inverter size is made according to the solar PV panel nominal power to be controlled. The size of the inverter is determined by the ratio between the active power supplied to the grid and the nominal power of the solar PV panel, and this ratio is between 0.8 and 0.9. When calculating this ratio, the power loss of solar PV panels and the efficiency of the inverter should be taken into account under real operating conditions. This ratio also depends on the installation methods that may cause a change in the power generated by the solar PV panels produced. In these installation methods the latitude, slope, and ambient temperature of solar PV panels should be considered.

For the correct sizing of the inverter, the characteristics of DC bus and AC bus voltage value, nominal power and maximum power value, nominal voltage and maximum voltage, maximum conversion efficiency, and range of maximum power point tracking voltage should be considered. In addition, the nominal values of the voltage and frequency at the inverter output must be evaluated. Voltage and frequency values at the output for grid-connected inverters are determined by the grid with defined tolerances. Regarding the voltage at the inverter input, the limit operating conditions of the solar PV panels should be evaluated to ensure safe and efficient operation of the inverter.

3.7 Grid-connected wind energy systems

Small power scale wind turbines connected to the grid used in distributed generation systems have been studied. Wind turbines have been used in many parts of the world recently. Such low-power capacity power generation systems are more sensitive to sudden changes in grid operating conditions. This operational sensitivity will lead to significant power fluctuations in the grid and a reduction in the quality of electrical power supply to end users. All of these reasons lead to frequency changes or sudden voltage fluctuations at the wind turbine output [80].

Energy storage and advanced control systems are required to eliminate these negative effects in low-power capacity wind turbines. This solution method may contain two energy storage units. The first is the rotating mechanical energy storage part that contains the wind turbine blades, gearbox, and rotor. A big and sudden change in the electrical output of the wind turbine can be eliminated with this mechanical part as in a machine rotating at a constant speed. The second energy storage element is the battery group. The battery group is located between the DC/DC converter and the inverter. The excess energy generated by the wind turbine can be stored in the battery group. When there is a decrease in wind speed, the energy in the battery group can be used. Thus power fluctuations in energy production can be prevented. In larger scale wind turbines, an auxiliary control unit is added to the inverter control to reduce ripple and increase the total output power. Thus the total output of the wind power system can be stabilized or leveled. The wind turbine controller must constantly monitor the output in order to obtain maximum power from the system outlet. The inverter output power should be adjusted by monitoring the wind turbine output. In this case the status of the battery group should also be monitored. The battery group should be kept full as much as possible in order to activate it in decreases in wind speed [81]. The wind turbine will track long-term changes in wind speed without the fluctuations caused by the wind. In high-power applications, power fluctuations at the wind turbine outlet can easily be absorbed by the grid. The aim of all the works carried out in this case is to obtain the maximum energy from wind energy.

The function of DC/DC converters at the wind turbine output is to adjust the turbine generator torque. At the same time, it is to obtain optimum power from the turbine blades by measuring the wind turbine shaft speed. The purpose of the inverter in wind turbines is to transfer the energy collected by the DC/DC converter to the grid. The need for a battery pack can be minimized by effective management between two power converter units. The wind turbine control algorithm needs to be fast enough for the inverter output power control to match the DC/DC converter output. Power electronics technology plays an important role in both wind turbine system configurations and control of offshore wind farms. Different generator types and control methods are used in wind farms. Wind farms can be connected to the local grid via a central AC bus [82].

Wind turbine system designs can be classified according to the types of wind energy generators, power electronics, speed controllability, and the way aerodynamic power is limited. These include fixed speed wind turbine systems, partial variable speed wind turbine systems with variable rotor resistance, variable speed wind turbine systems with partial scale frequency converter, and variable speed wind turbine systems with full-scale power converter.

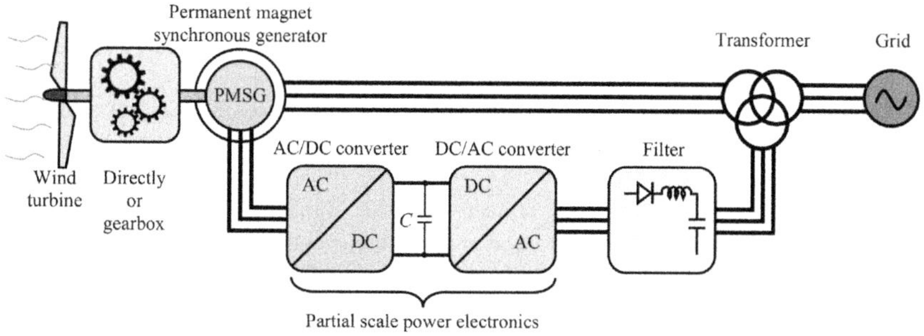

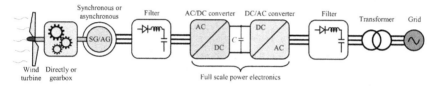

FIGURE 10.16   Wind turbine systems with partial scale power electronics.

FIGURE 10.17   Wind turbine systems with full scale power electronics.

Among these wind turbine systems, the topologies given in Figs. 10.16 and 10.17 are widely applied and used. Power electronic converters are very important in the topology structure given in Fig. 10.16 and they have different power ratios. The converter units were partially realized by power electronics converters in Fig. 10.16.

A double-fed induction generator was used as a wind turbine generator in this topology structure. However, in the topology structure given in Fig. 10.17, all power electronics converters realize the energy conversion. Synchronous generator or asynchronous generator structures are used in the topology structure. In today's wind turbine systems, structures with full-scale power electronic converters are used.

References

[1] Q. Wen, G. Liu, Z. Rao, S. Liao, Applications, evaluations and supportive strategies of distributed energy systems: a review, Energy Build 225 (2020) 1–19 https://doi.org/10.1016/j.enbuild.2020.110314.

[2] Q. Wen, G. Liu, W. Wu, S. Liao, Multicriteria comprehensive evaluation framework for industrial park-level distributed energy system considering weights uncertainties, J. Clean. Prod. 282 (2020) https://doi.org/10.1016/j.jclepro.2020.124530.

[3] M. Wang, H. Yu, X. Lin, R. Jing, F. He, C. Li, Comparing stochastic programming with posteriori approach for multi-objective optimization of distributed energy systems under uncertainty, Energy 210 (2020) 1–16 https://doi.org/10.1016/j.energy.2020.118571.

[4] J. Niu, Z. Tian, J. Zhu, L. Yue, Implementation of a price-driven demand response in a distributed energy system with multi-energy flexibility measures, Energy Convers. Manag. 208 (2020) 1–13 https://doi.org/10.1016/j.enconman.2020.112575.

[5] H. Li, S. Zhong, Y. Wang, J. Zhao, M. Li, F. Wang, et al. New understanding on information's role in the matching of supply and demand of distributed energy system, Energy 206 (2020) 1–15 https://doi.org/10.1016/j.energy.2020.118036.

[6] J. Yuan, Z. Xiao, C. Zhang, W. Gang, A control strategy for distributed energy system considering the state of thermal energy storage, Sustain. Cities Soc. 63 (2020) 1–11 https://doi.org/10.1016/j.scs.2020.102492.

[7] L. Li, S. Yu, Optimal management of multi-stakeholder distributed energy systems in low-carbon communities considering demand response resources and carbon tax, Sustain. Cities Soc. 61 (2020) 1–12 https://doi.org/10.1016/j.scs.2020.102230.

[8] B. Hong, Q. Li, W. Chen, B. Huang, H. Yan, K. Feng, Supply modes for renewable-based distributed energy systems and their applications: case studies in China, Glob. Energy Interconnect 3 (2020) 259–271 https://doi.org/10.1016/j.gloei.2020.07.007.

[9] X. Wu, J. Cao, C. Jiang, Y. Lou, S. Zhao, H. Madani, et al. Low carbon transition in climate policy linked distributed energy system, Glob. Transitions Proc. 1 (2020) https://doi.org/10.1016/j.gltp.2020.03.002.

[10] X. Wang, H. Tian, F. Yan, W. Feng, R. Wang, J. Pan, Optimization of a distributed energy system with multiple waste heat sources and heat storage of different temperatures based on the energy quality, Appl. Therm. Eng. 181 (2020) 1–10 https://doi.org/10.1016/j.applthermaleng.2020.115975.

[11] B. Yan, M. Di Somma, G. Graditi, P.B. Luh, Markovian-based stochastic operation optimization of multiple distributed energy systems with renewables in a local energy community, Electr. Power Syst. Res. 186 (2020) 1–9 https://doi.org/10.1016/j.epsr.2020.106364.

[12] Q. Wen, G. Liu, G. Wu, S. Liao, Genetic algorithm-based operation strategy optimization and multi-criteria evaluation of distributed energy system for commercial buildings, Energy Convers. Manag. 226 (2020) 1–16 https://doi.org/10.1016/j.enconman.2020.113529.

[13] J. Yuan, C. Cui, Z. Xiao, C. Zhang, W. Gang, Performance analysis of thermal energy storage in distributed energy system under different load profiles, Energy Convers. Manag. 208 (2020) 1–18 https://doi.org/10.1016/j.enconman.2020.112596.

[14] A. Marini, S.S. Mortazavi, L. Piegari, M.S. Ghazizadeh, An efficient graph-based power flow algorithm for electrical distribution systems with a comprehensive modeling of distributed generations, Electr. Power Syst. Res. 170 (2019) 229–243 https://doi.org/10.1016/j.epsr.2018.12.026.

[15] M. Tomasov, D. Motyka, M. Kajanova, P. Bracinik, Modelling effects of the distributed generation supporting e-mobility on the operation of the distribution power network, Transp. Res. Procedia 40 (2019) 556–563 https://doi.org/10.1016/j.trpro.2019.07.080.

[16] P.M. De Oliveira-De Jesus, C. Henggeler Antunes, A detailed network model for distribution systems with high penetration of renewable generation sources, Electr. Power Syst. Res. 161 (2018) 152–166 https://doi.org/10.1016/j.epsr.2018.04.005.

[17] S. Daiva, G. Saulius, A. Liudmila, Energy distribution planning models taxonomy and methods of distributed generation systems, Energy Procedia 107 (2017) 275–283 https://doi.org/10.1016/j.egypro.2016.12.150.

[18] A.A. Sadiq, S.S. Adamu, M. Buhari, Optimal distributed generation planning in distribution networks: a comparison of transmission network models with FACTS, Eng. Sci. Technol. Int. J. 22 (2019) 33–46 https://doi.org/10.1016/j.jestch.2018.09.013.

[19] A. Garry, F. Cadoux, M.C. Alvarez-Herault, N. Hadjsaid, Risk aversion model of distribution network planning rules considering distributed generation curtailment, Int. J. Electr. Power Energy Syst. 99 (2018) 385–393 https://doi.org/10.1016/j.ijepes.2018.01.035.

[20] R. Dashti, M. Ghasemi, M. Daisy, Fault location in power distribution network with presence of distributed generation resources using impedance based method and applying π line model, Energy 159 (2018) 344–360 https://doi.org/10.1016/j.energy.2018.06.111.

[21] E. Vaccariello, P. Leone, I.S. Stievano, Generation of synthetic models of gas distribution networks with spatial and multi-level features, Int. J. Electr. Power Energy Syst. 117 (2020) 1–19 https://doi.org/10.1016/j.ijepes.2019.105656.

[22] Z. Fang, Y. Lin, S. Song, C. Song, X. Lin, G. Cheng, Active distribution system state estimation incorporating photovoltaic generation system model, Electr. Power Syst. Res. 182 (2020) 1–14 https://doi.org/10.1016/j.epsr.2020.106247.

[23] A. Ahmed, M.F. Nadeem, I.A. Sajjad, R. Bo, I.A. Khan, A. Raza, Probabilistic generation model for optimal allocation of wind DG in distribution systems with time varying load models, Sustain. Energy Grids Networks 22 (2020) 1–13 https://doi.org/10.1016/j.segan.2020.100358.

[24] M.L. Kloubert, Assessment of generation adequacy by modeling a joint probability distribution model, Electr. Power Syst. Res. 189 (2020) 1–15 https://doi.org/10.1016/j.epsr.2020.106803.

[25] J. Bai, W. Wei, L. Chen, S. Mei, Modeling and dispatch of advanced adiabatic compressed air energy storage under wide operating range in distribution systems with renewable generation, Energy 206 (2020) 1–11 https://doi.org/10.1016/j.energy.2020.118051.

[26] F. Vazinram, M. Hedayati, R. Effatnejad, P. Hajihosseini, Self-healing model for gas-electricity distribution network with consideration of various types of generation units and demand response capability, Energy Convers. Manag. 206 (2020) 1–9 https://doi.org/10.1016/j.enconman.2020.112487.

[27] M. Tomasov, D. Motyka, M. Kajanova, P. Bracinik, Modelling effects of the distributed generation supporting e-mobility on the operation of the distribution power network, Transp. Res. Procedia 40 (2019) 556–563 https://doi.org/10.1016/j.trpro.2019.07.080.

[28] L. Luo, Z. Wu, W. Gu, H. Huang, S. Gao, J. Han, Coordinated allocation of distributed generation resources and electric vehicle charging stations in distribution systems with vehicle-to-grid interaction, Energy 192 (2020) 1–10 https://doi.org/10.1016/j.energy.2019.116631.

[29] D. Guillen, C. Salas, F. Trillaud, L.M. Castro, A.T. Queiroz, G.G. Sotelo, Impact of resistive superconducting fault current limiter and distributed generation on fault location in distribution networks, Electr. Power Syst. Res. 186 (2020) 1–14 https://doi.org/10.1016/j.epsr.2020.106419.

[30] Y. Xiang, Y. Wang, Y. Su, W. Sun, Y. Huang, J. Liu, Reliability correlated optimal planning of distribution network with distributed generation, Electr. Power Syst. Res. 186 (2020) 1–12 https://doi.org/10.1016/j.epsr.2020.106391.

[31] J. Snodgrass, L. Xie, Overvoltage analysis and protection of lightning arresters in distribution systems with distributed generation, Int. J. Electr. Power Energy Syst. 123 (2020) 1–15 https://doi.org/10.1016/j.ijepes.2020.106209.

[32] U.H. Ramadhani, M. Shepero, J. Munkhammar, J. Widén, N. Etherden, Review of probabilistic load flow approaches for power distribution systems with photovoltaic generation and electric vehicle charging, Int. J. Electr. Power Energy Syst. 120 (2020) 1–16 https://doi.org/10.1016/j.ijepes.2020.106003.

[33] R.W. Kenyon, B. Mather, B.M. Hodge, Coupled transmission and distribution simulations to assess distributed generation response to power system faults, Electr. Power Syst. Res. 189 (2020) 1–19 https://doi.org/10.1016/j.epsr.2020.106746.

[34] G. Gruosso, P. Maffezzoni, Data-driven uncertainty analysis of distribution networks including photovoltaic generation, Int. J. Electr. Power Energy Syst. 121 (2020) 1–17 https://doi.org/10.1016/j.ijepes.2020.106043.

[35] D. Strušnik, M. Golob, J. Avsec, Artificial neural networking model for the prediction of high efficiency boiler steam generation and distribution, Simul. Model. Pract. Theory 57 (2015) 58–70 https://doi.org/10.1016/j.simpat.2015.06.003.

[36] M. Koivisto, M. Degefa, M. Ali, J. Ekström, J. Millar, M. Lehtonen, Statistical modeling of aggregated electricity consumption and distributed wind generation in distribution systems using AMR data, Electr. Power Syst. Res. 129 (2015) 217–226 https://doi.org/10.1016/j.epsr.2015.08.008.

[37] K.M. Jagtap, D.K. Khatod, Loss allocation in radial distribution networks with various distributed generation and load models, Int. J. Electr. Power Energy Syst. 75 (2016) 173–186 https://doi.org/10.1016/j.ijepes.2015.07.042.

[38] K.N. Maya, E.A. Jasmin, A three phase power flow algorithm for distribution network incorporating the impact of distributed generation models, Procedia Technol. 21 (2015) 326–331 https://doi.org/10.1016/j.protcy.2015.10.040.

[39] A.S.N. Huda, R. Živanović, Large-scale integration of distributed generation into distribution networks: study objectives, review of models and computational tools, Renew. Sustain. Energy Rev. 76 (2017) 974–988 https://doi.org/10.1016/j.rser.2017.03.069.

[40] D. Santos-Martin, S. Lemon, SoL—a PV generation model for grid integration analysis in distribution networks, Sol. Energy 120 (2015) 549–564 https://doi.org/10.1016/j.solener.2015.07.052.

[41] J.C. Hernández, F.J. Ruiz-Rodriguez, F. Jurado, Modelling and assessment of the combined technical impact of electric vehicles and photovoltaic generation in radial distribution systems, Energy 141 (2017) 316–332 https://doi.org/10.1016/j.energy.2017.09.025.

[42] C. Ma, J.H. Menke, J. Dasenbrock, M. Braun, M. Haslbeck, K.H. Schmid, Evaluation of energy losses in low voltage distribution grids with high penetration of distributed generation, Appl. Energy 256 (2019) 1–15 https://doi.org/10.1016/j.apenergy.2019.113907.

[43] M.N. Faqiry, L. Edmonds, H. Wu, A. Pahwa, Distribution locational marginal price-based transactive day-ahead market with variable renewable generation, Appl. Energy 259 (2020) 1–17 https://doi.org/10.1016/j.apenergy.2019.114103.

[44] R. Alizadeh, R. Gharizadeh Beiragh, L. Soltanisehat, E. Soltanzadeh, P.D. Lund, Performance evaluation of complex electricity generation systems: a dynamic network-based data envelopment analysis approach, Energy Econ. 91 (2020) 1–14 https://doi.org/10.1016/j.eneco.2020.104894.

[45] C. Li, H. Zhou, J. Li, Z. Dong, Economic dispatching strategy of distributed energy storage for deferring substation expansion in the distribution network with distributed generation and electric vehicle, J. Clean. Prod. 253 (2020) 1–11 https://doi.org/10.1016/j.jclepro.2019.119862.

[46] J.D. McTigue, D. Wendt, K. Kitz, J. Gunderson, N. Kincaid, G. Zhu, Assessing geothermal/solar hybridization – integrating a solar thermal topping cycle into a geothermal bottoming cycle with energy storage, Appl. Therm. Eng. 171 (2020) 1–10 https://doi.org/10.1016/j.applthermaleng.2020.115121.

[47] T. Liu, Q. Liu, J. Lei, J. Sui, A new solar hybrid clean fuel-fired distributed energy system with solar thermochemical conversion, J. Clean. Prod. 213 (2019) 1011–1023 https://doi.org/10.1016/j.jclepro.2018.12.193.

[48] A. Rosato, A. Ciervo, G. Ciampi, M. Scorpio, F. Guarino, S. Sibilio, Energy, environmental and economic dynamic assessment of a solar hybrid heating network operating with a seasonal thermal energy storage serving an Italian small-scale residential district: influence of solar and back-up technologies, Therm. Sci. Eng. Prog. 19 (2020) 1–14 https://doi.org/10.1016/j.tsep.2020.100591.

[49] D. Yamashita, K. Tsuno, K. Koike, K. Fujii, S. Wada, M. Sugiyama, Distributed control of a user-on-demand renewable-energy power-source system using battery and hydrogen hybrid energy-storage devices, Int. J. Hydrog. Energy 44 (2019) 27542–27552 https://doi.org/10.1016/j.ijhydene.2019.08.234.

[50] D. Lau, N. Song, C. Hall, Y. Jiang, S. Lim, I. Perez-Wurfl, et al. Hybrid solar energy harvesting and storage devices: the promises and challenges, Mater. Today Energy 13 (2019) 22–44 https://doi.org/10.1016/j.mtener.2019.04.003.

[51] S. Varghese, R. Sioshansi, The price is right? How pricing and incentive mechanisms in California incentivize building distributed hybrid solar and energy-storage systems, Energy Policy 138 (2020) 1–16 https://doi.org/10.1016/j.enpol.2020.111242.

[52] P.S. Sikder, N. Pal, Modeling of an intelligent battery controller for standalone solar-wind hybrid distributed generation system, J. King Saud. Univ. Eng. Sci. 32 (2020) 368–377 https://doi.org/10.1016/j.jksues.2019.02.002.

[53] H. Mehrjerdi, Modeling and optimization of an island water-energy nexus powered by a hybrid solar-wind renewable system, Energy 197 (2020) 1–17 https://doi.org/10.1016/j.energy.2020.117217.

[54] M. Zareei, C. Vargas-Rosales, R. Villalpando-Hernandez, L. Azpilicueta, M.H. Anisi, M.H. Rehmani, The effects of an Adaptive and Distributed Transmission Power Control on the performance of energy harvesting sensor networks, Comput. Networks 137 (2018) 69–82 https://doi.org/10.1016/j.comnet.2018.03.016.

[55] M.W. Jeong, J.Y. Ryu, S.H. Kim, W. Lee, T.W. Ban, A completely distributed transmission algorithm for mobile device-to-device caching networks, Comput. Electr. Eng. 87 (2020) 1–18 https://doi.org/10.1016/j.compeleceng.2020.106803.

[56] J. Liu, H. Cheng, P. Zeng, L. Yao, C. Shang, Y. Tian, Decentralized stochastic optimization based planning of integrated transmission and distribution networks with distributed generation penetration, Appl. Energy 220 (2018) 800–813 https://doi.org/10.1016/j.apenergy.2018.03.016.

[57] C.K. Das, T.S. Mahmoud, O. Bass, S.M. Muyeen, G. Kothapalli, A. Baniasadi, et al. Optimal sizing of a utility-scale energy storage system in transmission networks to improve frequency response, J. Energy Storage 29 (2020) 1–12 https://doi.org/10.1016/j.est.2020.101315.

[58] Y. Chen, T. Koch, N. Zakiyeva, B. Zhu, Modeling and forecasting the dynamics of the natural gas transmission network in Germany with the demand and supply balance constraint, Appl. Energy 278 (2020) 1–11 https://doi.org/10.1016/j.apenergy.2020.115597.

[59] S. Das, A. Verma, P.R. Bijwe, Efficient multi-year security constrained ac transmission network expansion planning, Electr. Power Syst. Res. 187 (2020) 1–18 https://doi.org/10.1016/j.epsr.2020.106507.

[60] E. da S. Oliveira, I.C. Silva Junior, L.W. de Oliveira, I.M. de Mendonça, P. Vilaça, J.T. Saraiva, A two-stage constructive heuristic algorithm to handle integer investment variables in transmission network expansion planning, Electr. Power Syst. Res. 192 (2020) https://doi.org/10.1016/j.epsr.2020.106905.

[61] S. Sayed, A. Massoud, Minimum transmission power loss in multi-terminal HVDC systems: a general methodology for radial and mesh networks, Alexandria Eng. J. 58 (2019) 115–125 https://doi.org/10.1016/j.aej.2018.12.007.

[62] C.E. Bruzek, A. Allais, K. Allweins, D. Dickson, N. Lallouet, E. Marzahn, Using superconducting DC cables to improve the efficiency of electricity transmission and distribution (T&D) networks: an overview, Supercond. Power Grid Mater. Appl. 1 (2015) 189–224 https://doi.org/10.1016/B978-1-78242-029-3.00006-6.

[63] M. Isuru, M. Hotz, H.B. Gooi, W. Utschick, Network-constrained thermal unit commitment fortexhybrid AC/DC transmission grids under wind power uncertainty, Appl. Energy 258 (2020) 1–16 https://doi.org/10.1016/j.apenergy.2019.114031.

[64] J. Liao, N. Zhou, Q. Wang, DC-side harmonic analysis and DC filter design in hybrid HVDC transmission systems, Int. J. Electr. Power Energy Syst. 113 (2019) 861–873 https://doi.org/10.1016/j.ijepes.2019.06.013.

[65] Z. Cen, P. Kubiak, C.M. López, I. Belharouak, Demonstration study of hybrid solar power generation/storage micro-grid system under Qatar climate conditions, Sol. Energy Mater. Sol. Cells 180 (2018) 280–288 https://doi.org/10.1016/j.solmat.2017.06.053.

[66] M. Jahangiri, M.H. Soulouknga, F.K. Bardei, A.A. Shamsabadi, E.T. Akinlabi, S.M. Sichilalu, et al. Techno-econo-environmental optimal operation of grid-wind-solar electricity generation with hydrogen storage system for domestic scale, case study in Chad, Int. J. Hydrog. Energy 44 (2019) 28613–28628 https://doi.org/10.1016/j.ijhydene.2019.09.130.

[67] D.S. Mallapragada, N.A. Sepulveda, J.D. Jenkins, Long-run system value of battery energy storage in future grids with increasing wind and solar generation, Appl. Energy 275 (2020) 1–14 https://doi.org/10.1016/j.apenergy.2020.115390.

[68] A. Mérida García, J. Gallagher, A. McNabola, E. Camacho Poyato, P. Montesinos Barrios, J.A. Rodríguez Díaz, Comparing the environmental and economic impacts of on- or off-grid solar photovoltaics with traditional energy sources for rural irrigation systems, Renew. Energy 140 (2019) 895–904 https://doi.org/10.1016/j.renene.2019.03.122.

[69] S. Buragohain, K. Mohanty, P. Mahanta, Experimental investigations of a 1 kW solar photovoltaic plant in standalone and grid mode at different loading conditions, Sustain. Energy Technol. Assess. 41 (2020) 1–12 https://doi.org/10.1016/j.seta.2020.100796.

[70] J. Li, P. Liu, Z. Li, Optimal design and techno-economic analysis of a solar-wind-biomass off-grid hybrid power system for remote rural electrification: a case study of west China, Energy 208 (2020) 1–17 https://doi.org/10.1016/j.energy.2020.118387.

[71] J.S. Jagabar, V. Krishnaswamy, An assessment of recent multilevel inverter topologies with reduced power electronics components for renewable applications, Renew. Sustain. Energy Rev. 82 (2018) 3379–3399 https://doi.org/10.1016/j.rser.2017.10.052.

[72] B. Vidales, J.L. Monroy-Morales, J.R. Rodríguez-Rodriguez, M. Madrigal, D. Torres-Lucio, A transformerless topology for a micro inverter with elevation factor of 1:10 for photovoltaic applications, Int. J. Electr. Power Energy Syst. 109 (2019) 504–512 https://doi.org/10.1016/j.ijepes.2019.02.006.

[73] R. Dogga, M.K. Pathak, Recent trends in solar PV inverter topologies, Sol. Energy 183 (2019) 57–73 https://doi.org/10.1016/j.solener.2019.02.065.

[74] E. Kabalcı, Review on novel single-phase grid-connected solar inverters: circuits and control methods, Sol. Energy 198 (2020) 247–274 https://doi.org/10.1016/j.solener.2020.01.063.

[75] M.P. Viswanathan, B. Anand, Particle swarm optimization technique for multilevel inverters in solar harvesting micro grid system, Microprocess. Microsyst. 79 (2020) 1–19 https://doi.org/10.1016/j.micpro.2020.103288.

[76] K. Zeb, W. Uddin, M.A. Khan, Z. Ali, M.U. Ali, N. Christofides, et al. A comprehensive review on inverter topologies and control strategies for grid connected photovoltaic system, Renew. Sustain. Energy Rev. 94 (2018) 1120–1141 https://doi.org/10.1016/j.rser.2018.06.053.

[77] Z. Liao, C. Cao, D. Qiu, Analysis on topology derivation of single-phase transformerless photovoltaic grid-connect inverters, Optik (Stuttg) 182 (2019) 50–57 https://doi.org/10.1016/j.ijleo.2018.12.169.

[78] Z. Ahmad, S.N. Singh, Improved modulation strategy for single phase grid connected transformerless PV inverter topologies with reactive power generation capability, Sol. Energy 163 (2018) 356–375 https://doi.org/10.1016/j.solener.2018.01.039.

[79] R.M. Silva, A.F. Cupertino, G.M. Rezende, C.V. Sousa, V.F. Mendes, Power control strategies for grid connected converters applied to full-scale wind energy conversion systems during LVRT operation, Electr. Power Syst. Res. 184 (2020) 1–14 https://doi.org/10.1016/j.epsr.2020.106279.

[80] A. Tahir, M. EL-Mukhtar, A. EL-Faituri, F. Mohamed, Grid connected wind energy system through a back-to-back converter, Comput. Electr. Eng. 85 (2020) 1–16 https://doi.org/10.1016/j.compeleceng.2020.106660.

[81] K.A. Chinmaya, G.K. Singh, Modeling and experimental analysis of grid-connected six-phase induction generator for variable speed wind energy conversion system, Electr. Power Syst. Res. 166 (2019) 151–162 https://doi.org/10.1016/j.epsr.2018.10.007.

[82] M. Mansour, M.N. Mansouri, S. Bendoukha, M.F. Mimouni, A grid-connected variable-speed wind generator driving a fuzzy-controlled PMSG and associated to a flywheel energy storage system, Electr. Power Syst. Res. 180 (2020) 1–15 https://doi.org/10.1016/j.epsr.2019.106137.

Chapter 11

Future of Electric Vehicles in Solar Hybrid Systems

1 Electric vehicles and solar hybrid systems

Today, fossil fuels are widely used in transportation and electricity generation. The concern about the decrease in fossil fuels has increased the search for alternative energy sources. The existence and damages of fossil fuels in the future are not a sustainable situation. In addition, oil prices are expected to increase gradually in the coming years. For all these reasons the development of electric vehicles has accelerated, especially in transportation. The sale and production of vehicles with internal combustion engines have been banned by many countries. The use of electric vehicles with such regulations will continue to spread. The acceleration of electric vehicle sales is increasing with the developments made on this technology. It is predicted that concerns about global climate change will be reduced, thanks to electric vehicles. Thanks to the high battery capacities of electric vehicles, it can be used as a backup energy source in a distributed grid structure. Electric vehicles can stay connected to the grid and stand ready as a backup source according to the vehicle-to-grid (V2G) topology [1–4].

With this technological feature, electric vehicles can provide services for load trimming, spinning reserve, voltage, and frequency regulation in the distributed grid. The use of solar and wind energies has become widespread recently. The increase of renewable resources in distributed power generation systems brings up the issue of continuity of energy. By their nature, renewable energy sources have an intermittent and unstable structure. These interruptions have an unpredictable output characteristic. Thanks to the energy storage feature of electric vehicles, these interruptions can be prevented in residential applications. Thus the electric vehicle functions as a backup power source both for transportation purposes and for domestic ones. Wind power installations have reached high power capacities in developed countries such as Europe and China. This renewable energy source has created the need for energy storage system. In this case, electric vehicles come into play and become a solution source [5–8]. The V2G structure takes on duty as a dynamic energy storage system.

A separate power plant foundation is formed with electric vehicles and distributed grid structure. The charging/discharging process of electric vehicle batteries in the V2G structure requires a superior control algorithm. A number of

Solar Hybrid Systems. http://dx.doi.org/10.1016/B978-0-323-88499-0.00011-2

difficulties arise in this situation. These control difficulties are the optimization and operation of the grid. The electric vehicles are seen as both an energy storage unit and a load for the grid. As an electric vehicle user, when the vehicle serves as an energy storage unit in the grid, it raises tariff concerns. In this case a legal base should be established to satisfy an electric vehicle user. In case electric vehicles are charged over the grid, the charging tariff is also important for the user [9,10].

For electric vehicles the energy flow should be measured with a monitoring system in the case of receiving or giving power from the grid. In addition, the control unit to which the electric vehicle is connected must support bidirectional power flow. While this charging unit is charging the electric vehicle battery, it must also provide the flow of energy from the electric V2G, if needed. All these features must be visible and controlled by both the vehicle user and the grid operator. Another problem is that the vehicle user wants the electric vehicle battery to be constantly charged. In this case the grid operator may encounter an incomplete energy storage unit, when needed. These developments will reveal our way of using electric vehicles with legal regulations over time [11,12].

The energy flow between the electric vehicle and the grid should be carried out with advanced sensor and communication infrastructure. This communication should be shared with the vehicle user and the grid operator. In this case a strong communication network is needed. This energy transfer between the electric vehicle and the grid forms the basis of the complex grid structure. This grid structure is a smart grid that will emerge in the future. It reveals the concepts of communication, information sharing, smart sensors, and real-time tariffs within the smart grid structure. Voltage fluctuations that may occur in the grid can be prevented by electric vehicles. In order to realize all these situations, a modern power system model must exist between electric vehicles and smart grid. In order to have all these features in the grid structure, there is a need for the widespread use of electric vehicles, the increase of renewable energy sources in the distributed grid, and a strong control mechanism [13].

1.1 Integration of electric vehicles to the electricity grid

The increase in the number of electric vehicles means an overload for the existing grid structure. This is because the existing grid structure works unidirectionally with the power flow from the grid to the loads. If the existing grid is not transformed into a distributed power generation structure, it will reveal the problems of work, economic factors, and capacity. Due to the increase in the number of electric vehicles, the power capacity of the existing grid should be increased. In addition to this investment, the use of renewable energy resources should be supported and encouraged to its users. In addition, all users' existing electricity meters should be replaced with smart and bidirectional ones. The regulations on tariffs should be made and a legal base should be established with this meter change [14–16]. At least as a starting point, steps should be taken to reduce the energy value transferred to the grid by the user from the energy consumed by the user. In addition, charging stations must be established in commercial, public, or workplaces for fast charging of electric vehicles. A uniform easy-to-use payment method should be developed for these charging stations. In the current situation, increasing widespread use of electric vehicles and serious problems await the grid infrastructure. The most important of these effects are overheating of distribution transformers, increase in thermal effects in transmission lines, and voltage fluctuations. On the other hand, the adoption and implementation of smart grid structure will be seen as improvements in grid performance, efficiency, and power quality. For this the existing grid structure should be planned and rearrangements should be made within the framework of standards [17,18].

Large electric vehicle fleets will be able to operate as a backup energy source when not in use. The realization of this scenario depends on the agreement between the electric vehicle owners and the distribution company. This situation should create an attractive situation for the electric vehicle user as well. Thus a coordinated electricity generation and a distribution process are realized between the distribution system operators, electric vehicle user, and energy service providers. Electric vehicles take the role of a virtual power plant for the grid structure [19]. The distribution operator always knows that there is an energy storage unit in the grid and puts these units into use when needed. Such a grid structure will prevent unpredictable electrical events such as blackout.

Battery technology is one of the most important factors for electric vehicles. Today, lithium ion–based batteries are used in electric vehicles. When a user compares a conventional vehicle to an electric vehicle, one of the first concerns is range. Improvement in battery technology will primarily increase vehicle range. On the other hand, the charging time and power of a high-capacity battery will cause another problem on the grid side. Continuous charging/ discharging of electric vehicles in the V2G structure in the grid makes the battery cycle life question. This situation creates a negative situation for the electric vehicle user [20,21]. If lithium ion–based batteries are operated in high charge/discharge cycles, their cycle life decreases. High discharge currents will affect the lifetime of the batteries in the case of connecting electric vehicles with the V2G structure, especially in the case of backup energy source behavior. These high discharge rates should be adjusted by both the user and the distribution operator at a rate that will not damage the batteries. Thanks to smart charge and discharge algorithms, the number of cycles of electric vehicle batteries can be extended. The smart charge/discharge algorithm should also take into account many parameters of the battery group such as aging, self-discharge, and temperature. This type of controls and algorithms will improve, thanks to the electric vehicle and V2G integration. High power density, small size/volume, and low cost will be important development issues for batteries in the future. In addition to all these developments, energy estimation, range estimation, tariff rates/pricing, and vehicle driving features should be monitored in real time with an advanced communication [22–24]. As a result, the structure that can realize all these features is the smart grid.

2 Electric vehicle charge levels

2.1 Electric vehicle charging methods and V2G

Electric vehicle batteries have high power capacities. They can meet the long-term demand for medium- and high power. Charging power levels of electric vehicles are examined in the following three groups [25,26]:

- Level 1: It is the power level that can be obtained from household sockets. The charging capacity of this level is approximately 3 kW.
- Level 2: It is the power level that can be obtained through the special charging socket and wiring. The charging capacity of this level is approximately 10–20 kW.
- Level 3: It is the power level that can be obtained by using a special charging socket, wiring, and a special external direct current (DC) fast charging converter. The charging capacity of this level is approximately 40 kW.

At Level 1 the electric vehicle usually charges slowly, at Level 2 at a medium level, and at Level 3 fast. According to the preference of the electric vehicle user, the charge level can be provided by the vehicle manufacturer. Different solutions are offered according to the electric vehicle user needs. It is generally applied in areas where the vehicle can stop for a long time for slow charging. These are home, parking, and individual charging stations. Slow charging stations or charging points are the Level 1 power level. Level 2 or Level 3 power levels are offered in the unit where bus and trucks are located at special charging stations reserved for electric vehicle fleets. There are individual charging stations open for medium charge levels. This charge is not suitable for level-controlled charging. The power levels of these charging stations are Level 2 because these station locations are short-term parking spaces. Another unit is battery exchange stations. This application may not be suitable for all electric vehicles. According to the state of stock in charge levels, Level 1 is used for slow charging and Level 2 or Level 3 power levels are used for fast charging. Fast charging stations are usually commercial stations with dedicated power conversion units. The power level at this station is Level 2 or Level 3 and charging is carried out in short periods.

When charging electric vehicles, one of the methods adopted in terms of grid is the free charging method. The electric vehicle battery is charged in an uncontrolled time period in this charge acceptance method. In this case the electric vehicle is fully charged freely at any time and is classified as a normal grid load. Thereafter the electric vehicle is connected to the grid by the user, the charging process starts, and it continues until the battery is fully charged or the user stops charging. In this charging method, it is assumed that the electricity fee tariff will be fixed throughout the day. There is no economic incentive method in this charging case. In this case, voltage fluctuations may occur due to sudden loads on the grid side. The grid infrastructure should be strengthened to solve the anticipated problems, but this solution is very costly [27].

In the multitariff, as in the previous charging approach, electric vehicles are charged at any time. However, electric vehicle users are subject to double tariffs on electricity prices. The electric vehicle user will charge his vehicle anytime he wishes only to be evaluated on a more favorable tariff in this method. Since this charging method does not contain an active management, it is in the same class with other loads in terms of grid structure. This charging method is also accepted as an uncontrolled charging approach. It is a charging method that may cause technical problems in terms of the grid [28,29].

Smart charging method is one of the suitable methods for electric vehicles and grid infrastructure. The electric vehicle charging rates are actively monitored from the control center in this method. Electric vehicles will be charged at certain rates by the control center while the grid is operating under normal conditions. This situation may not be suitable for the vehicle user in some cases. In case the electric vehicle user needs high charging current for a short time, this is reflected as a disadvantage. The control center will usually request charging of electric vehicles during periods of low power demand of the grid. This usually happens at night. In this case the grid will not be overloaded and the vehicle user will be charged at a low electricity tariff.

This charging method explains the future position of electric vehicles. While the grid charges electric vehicles, it will be able to demand energy from electric vehicles. With the electric vehicle V2G operating method, both the load demand powers will be met and will be used from the energy storage capacity. From the grid perspective, electric vehicles will be used to solve problems such as voltage and power fluctuations in some problematic areas. In addition to all these, operating the electric vehicle battery in V2G mode increases the charge/discharge cycle. This situation affects the battery cycle life. Electric vehicle users should be encouraged about charge/discharge tariff fees and about widespread use of this method with smart energy management in V2G mode [30–32].

2.2   Electric vehicle charging modes

In Mode 1 the electric vehicle battery is charged with slow charging through a normal electrical outlet. There is no communication between the electric vehicle and the charging point in Mode 1. It is necessary to provide protection against faults with a ground line between the electric vehicle and the charging point. Mode 1 is an undesirable method of charging. In Mode 2 the electric vehicle battery is charged from the alternative current (AC) line with slow charging. The charging cable includes control, communication, and leakage current protection units in Mode 2. In Mode 3 the electric vehicle battery is charged over the AC line with slow or semi-fast charging with a special charging socket. It is compatible with the socket structure in Mode 1 and Mode 2. The electric vehicle has the same feature on both sides of the charging cable, so there is no direction. In Mode 3 the charging station is responsible for charging, communication, and protection of the battery. Mode 3 is often used in public charging stations.

Mode 4 has a special socket to charge the battery group, and the electric vehicle is charged with fast charging. The battery is charged with high current by using the DC power line in Mode 4. The AC/DC converter unit due to its high power capacity is located in the charging station. This charge control unit includes control, communication, and protection features [33]. Four different charging modes for electric vehicles are given in Fig. 11.1.

The charging topology of electric vehicles is an interactive phenomenon for the electric vehicle and the grid. There are classifications regarding different power and charging times for charging electric vehicles. These classifications are slow and fast charging scenarios. These charging scenarios are standardized. While the slow charging scenario is usually done at home and office areas, fast charging is carried out at commercial or public charging stations. AC and DC charge levels for electric vehicles according to the Society of Automotive Engineers (SAE) J1772 standard are given in Table 11.1. AC Level 1 (120 V 12 A) offers charging at home, while AC Level 2 (240 V up to 80 A) takes place at workplaces, public, and commercial charging stations. DC Levels 1, 2, and

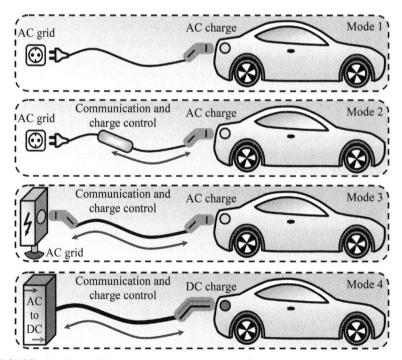

FIGURE 11.1 Four different charging modes for electric vehicles.

TABLE 11.1 AC and DC charge levels for electric vehicles according to SAE J1772 standard.

Power levels	Voltage	Current capacity	Power capacity
AC Level 1	120 V AC	12 A	1.4 kW
		16 A	1.9 kW
AC Level 2	240 V AC	Up to 80 A	19.2 kW
AC Level 3	Under development (over 20 kW)		
DC Level 1	200–500 V DC	<80 A	Up to 40 kW
DC Level 2	200–500 V DC	<200 A	Up to 100 kW
DC Level 3	200–600 V DC	<400 A	Up to 240 kW

3 are in the fast-charging class and are carried out only in commercial and public charging stations [34].

Since the subject of electric vehicles is a developing and maturing technology, there are many standards on this subject. One of these is that the DC bus voltage to be used in the electric vehicle is limited to 400 V. Battery capacities of electric vehicles vary according to the demands of the manufacturer and the user. Accordingly, charging time, capacity, and current values vary for certain electric vehicles. The standards regarding fast charging are still under development. One of the biggest reasons for this is the advances in battery and semiconductor technology.

These standards are created according to the limits allowed by technologies. The SAE carries out studies on AC and DC fast charging solutions. One of these fast-charging studies is to establish a safe and efficient connector standard for electric vehicles. In this context, AC single and three-phase charging and DC fast charging will be carried out with a single connector (Hybrid Combo). They developed the CHAdeMO standard connector for fast charging of electric vehicles by Tokyo Electric Power Company. Fig. 11.1 shows the AC/DC standard charging connectors developed for electric vehicles. DC fast charging in electric vehicles Level 1 supports up to 36 kW Level 2 power transfer to 90 kW [35–37].

The existing grid structure is AC-based and designed for domestic load groups. In order to charge an electric vehicle at high powers, a recovered power circuit is required. However, high powers cause cost and thermal problems in the existing grid infrastructure. Fast charging in electric vehicles is carried out with DC voltage. In this case the DC fast charging infrastructure must have high power capacities. There are many studies in this area and high power capacity DC fast charging stations will be widely used in the future [38].

The biggest challenge in these fast-charging stations is power converters. Another challenge is the grid infrastructure that can meet the power demand

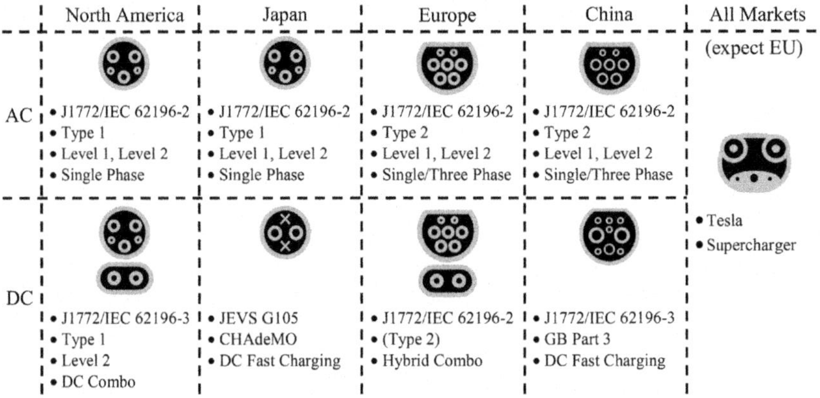

FIGURE 11.2 The AC and DC charging socket types produced for electric vehicles.

of these converters. The grid and converter models and feasibility studies are required for all these. SAE J1772 standard has been revised for bidirectional energy flow in order to reach DC fast charge levels regarding V2G integration. Bidirectional chargers with low electromagnetic interference are being developed for V2G and electric vehicle compatibility [39,40]. The AC and DC charging socket types produced for electric vehicles are given in Fig. 11.2.

2.3 Electric vehicle charging methods

Electric vehicles have the potential to diversify the energy source used for transportation while reducing carbon emissions. While the transition to electric vehicles is beneficial in several ways, range, concern, and long charging times remain critical challenges for collective adaptation of electric vehicles. Range concern can be reduced by having higher energy battery packs (> 100 kWh). However, higher power battery packs require longer charging times.

The cable charging system has AC and DC charging technologies. AC charging systems have a maximum charging capacity of 43 kW according to the IEC TS 61980-3 standard [41]. DC fast-charging systems increase charging power levels (> 100 kW) and reduce the charging time to less than an hour. As the charging power level increases, the cable conductive cross section increases and the cable structure becomes bulky and heavy. It also requires active thermal management of the charging cable and makes the charging system susceptible to hazards. Due to the space constraints on the electric vehicle, AC charging technology is used for powers below 43 kW, while DC charging technology is used for powers above this level [42]. Fig. 11.3 shows the present electric vehicle charging system topology.

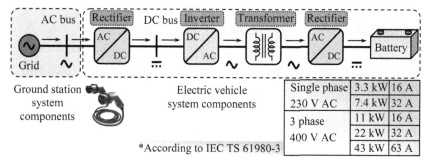

Single phase	3.3 kW	16 A
230 V AC	7.4 kW	32 A
3 phase	11 kW	16 A
400 V AC	22 kW	32 A
*According to IEC TS 61980-3	43 kW	63 A

FIGURE 11.3   The present electric vehicle charging system topology.

2.3.1   On-board charging systems

There are two different methods for charging electric vehicles; it is an onboard and off-board charging system. In the onboard charging system, the charging power circuit is in the electric vehicle. The onboard charging system is low in cost and has a compact structure in small dimensions. It is possible to charge wherever there is an AC line with the onboard charging system. In addition, the effects of this charging system on the grid are very low. It is generally suitable for domestic use and the battery is charged overnight. Thus the load level on the grid is low and the electric vehicle is charged on a low tariff. The disadvantage of this charging system is its slow charging current (Level 1 and Level 2). Therefore the electric vehicle charging time is long (6–8 hours). The power capacity of the onboard charging system is low [43]. Fig. 11.4 is given to show the onboard charging system topology.

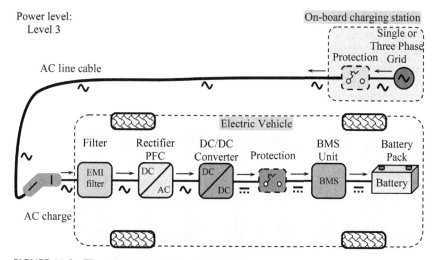

FIGURE 11.4   The onboard charging system topology.

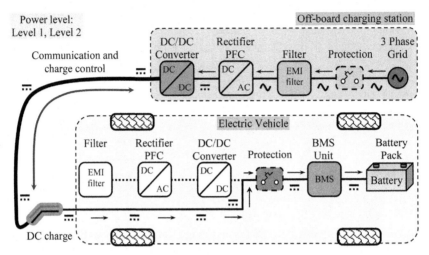

FIGURE 11.5 **The off-board charging system topology.**

2.3.2 Off-board charging systems

In the off-board charging system, the charging power circuit is outside the vehicle. The charging is carried out with a DC bus in this charging method. Since the charging power circuit is not on the electric vehicle, high current and power values (Level 2 and Level 3) are reached. This charging method can usually be done with commercial or public charging stations. Thanks to the off-board charging method, long distance travels can be made and charging is performed in short time (less than 1 hour). Thus efficiency on the electric vehicle is increased in terms of weight, volume, and cost. The charging power circuit is bidirectional and suitable for V2G structure in this charging method. Off-board charging power circuitry is complex and costly. During this charge management work, it has a high impact on the grid and requires security measures [43]. Fig. 11.5 is given to show the off-board charging system topology.

2.3.3 Wireless charging systems

There is no physical connection between the ground station and the electric vehicle in the wireless electric vehicle charging system. Energy transfer is provided by the magnetic connection between the receiver and transmitter coils in wireless energy transfer, and electrical isolation is also provided by the air gap between the coils. There is no connection between the electric vehicle and the charging system by a conductive. Therefore it is safer as the risk of electric shock is lower than wired systems. Fig. 11.6 shows the wireless electric vehicle charging system topology.

Thanks to the wireless energy transfer for electric vehicles, charging in harsh weather conditions without getting wet and contaminated brings ease of

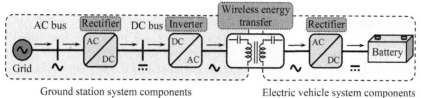

Ground station system components   Electric vehicle system components

FIGURE 11.6   **The wireless electric vehicle charging system topology.**

use for the elderly and the disabled users. In the future, electric vehicles will be able to be charged with wireless energy transfer technology while stopping at intersections or while driving [44,45].

When wired and wireless charging systems are compared, while the connection cable used in the cable charging system causes problems such as complexity, fault, electric shock, disconnection, and power failure, wireless charging systems do not have any connectors that cause these problems. In cases of harsh weather conditions such as rain, snow, or wind, plugging the cable into the charging socket is both dangerous and difficult. In this respect, wireless charging systems are easy and practical [46].

In the primary coil side control in the wireless charging system, the system is controlled as closed loop by using parameters such as inverter output current, transmitter coil current [47] and resonance capacitor voltage. By using these values, voltage amplitude, current amplitude, and frequency at DC/AC inverter circuit output are controlled [48]. The purpose is to operate the system at resonant frequency and to control the magnitude of the power to be transferred. In case the system operates at resonance frequency, the switching losses are decreased occurring when the semiconductor switches turn on or turn off and the power transfer efficiency is increased. The physical basis of frequency control is that the apparent input impedance from the input of the coils exhibits two more resonance behaviors around the resonant frequency due to the physical effect called frequency separation [49]. Since the values of these frequencies depend on the coupling constant between the coils, the system frequency is controlled to correspond to one of these frequencies in frequency control.

Wireless energy transfer system consists of two important units. These are coil group and power circuit. In order for the wireless energy transfer system to work with high efficiency, the coil group must be operated at active power and resonance. The resonance frequency (F_R) relationship of an electrical system is given in Eq. (11.1) [50]. Eq. (11.1) shows the relationship between coil inductance and switching frequency used in wireless charging systems. The inductance value of the coil is inversely proportional to the frequency and it must be operated at high frequencies in order for the inductance value to be small.

$$F_R = \frac{1}{2\pi\sqrt{LC}}$$

(11.1)

High frequencies will decrease the inductance (L) value of the coil in the wireless energy transfer system. The decrease of the coil structure (weight and volumetric) will reduce the weight of the coil that will be located especially in the electric vehicle part. Thus the weight of the wireless charging coil on the electric vehicle is reduced. This will have contributed to the range of electric vehicles. Therefore the operation of wireless charging circuit topologies at high frequencies is important for high power transmission, vehicle weight, and aerodynamics. Wireless charging technology will soon become the standard equipment in every vehicle with the development and the widespread use of electric vehicles.

Inductive coils are one of the important components affecting the amount and the efficiency of power transfer in wireless energy transfer systems. The ultimate goal in coil designs is to produce coils exhibiting high magnetic coupling constant as small as possible, high efficiency at desired power levels, and high tolerance to misalignment distortion [51]. Circular and double-D (DD) coil-type approaching back-to-back coils have been important coil geometries used for high power wireless electric vehicle charging applications [52]. The inductive coils in wireless energy transfer are produced from Litz wire to provide low conduction losses. The noncontact power transmission performance between the coils in the transmitter and receiver parts is highly dependent on the intensity distribution of the magnetic field generated by the transmitter coil and the level of current this magnetic field induces in the receiver coil. The DD-type coil structure section view is given in Fig. 11.7.

2.3.3.1 Coil design in wireless charging systems

Inductive coils are passive components that perform noncontact power transfer at high frequency in accordance with the principle of Ampere and Faraday laws (magnetic field coupling). These coils are produced by winding the conductive wire in certain geometry of air or a ferrite material on the pole in the structure or around it. Due to the skin effect and proximity effect on the conductive wires, the resistance values and therefore the losses at high frequencies increase. The ferrite structures are generally required in coils in high power systems, because

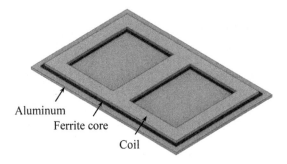

Aluminum
Ferrite core
Coil

FIGURE 11.7 The DD type coil structure section view.

the strong magnetic field formed around the coils should be prevented from forming in undesirable areas. Losses due to eddy currents should be reduced. Unnecessary distribution of these areas should be avoided in terms of EMI–EMC. Moreover, the use of ferrite makes a significant contribution to increasing the inter-coil coupling. The disadvantage of ferrites is that they increase the weight of the coils considerably. In this context the amount of ferrite required to be used by simulations should be used and its overuse should be avoided both in terms of cost and weight gain. Ferrites are mostly produced by bonding I-type ferrite bars or other geometrical ferrite materials to one another [53].

Resonant circuits are used to compensate the reactive value of the impedance that appears when looking toward the coils. The compensation circuit is most commonly used to compensate the reactive value of the impedance visible when looking toward the coils. Compensation circuit basically consists of capacitors connected in series or parallel to the coils. These structures are realized in four different topologies that are serial–serial (SS), serial–parallel (SP), parallel–serial (PS), and parallel–parallel (PP) as in Fig. 11.8 [54].

For example, in a system with PP compensation, the amount of power transferred to the load connected to the coil on the compensated receiver side, depending on the current entering the transmitter coil, can be found by the following equation [55]:

$$P_{out} = \omega I_P^2 \frac{M^2}{L} Q_L \tag{11.2}$$

On the input side the resistance compensated toward the coil side is calculated by the following equation:

$$R_{in} = \frac{M^2 R_L}{L^2} \tag{11.3}$$

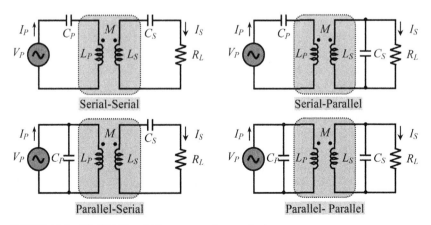

FIGURE 11.8 SS, SP, PS, and PP compensation topologies.

Similar equations are widely available in the literature for all compensation schemes. As can be seen from Eq. (11.3), the power level at the output is directly dependent on the coupling constant ($k = M/L$), frequency, and the quality factor (Q_L) of the receiver coil under load. The voltage level to occur at the input depends on the impedance seen from the input and the input current to be applied, as seen in the relation in Eq. (11.3). In short the voltage and current levels that will occur at the inputs and outputs of the system and the coil parameters that should be designed according to the amount of power to be transferred to the conjugate load resistance (RL) visible at the receiver coil output should be determined approximately.

2.3.3.2 Semiconductor technology used in wireless charging systems

Today, semiconductor technologies with high-frequency switching are available. The transistor has dominated the silicon world for many years, but this is changing with the increase of technology. Recently, compound semiconductors made of two or three materials have been developed in silicon technology, thus offering many superior properties. For example, it led to the emergence of compound semiconductors and light-emitting diodes. In addition to these technological developments, the production of compound semiconductors reveals the difficulty and the cost of manufacturing these materials. Despite all the factors, silicone technology provides very important benefits. More demanding applications such as industrial products, military application areas, electrical systems, and electric vehicles require the use of compound semiconductors. Commercially available products that emerge using two compound semiconductor technologies are Gallium Nitride Field Effect Transistor (GaN) and Silicon Carbide (SiC) power transistors. These semiconductor materials can compete with the long-life silicon metal–oxide–semiconductor field-effect transistor (MOSFET). Although GaN FET and SiC MOSFET semiconductor technologies are similar in some respects, they also have important differences. The most important feature that distinguishes GaN FETs from other semiconductor switching elements is the high-speed turn-on and turn-off time it has gained with high-speed electron carrying capacity [56]. The turn-on time's graph of GaN FET and SiC MOSFET semiconductors is given in Fig. 11.9. Thanks to the high switching speed of GaN FETs compared to SiC MOSFETs, it ensures higher dV/dt or dI/dt ratios.

SiC MOSFETs up to 1700 V are available around 70 A and 45 mΩ, but internal reverse bias diodes are slow. GaN FET semiconductor switching elements are manufactured commercially by GaN Systems, Texas Instruments, Infineon, and Transphorm. Currently, Transphorm's TP65H015G5WS model, in TO-247 package with 650 V, 95 A, and 15 mΩ values, has 78 ns turn-on and 132 ns turn-off time. GaN FETs are semiconductors that can switch faster than most SiC MOSFETs. The switch-on resistance (R_{DSON}) of existing GaN FETs has very lower values than traditional silicon MOSFETs [57]. The switching frequency and power ranges of MOSFET, IGBT, SiC MOSFET, and GaN FET semiconductor technologies are given in detail in Fig. 11.10.

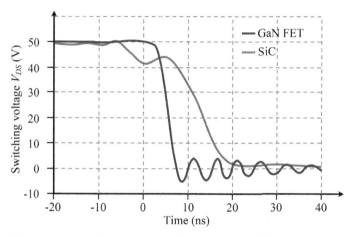

FIGURE 11.9 The turn-on times graph of GaN FET and SiC MOSFET.

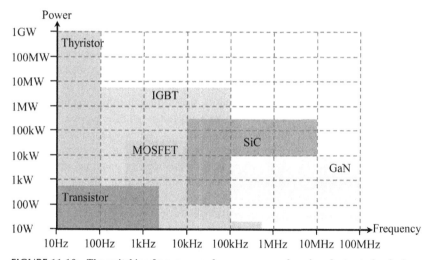

FIGURE 11.10 The switching frequency and power ranges of semiconductor technologies.

GaN FET and SiC MOSFETs are getting ready to solve the challenges of electric vehicles. General specifications for GaN FET and SiC MOSFETs are operating voltage ranges of 650, 900, and 1200 V, high-speed switching frequency (up to 10 MHz for SiC MOSFET and up to 100 MHz for GaN FET), higher operating temperature (<200 °C), minimum power loss, higher efficiency, and lower conduction resistance (<10 mΩ) [58].

GaN FETs are widely used in wireless devices as power amplifiers at frequencies up to 100 MHz. Some of its primary uses are cellular base station power amplifiers, military radar, satellite transmitters, and general radio frequency amplification. However, due to their high voltages (up to 1000 V), high temperatures, and high switching frequencies, they are used in various power supply applications such as DC/DC converters, inverters, and battery chargers [58].

On wireless energy transfer systems for electric vehicles, a 20-kW wireless energy transfer system at 180 Hz was proposed by Bolger and Kristen in 1978 for intercity roads and tested in a laboratory environment with the produced transmitter–receiver prototypes [59]. This study is one of the first examples of dynamic charging systems proposed for charging vehicles while traveling. In the study, wireless energy transfer was carried out in a very short distance of only 2.5 cm of a very large magnetic coil structure with an inductive coil width of 60 cm and a length of 153 cm. The reason why the distance is so short is that the operating frequency is low and the current on the transmitter coil at these low frequencies creates low levels of electromotive force on the receiver coil. Therefore the lower the operating frequency, the higher the magnetic coupling between the coils is required. This leads to the conclusion that in addition to the short distance between the coils, the tolerance of the system against the alignment between the coils at low frequencies is quite low [60].

In the recent years, studies of wireless energy transfer systems targeting electric vehicle charging systems have gained speed again. One of the reasons for this is the widespread use of electric vehicles, and the other reason is the developments in power electronic circuits that can switch at high speed depending on the developments in semiconductor technology. As the cost of reaching high powers at high frequencies decreases, the minimum limits required in the magnetic coupling constant between coils decrease, enabling power transfer over longer distances (15–30 cm) [61]. Adaptive frequency tuned power transmission systems have been presented to keep the changing power transfer efficiency constant due to the distance between the coils or misalignment in low power high-frequency systems [62]. In these studies, solutions are presented to find the minimum point of the reflection coefficient by the transmitter and to follow the resonance frequency that changes due to the distance change. In power transfer systems for electric vehicles, the change in resonant frequency is performed by methods such as finding the phase difference between the input voltage and current or the point where the current is maximum, since coupling elements cannot be found at low frequencies and high powers [63].

Wireless energy transfer for electric vehicles is an emerging technology and is under development. A second coil is placed under the electric vehicle that will receive the magnetic field produced by the primer coil on the ground. Electric vehicle charging current, voltage, and power data are transferred to the other party via wireless communication and the primary coil switching on the transmitter part is performed by taking this data as a reference. According to the IEC TS 61980-3 standard, it is possible to charge wirelessly up to 11.1 kW for electric vehicles using a wireless energy transfer system. Power values in the standard are updated and changed by research and development studies [64]. Wireless inductive charging based on wireless energy transfer is a safe and convenient form of electric vehicle battery charging. It does not require manual connection of charging cables, provides galvanic isolation from the grid to the vehicle, is suitable for autonomous charging, and is more resistant to harsh weather conditions. The SAE published the TIR J2954 technical information report on wireless energy transfer for light commercial and passenger electric vehicles in 2017. According to the TIR J2954 report, wireless charging systems are designed to increase up to 22 kW [65]. Although there has been a considerable amount of research and development focusing on various aspects of wireless charging of electric vehicles below 22 kW power levels, there are only a few studies focusing on wireless charging at power levels above 100 kW [66].

While examining the wireless energy transfer studies, both grid-to-vehicle and V2G power transfer mode experiments were conducted in the study on the integration of electric vehicles with the grid. A 20 kW wireless charging system has been developed by using bidirectional inductance-capacitance-capacitance (LCC)–LCC resonance structure in wireless energy transfer coils. Both analytical and experimental results show that the frequency response of input and output impedance and the voltage gain vary significantly with the charging power of the battery. It has been shown that power circuit operating frequency increases efficiency by reducing system losses when operated with zero voltage switching technique [67].

In the high-frequency inverter structure used in wireless energy transfer, a new control strategy has been proposed to reduce the voltage notches in the inverter output. Voltage notches have been shown to increase inverter switching losses and to negatively affect the overall system efficiency. It was determined that by reducing the voltage notches in the inverter, the total losses decreased by 17 %. In the high-frequency inverter, a trigger signaling at 83.24 kHz was applied by using SiC MOSFET. A wireless energy transfer of 1 kW was realized at 150 mm of air gap between the coils [68].

Single-phase 100-kW energy transfer was performed wirelessly for electric vehicle battery charging at Oak Ridge National Laboratory (ORNL). In the study, SiC MOSFETs are used in high-frequency single-phase inverter structure. The switching frequency of the high-frequency inverter is 22 kHz. The coil type used in wireless energy transfer has DD geometry. Series–series resonant circuit is used as a resonant circuit. The distance between the coils in the wireless energy transfer system is 127 mm and the efficiency of all power circuits is 96 % [69].

A three-phase bidirectional 20-kW LCC wireless charging system with resonance structure was tested in the study carried out in ORNL. In the study, experimental studies were carried out in 280 mm air gap using the DD structure in the coils. SiC MOSFET semiconductor switching elements are used in high-frequency three-phase inverter structure [70].

In the laboratory environment at ORNL, 14-kW wireless energy transfer for Toyota RAV4 electric vehicle has been tested. In this study conducted on a real electric vehicle, integration among battery charging communication, wireless charging system with J1772 protocol, and electric vehicle was achieved. In the study a three-phase grid and a high-frequency SiC MOSFET inverter structure are used. Toyota RAV4 electric vehicle has been tested up to 14 kW in 160 mm magnetic air gap with 95.16 % DC/DC efficiency [71].

In the other study carried out in ORNL, 18 kW wireless energy transfer was performed by using three-phase high-frequency SiC MOSFET inverter structure. SiC MOSFET inverter was triggered at 83.5 kHz and a coil structure with circular geometry was used. The coils were tested with a series–series resonant circuit at 127-mm air gap [72].

Tests were carried out in ORNL with a three-phase SiC MOSFET inverter-based 50 kW power and 150 mm of air gap between the coils for wireless energy transfer. Series–series resonance structure is preferred as the resonant circuit of the coils. By using the zero-voltage switching method as a high-frequency inverter control structure, 95 % DC/DC efficiency has been obtained [73].

A 1.2-kW study was carried out using SiC MOSFET as a single-phase high-frequency inverter structure for wireless energy transfer. In the study the inverter is triggered at 85 kHz and a control algorithm compatible with SAE J2954 standard has been developed [74].

According to the IEC TS 61980-3 standard, single-phase systems can be charged wirelessly with a power range of 3.7–7.4 kW. A power value of 7.4 kW is in the fast-charging class. However, it is possible to increase this power value to much higher values with new-generation high-frequency switching elements. In ORNL, 50-kW power transfer has been carried out for the electric vehicle with a single-phase inverter structure wirelessly.

2.3.4 Electric vehicle charging method in the future

Wireless energy transfer systems are in the research phase for charging electric vehicles on the future highways. Energy transfer optimization between the moving vehicle and the coils on the highway is one of the important issues in these systems. The wireless charging system located on the highways communicates wirelessly when the electric vehicle enters the charging road and the energy transfer starts. While there is one coil on the electric vehicle, there are coil groups extending over multiple kilometers on the highway. The biggest problem is the low magnetic coupling of the coils while the electric vehicle is under way. As the electric vehicle starts moving, each time the coils transfer high energy initially and then low energy. The solution to this is to increase the number of coils on the highway floor. Electrically, coil groups can be multiplied by connecting in parallel over a DC bus. Thus the electric vehicle in motion can be charged at high power values. Fig. 11.11 shows the topology of wireless charging with more than one parallel-connected coil for electric vehicles in motion.

Researchers from the Korea Advanced Institute of Science and Technology performed an E and U core–based wireless energy transfer by connecting coil groups in parallel. In the system, 100 kW of power has been transferred with

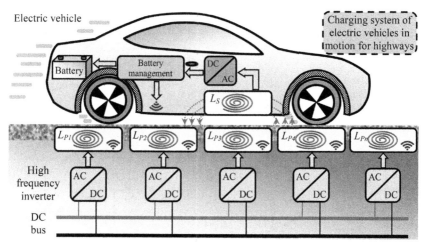

FIGURE 11.11 Wireless charging topology of electric vehicle in motion.

five 20-kW coil groups in total. Five high-frequency inverter architecture were used in the wireless energy transfer performed [75].

Another study was carried out for electric rail vehicles in motion. The system realizes a total of 1-MW wireless energy transfer, and four coil groups are connected in parallel with 250 kW [76]. It is necessary to use parallel converter topologies to reach high charging current and power values in electric vehicles.

3 Configuration electric vehicle and solar hybrid systems

3.1 AC bus–connected systems

AC bus–connected systems are one of the connection methods used in fast charging of electric vehicles. The topology of common AC bus–connected electric vehicle and solar hybrid system is given in Fig. 11.12. In this system structure the grid voltage is reduced with a low-frequency step-down type transformer. The primary side of this transformer is connected to the AC grid line and the secondary side to the electric vehicle and solar hybrid system structure. Solar photovoltaic (PV) panel, wind energy, high-capacity energy storage unit, and electric vehicle charging units are collected individually in a common AC bus. This system is very simple to use and apply. It is similar to the transformer centers in the existing grid structure. The fast charging of electric vehicles can be achieved with this common AC bus bar structure. The charging station installed for each electric vehicle in the AC common bus system contains factors that disrupt the power factor on the grid side individually and the grid voltage may be distorted. Increasing the charging stations in this AC bus reduces the overall efficiency of the line. In addition, the transducer in each unit increases the complexity of the system since it contains its own filter sensor and control

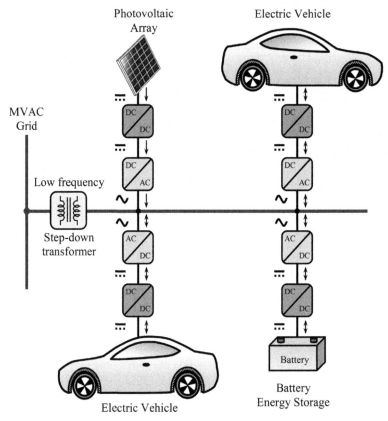

FIGURE 11.12 The topology of common AC bus–connected system.

algorithm. The increase in renewable energy sources and energy storage units increases the transformation phases and the system cost increases. In AC bus–connected systems, the control of AC systems, their integration into the grid, and the improvement of the power factor are quite complex compared to other topologies [77].

3.2 DC bus–connected systems

Common DC bus–connected systems are structures that allow electric vehicles to be charged with DC lines at high powers. The topology of common DC bus–connected electric vehicle and solar hybrid system is given in Fig. 11.13. In this structure, there is a step-down transformer and AC/DC converter unit with high power capacity between the mains and hybrid systems. This structure is highly flexible for a distributed renewable energy grid structure. It can easily be combined with the common DC bus. Solar PV panel, energy storage unit, and electric vehicles are connected to a DC bus with DC/DC converter circuit. Since

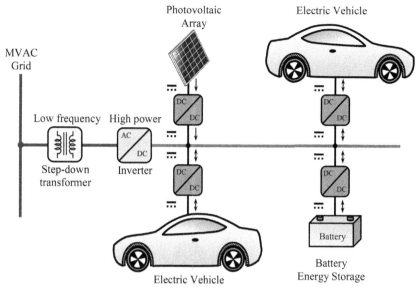

FIGURE 11.13   The topology of common DC bus–connected system.

the common transmission line is a DC bus, problems such as synchronization are eliminated and no reactive power occurs. A part of the smart grid structure is formed with this one. System costs are reduced and system efficiency increases as the number of conversions between units decreases. The system supports the operation of DC loads as the power line is carried by a DC bus. Especially since the fast-charging process of electric vehicles is done with a DC bus, the battery group is charged easily and quickly. Since the high-power AC/DC converter at this common DC bus system input operates at high frequency, its impact on the grid is high. Since high power transfer is realized with a DC bus, cutting operations are difficult since there are no zero crossing points. For this reason, it requires additional protection devices [78].

3.3   DC bus–connected systems with high-frequency transformers

The easiest way to charge electric vehicles fast can be done with a DC bus. In conventional converter structures, the AC voltage is primarily a step down by a low-frequency transformer. Achieving high powers will increase transformer size, weight, and cost. In addition, a large conductive cross section is required in this type of fast charging system. In order to overcome all these problems, the converter structure, including high-frequency transformer, is used. The low-frequency bulky transformers are not used with this approach. The topology of high-frequency transformer, DC bus–connected electric vehicle, and solar hybrid system are given in Fig. 11.14. Bidirectional energy flow can be

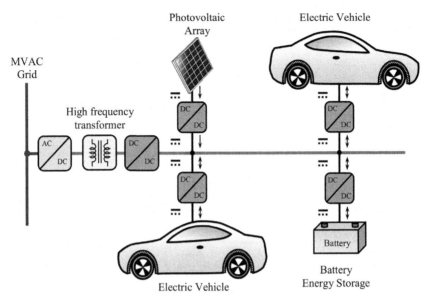

FIGURE 11.14 **The topology of DC bus–connected systems with high-frequency trans-formers.**

provided with a high-frequency AC/DC converter and electrical isolation is also provided. Thus the number of converters used in renewable energy sources such as common DC bus connection and electric vehicle charging stations is reduced. The electric vehicles are provided to be charged quickly and the charging station installation cost is reduced with this structure [79].

3.4 Role of electric vehicles in the grid in future

While electric vehicles demand power from the grid during charging, they can operate as dynamic energy storage, when needed, and support the grid (V2G structure). While a single electric vehicle will not be a sufficient energy storage source for a zone, multiple electric vehicle batteries will provide high capacity. In this case, for the realization of the V2G structure, support can be provided to the grid by commissioning it through communication between the electric vehicles and the grid operator. Thus electric vehicles will take the role of energy storage in distributed power generation systems.

Electric vehicles need control units that provide bidirectional power flow in order to be able to interact with the smart grid. Clustered electric vehicles are a challenge for the V2G structure. Some electric vehicles need to be charged, while others need to support the grid. In this case the control algorithm has to predict all these parameters. Thanks to electric vehicles, it can eliminate the voltage fluctuation and frequency shift that may occur in the grid. In the bidirectional power flow system of electric vehicles, the battery cycle life decreases because the electric vehicle battery group is constantly exposed to the charge/discharge cycle. In addition, protection and measurement systems in bi-directional power flow systems bring extra costs.

There is a need for communication between the V2G structure and the electric vehicle fleets, which is prevented by the public institution. When considered on a regional and country base, high-capacity energy control is mentioned. This raises the issue of energy security. This control structure can be centralized, hierarchical, or regional. Decision-making and information exchange are controlled from a single point in a central control structure. The hierarchical control structure performs spatial-based decision-making and subinformation exchange. Then this flow of information takes place toward the regions. This information includes power data received from smart meters, power data provided to the grid by power generation systems, and power data of all load demand in the grid. After the control center processes all the data, it calculates the analysis of the electric vehicles connected to the system and the total energy storage capacity. It then uses electric vehicles to compensate for regional or residential consumption deviations. Electric vehicles will serve as virtual power plants within the V2G structure. Thus electric vehicles will play a role in the electricity grid and energy market [80–84].

4   Electric vehicles and solar PV panel energy

Electricity production from solar PV panels is carried out at high capacities and makes a significant contribution to the energy production share. Solar PV panels provide a cumulative power flow to the electricity grid. Since electric vehicles have an increasing power demand share from the grid, solar PV panels have started to be used as a support to the grid. Electric vehicles usually stop in parking areas during working hours. It is observed that solar PV panel roof systems are becoming widespread especially in open parking areas [85]. The application topology of solar PV panel and electric vehicle charging station is given in Fig. 11.15. It is also possible to apply V2G structure with solar PV pan-

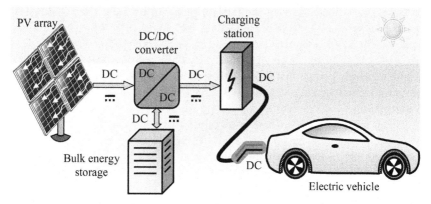

FIGURE 11.15   The application topology of solar PV panel and electric vehicle charging station.

els. Thus operating and charging costs will be reduced. The advantages of an electric vehicle charged with a renewable energy source will increase in terms of carbon emissions and fuel costs. The charging of electric vehicles from solar PV panels to the grid will be reduced. Solar PV panel–supported electric vehicle charging stations provide gains and benefits in many ways. In this charging system, after the electric vehicles are fully charged, the remaining energy is transferred and support is provided to the grid.

The investment cost of the solar PV panel–supported electric vehicle charging station can be met with a long-term payback. Thus the charging system will provide profit after the amortization period. Like solar systems, this charging system is also affected by seasonal conditions. An energy storage unit is also added to high-power electric vehicle charging systems. Thus the energy that is not transferred to the grid is stored in the high-capacity energy storage unit. Later, the electric vehicle can be charged even in harsh weather conditions. In such a structure, it is necessary to use a bidirectional converter. In addition, since DC energy storage will occur, the Level 3 charging method can be used to charge the electric vehicle at high charging currents. In addition, this high-power capacity energy storage unit can also be used for the V2G structure [86]. It can provide energy flow in peak load time as a support to the grid.

This bidirectional energy flow relationship is complementary to system deficits. Solar PV panel–supported electric vehicle charging station will take its place in the smart grid part by supporting the V2G structure. Thus solar PV panel–supported electric vehicle charging stations will meet the charge demands of vehicles in the future and contribute to the load reduction in the present grid.

Solar PV panels have gained extra popularity over the past decade as a means of distributed generation. This is mainly due to two reasons. The first is to increase the efficiency of solar PV panels. The second is that solar PV panel costs decrease every year. This has made solar energy not only a clean but also a cheap energy source. Distribution system operators around the world are expected to reduce their supply tariffs for solar energy in the next few years to match wholesale electricity prices. Installation of solar PV panels as large power plants is increasing. In addition the potential for solar PV panel installation of workplace roofs and car parks is increasing [87]. All of these cases indicate a huge PV potential for charging electric vehicles. The energy production of solar PV panels is greatly affected by daily and seasonal changes. This requires energy storage that is an expensive component. Electric vehicles can be considered as a great potential to meet the storage needs without the need for a storage unit.

Using solar PV panels, some or all of the energies required for electric vehicle battery charging can be provided by renewable energy sources. One of the most important factors in increasing the life of the battery group in electric vehicles is battery energy management. Energy management should be carried out by considering a number of factors such as charge/discharge cycle and thermal effect of lithium-based batteries used in electric vehicles. Thus optimum use of the electric vehicle battery group is provided. A hybrid energy storage structure can be obtained by using fuel cell, battery group, and ultracapacitor together in hybrid electric vehicles. In this case an energy management algorithm should be written for the optimum use of energy from three sources [88].

Economic applicability can be increased by centrally coordinating energy management between renewable energy–based buildings and electric vehicles. On the other hand, monitoring the battery group is critical in terms of evaluating the technological-economic performance and energy flexibility. For flexible energy management, an advanced energy control strategy that fully utilizes the amortization characteristics of the battery group should be developed. The purpose of the energy management algorithm is to increase the number of cycles of the battery group, to shorten the amortization period, and to prevent deep discharge. In addition, the energy stored in the electric vehicle battery in these systems will be able to operate in V2G mode, thanks to the bidirectional converter circuits. While all these processes are taking place, an energy requirement will be provided with a continuous electrical energy on the battery side and preventing voltage fluctuations [89].

The electric vehicles can be charged wirelessly with a solar PV panel–based structure. An example of solar PV panel and wireless electric vehicle charging topology and the control diagram of the whole system are shown in Fig. 11.16. In this system the electric vehicle battery is charged by renewable energy sources in an uninterrupted and high-quality method. In general, solar PV panel and wireless charging system, wireless energy transfer system, electric vehicle bat-

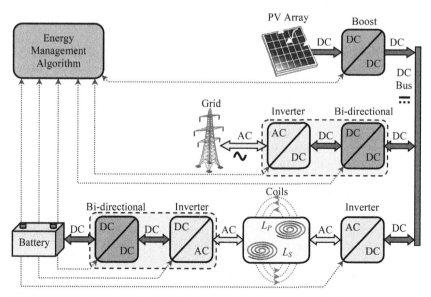

FIGURE 11.16 Example of solar PV panel and wireless electric vehicle charging topology and the control diagram.

tery group, power electronic converter circuits (DC/DC boost converter, DC/DC bidirectional converter, AC/DC bidirectional converter), and dynamic energy management algorithm. The energy management algorithm in the system provides two energy flow directions, in the direction of the electric vehicle battery and in the direction of the grid. In this system, the energy required for electric vehicle battery charging is generally met from solar PV panels. If the power is not sufficient, energy support is provided from the grid to the electric vehicle. In addition, the energy produced by solar PV panels and the energy stored in the electric vehicle battery are transferred to the grid direction when needed.

4.1 Energy management system in the V2G

In systems with solar hybrid units and electric vehicles, energy management systems are needed to make the power flow between the grid and the units at an optimum level. The energy management system carries out electric vehicle charging operations, taking into account the solar PV panel power generation [90]. Figs. 11.17 and 11.18 give an example of solar PV panel, electric vehicle, and grid power flow scenarios. The energy flow directions determined according to the scenarios in the solar hybrid system are shown with arrows. In the solar hybrid system, there is a bidirectional energy flow between the grid and the DC bus and between the electric vehicle battery and the DC bus. There is unidirectional energy flow between the solar PV panel and the DC bus. While the energy needed at the DC bus point can be met from the solar PV panel system or the grid, the remaining energy can be transferred to the direction of the grid. The system works depending on the occupancy of the electric vehicle battery and solar PV panel production. The energy management algorithm determines these study scenarios.

In Scenario 1, all energy flow in the DC bus is in the direction of the electric vehicle battery and the electric vehicle is charged. In this scenario, if the energy generated from solar PV panels is sufficient for electric vehicle battery charging, the electric vehicle battery can be charged by solar PV panels (Scenario 1a). If solar PV panels do not generate energy, all of the energy can be met from the grid (Scenario 1b). If the energy produced from the solar PV panels is lower than the energy demanded by the electric vehicle charger, the remaining part of the required energy can be met from the grid (Scenario 1c).

In Scenario 2, all energy flow in the DC bus is toward the grid direction. If the electric vehicle battery is not charged, only energy is transferred from solar PV panels in the direction of the grid (Scenario 2a). In cases where there is a need for energy, the need for grid can be met with the remaining energy in the electric vehicle battery (Scenario 2b). If there is energy remaining in the electric vehicle battery and there is power generation by solar PV panels, the energy can be transferred to the grid direction (Scenario 2c).

Thanks to the solar hybrid energy management system, the following charging scenarios can be designed:

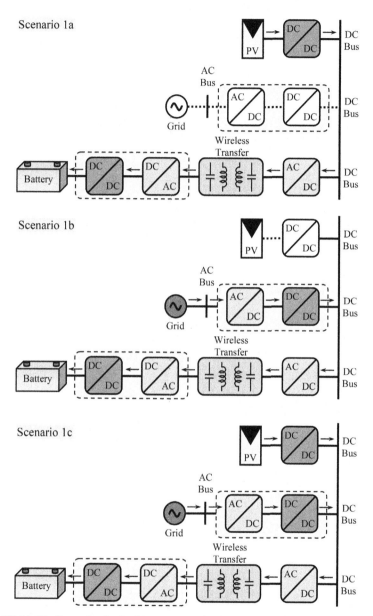

FIGURE 11.17 **Example of solar PV panel, electric vehicle, and grid power flow scenarios (Scenario 1a, Scenario 1b, and Scenario 1c).**

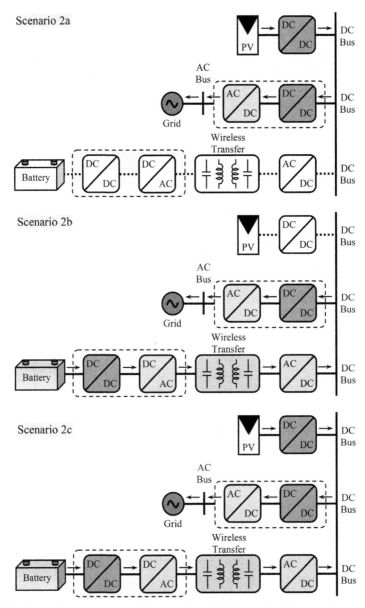

FIGURE 11.18 Example of solar PV panel, electric vehicle, and grid power flow scenarios (Scenario 2a, Scenario 2b, and Scenario 2c).

- If the energy of solar PV panels is lower than the energy required for electric vehicle battery charging, the remaining energy can be provided from the grid. If there is enough time, electric vehicles with variable power can be charged only with the energy generated from solar PV panels. If there is a

time constraint, electric vehicle charging scenarios with maximum power can be created from both solar PV panels and the grid.
- If the energy of solar PV panels is higher than the energy required for electric vehicle battery charging, the excess energy can be transferred to the grid.
- If the energy stored in the electric vehicle battery is excessive, it can be transferred to the grid. For example, the energy in the electric vehicle battery that is charged at the workplace can be used to meet the energy needs in the case of a power cut or when energy prices are high.

Solar PV panel, electric vehicle, and grid structure can be modeled as a linear programming–based optimization problem. The following equation states that the charging power of each electric vehicle cannot be greater than the charging capacity of the respective charging station:

$$P_{m,t}^{EV-char} \leq CR_m^{EV}, \forall m, t \in [T_m^{in}, T_\ddot{o}^{out}] \tag{11.4}$$

The change in the state of charge of electric vehicles is modeled by Eq. (11.5) that expresses the battery energy state when the electric vehicle arrives and connects to the station. The battery energy state $\left(SoE_{m,t}^{EV}\right)$ in period t must equal the sum of the energy state $\left(SoE_{m,t-1}^{EV}\right)$ in the previous period and the charge energy $\left(P_{m,t}^{EV-char} \cdot CE_m^{EV} \cdot \Delta T\right)$ supplied from the charging station. Eq. (11.6) indicates the remaining energy state in the battery when the electric vehicle arrives at the relevant charging station.

$$SoE_{m,t}^{EV} = SoE_{m,t-1}^{EV} + P_{m,t}^{EV-char} \cdot CE_m^{EV} \cdot \Delta T, \forall m, t > T_m^{in} \tag{11.5}$$

$$SoE_{m,t}^{EV} = SoE_m^{EV,start}, \forall m, t = T_m^{in} \tag{11.6}$$

In line with the request of the electric vehicle owner, the condition that the battery energy state of the electric vehicle at the time of leaving the charging station is equal to the charge level $\left(SoE_m^{EV,demand}\right)$ as requested by the user is provided by Eq. (11.7). Thus the comfort of the end user, the owner of the electric vehicle, is also taken into consideration. Eq. (11.8) models the battery limits of the relevant electric vehicle and provides the condition that it should be between the maximum battery energy capacity $\left(SoE_m^{EV,max}\right)$ and the minimum battery energy level $\left(SoE_m^{EV,min}\right)$. It should be kept in mind that if the electric vehicles are not connected to any charging station, the battery energy level and battery charge power of the electric vehicle should not get any value. This condition is provided with the help of Eq. (11.9).

$$SoE_{m,t}^{EV} = SoE_m^{EV,demand}, \forall m, t = T_m^{out} \tag{11.7}$$

$$SoE_m^{EV,min} \leq SoE_{m,t}^{EV} \leq SoE_m^{EV,max}, \forall m, \forall t = T_m^{out} \tag{11.8}$$

$$SoE_m^{EV} = 0, P_{m,t}^{EV,char} = 0, \forall m, \notin \left[T_m^{in}, T_m^{out}\right] \tag{11.9}$$

References

[1] A. Goswami, P. Kumar Sadhu, Stochastic firefly algorithm enabled fast charging of solar hybrid electric vehicles, Ain Shams Eng. J. (2020) https://doi.org/10.1016/j.asej.2020.08.016. In press.

[2] M. Schwarz, Q. Auzepy, C. Knoeri, Can electricity pricing leverage electric vehicles and battery storage to integrate high shares of solar photovoltaics?, Appl. Energy 277 (2020) 1–19 https://doi.org/10.1016/j.apenergy.2020.115548.

[3] M. Alilou, B. Tousi, H. Shayeghi, Home energy management in a residential smart micro grid under stochastic penetration of solar panels and electric vehicles, Sol. Energy 212 (2020) 6–18 https://doi.org/10.1016/j.solener.2020.10.063.

[4] J.H. Angelim, C. de Mattos Affonso, Probabilistic assessment of voltage quality on solar-powered electric vehicle charging station, Electr. Power Syst. Res. 189 (2020) 1–16 https://doi.org/10.1016/j.epsr.2020.106655.

[5] A. Rosato, A. Ciervo, G. Ciampi, M. Scorpio, F. Guarino, S. Sibilio, Impact of solar field design and back-up technology on dynamic performance of a solar hybrid heating network integrated with a seasonal borehole thermal energy storage serving a small-scale residential district including plug-in electric vehicles, Renew. Energy 154 (2020) 684–703 https://doi.org/10.1016/j.renene.2020.03.053.

[6] R. Maruthi Prasad, A. Krishnamoorthy, Design validation and analysis of the drive range enhancement and battery bank deration in electric vehicle integrated with split power solar source, Energy 172 (2019) 106–116 https://doi.org/10.1016/j.energy.2019.01.116.

[7] H. Mehrjerdi, M. Bornapour, R. Hemmati, S.M.S. Ghiasi, Unified energy management and load control in building equipped with wind-solar-battery incorporating electric and hydrogen vehicles under both connected to the grid and islanding modes, Energy 168 (2019) 919–930 https://doi.org/10.1016/j.energy.2018.11.131.

[8] F. He, H. Fathabadi, Novel standalone plug-in hybrid electric vehicle charging station fed by solar energy in presence of a fuel cell system used as supporting power source, Renew. Energy 156 (2020) 964–974 https://doi.org/10.1016/j.renene.2020.04.141.

[9] Z. Cabrane, D. Batool, J. Kim, K. Yoo, Design and simulation studies of battery-supercapacitor hybrid energy storage system for improved performances of traction system of solar vehicle, J. Energy Storage 32 (2020) 1–13 https://doi.org/10.1016/j.est.2020.101943.

[10] F.A. Tiano, G. Rizzo, M. Marino, A. Monetti, Evaluation of the potential of solar photovoltaic panels installed on vehicle body including temperature effect on efficiency, eTransportation 5 (2020) 100067 https://doi.org/10.1016/j.etran.2020.100067.

[11] D. Gudmunds, E. Nyholm, M. Taljegard, M. Odenberger, Self-consumption and self-sufficiency for household solar producers when introducing an electric vehicle, Renew. Energy 148 (2020) 1200–1215 https://doi.org/10.1016/j.renene.2019.10.030.

[12] H. Mehrjerdi, R. Hemmati, Electric vehicle charging station with multilevel charging infrastructure and hybrid solar-battery-diesel generation incorporating comfort of drivers, J. Energy Storage 26 (2019) 1–15 https://doi.org/10.1016/j.est.2019.100924.

[13] K. Verclas, Electric vehicle and solar energy pilot: opportunity to address suburban energy challenges, Electr. J. 31 (2018) 48–56 https://doi.org/10.1016/j.tej.2018.06.007.

[14] Y. Wang, M. Kazemi, S. Nojavan, K. Jermsittiparsert, Robust design of off-grid solar-powered charging station for hydrogen and electric vehicles via robust optimization approach, Int. J. Hydrogen Energy 45 (2020) 18995–19006 https://doi.org/10.1016/j.ijhydene.2020.05.098.

[15] O. Olatunde, M.Y. Hassan, M.P. Abdullah, H.A. Rahman, Hybrid photovoltaic/small-hydropower microgrid in smart distribution network with grid isolated electric vehicle charging system, J. Energy Storage 31 (2020) 1–11 https://doi.org/10.1016/j.est.2020.101673.

[16] S. Bairabathina, S. Balamurugan, Review on non-isolated multi-input step-up converters for grid-independent hybrid electric vehicles, Int. J. Hydrogen Energy 45 (2020) 21687–21713 https://doi.org/10.1016/j.ijhydene.2020.05.277.

[17] Y. Zou, J. Zhao, X. Gao, Y. Chen, A. Tohidi, Experimental results of electric vehicles effects on low voltage grids, J. Cleaner Prod. 255 (2020) 1–12 https://doi.org/10.1016/j.jclepro.2020.120270.

[18] C. Zhang, J.B. Greenblatt, P. MacDougall, S. Saxena, A. Jayam Prabhakar, Quantifying the benefits of electric vehicles on the future electricity grid in the midwestern United States, Appl. Energy 270 (2020) 1–9 https://doi.org/10.1016/j.apenergy.2020.115174.

[19] S.R. Etesami, W. Saad, N.B. Mandayam, H.V. Poor, Smart routing of electric vehicles for load balancing in smart grids, Automatica 120 (2020) 1–16 https://doi.org/10.1016/j.automatica.2020.109148.

[20] L. Luo, Z. Wu, W. Gu, H. Huang, S. Gao, J. Han, Coordinated allocation of distributed generation resources and electric vehicle charging stations in distribution systems with vehicle-to-grid interaction, Energy 192 (2020) 1–18 https://doi.org/10.1016/j.energy.2019.116631.

[21] J.K. Szinai, C.J.R. Sheppard, N. Abhyankar, A.R. Gopal, Reduced grid operating costs and renewable energy curtailment with electric vehicle charge management, Energy Policy 136 (2020) 1–19 https://doi.org/10.1016/j.enpol.2019.111051.

[22] M.J. Salehpour, S.M.M. Tafreshi, Contract-based utilization of plug-in electric vehicle batteries for day-ahead optimal operation of a smart micro-grid, J. Energy Storage 27 (2020) 1–14 https://doi.org/10.1016/j.est.2019.101157.

[23] T.U. Solanke, V.K. Ramachandaramurthy, J.Y. Yong, J. Pasupuleti, P. Kasinathan, A. Rajagopalan, A review of strategic charging–discharging control of grid-connected electric vehicles, J. Energy Storage 28 (2020) 1–13 https://doi.org/10.1016/j.est.2020.101193.

[24] B.K. Sovacool, J. Kester, L. Noel, G.Z. de Rubens, Energy injustice and Nordic electric mobility: inequality, elitism, and externalities in the electrification of vehicle-to-grid (V2G) transport, Ecol. Econ. 157 (2019) 205–217 https://doi.org/10.1016/j.ecolecon.2018.11.013.

[25] J. Kester, G. Zarazua de Rubens, B.K. Sovacool, L. Noel, Public perceptions of electric vehicles and vehicle-to-grid (V2G): insights from a Nordic focus group study, Transp. Res. D Transp. Environ. 74 (2019) 277–293 https://doi.org/10.1016/j.trd.2019.08.006.

[26] R. Shi, S. Li, P. Zhang, K.Y. Lee, Integration of renewable energy sources and electric vehicles in V2G network with adjustable robust optimization, Renew. Energy 153 (2020) 1067–1080 https://doi.org/10.1016/j.renene.2020.02.027.

[27] X. Li, Y. Tan, X. Liu, Q. Liao, B. Sun, G. Cao, et al. A cost-benefit analysis of V2G electric vehicles supporting peak shaving in Shanghai, Electr. Power Syst. Res. 179 (2020) 1–15 https://doi.org/10.1016/j.epsr.2019.106058.

[28] S. Cheng, Z. Li, Multi-objective network reconfiguration considering V2G of electric vehicles in distribution system with renewable energy, Energy Procedia 158 (2019) 278–283 https://doi.org/10.1016/j.egypro.2019.01.089.

[29] B.K. Sovacool, J. Kester, L. Noel, G. Zarazua de Rubens, Are electric vehicles masculinized? Gender, identity, and environmental values in Nordic transport practices and vehicle-to-grid (V2G) preferences, Transp. Res. D Transp. Environ. 72 (2019) 187–202 https://doi.org/10.1016/j.trd.2019.04.013.

[30] M. Sufyan, N.A. Rahim, M.A. Muhammad, C.K. Tan, S.R.S. Raihan, A.H.A. Bakar, Charge coordination and battery lifecycle analysis of electric vehicles with V2G implementation, Electr. Power Syst. Res. 184 (2020) 1–11 https://doi.org/10.1016/j.epsr.2020.106307.

[31] W. Meesenburg, A. Thingvad, B. Elmegaard, M. Marinelli, Combined provision of primary frequency regulation from vehicle-to-grid (V2G) capable electric vehicles and community-scale heat pump, Sustain. Energy Grids Networks 23 (2020) 1–9 https://doi.org/10.1016/j.segan.2020.100382.

[32] R.J. Flores, B.P. Shaffer, J. Brouwer, Electricity costs for an electric vehicle fueling station with Level 3 charging, Appl. Energy 169 (2016) 813–830 https://doi.org/10.1016/j.apenergy.2016.02.071.

[33] H.S. Das, M.M. Rahman, S. Li, C.W. Tan, Electric vehicles standards, charging infrastructure, and impact on grid integration: a technological review, Renew. Sustain. Energy Rev. 120 (2020) 1–10 https://doi.org/10.1016/j.rser.2019.109618.

[34] M. Fritz, P. Plötz, S.A. Funke, The impact of ambitious fuel economy standards on the market uptake of electric vehicles and specific CO_2 emissions, Energy Policy 135 (2019) 1–14 https://doi.org/10.1016/j.enpol.2019.111006.

[35] J. Kester, L. Noel, X. Lin, G. Zarazua de Rubens, B.K. Sovacool, The coproduction of electric mobility: selectivity, conformity and fragmentation in the sociotechnical acceptance of vehicle-to-grid (V2G) standards, J. Cleaner Prod. 207 (2019) 400–410 https://doi.org/10.1016/j.jclepro.2018.10.018.

[36] D.J. Swart, A. Bekker, J. Bienert, The subjective dimensions of sound quality of standard production electric vehicles, Appl. Acoust. 129 (2018) 354–364 https://doi.org/10.1016/j.apacoust.2017.08.012.

[37] V. Ruiz, A. Pfrang, A. Kriston, N. Omar, P. Van den Bossche, L. Boon-Brett, A review of international abuse testing standards and regulations for lithium ion batteries in electric and hybrid electric vehicles, Renew. Sustain. Energy Rev. 81 (2018) 1427–1452 https://doi.org/10.1016/j.rser.2017.05.195.

[38] Z. Wang, J. Hong, L. Zhang, P. Liu, Voltage fault detection and precaution of batteries based on entropy and standard deviation for electric vehicles, Energy Procedia 105 (2017) 2163–2168 https://doi.org/10.1016/j.egypro.2017.03.611.

[39] S. Niu, H. Xu, Z. Sun, Z.Y. Shao, L. Jian, The state-of-the-arts of wireless electric vehicle charging via magnetic resonance: principles, standards and core technologies, Renew. Sustain. Energy Rev. 114 (2019) 1–12 https://doi.org/10.1016/j.rser.2019.109302.

[40] P.M. Sneha Angeline, Evolution of electric vehicle and its future scope, Mater. Today Proc. 33 (7) (2020) 3930–3936 https://doi.org/10.1016/j.matpr.2020.06.266.

[41] P. Morrissey, P. Weldon, M. O'Mahony, Future standard and fast charging infrastructure planning: an analysis of electric vehicle charging behaviour, Energy Policy 89 (2016) 257–270 https://doi.org/10.1016/j.enpol.2015.12.001.

[42] J.Y. Yong, S.M. Fazeli, V.K. Ramachandaramurthy, K.M. Tan, Design and development of a three-phase off-board electric vehicle charger prototype for power grid voltage regulation, Energy 133 (2017) 128–141 https://doi.org/10.1016/j.energy.2017.05.108.

[43] H. Ngo, A. Kumar, S. Mishra, Optimal positioning of dynamic wireless charging infrastructure in a road network for battery electric vehicles, Transp. Res. D Transp. Environ. 85 (2020) 1–15 https://doi.org/10.1016/j.trd.2020.102385.

[44] P. Machura, Q. Li, A critical review on wireless charging for electric vehicles, Renew. Sustain. Energy Rev. 104 (2019) 209–234 https://doi.org/10.1016/j.rser.2019.01.027.

[45] S. Niu, H. Yu, S. Niu, L. Jian, Power loss analysis and thermal assessment on wireless electric vehicle charging technology: the over-temperature risk of ground assembly needs attention, Appl. Energy 275 (2020) 1–16 https://doi.org/10.1016/j.apenergy.2020.115344.

[46] A. Ahmad, M.S. Alam, Y. Rafat, S. Shariff, Designing and demonstration of misalignment reduction for wireless charging of autonomous electric vehicle, eTransportation 4 (2020) 100052 https://doi.org/10.1016/j.etran.2020.100052.

[47] C. Panchal, S. Stegen, J. Lu, Review of static and dynamic wireless electric vehicle charging system, Eng. Sci. Technol. Int. J. 21 (2018) 922–937 https://doi.org/10.1016/j.jestch.2018.06.015.

[48] Y.J. Jang, Survey of the operation and system study on wireless charging electric vehicle systems, Transp. Res. C Emerg. Technol. 95 (2018) 844–866 https://doi.org/10.1016/j.trc.2018.04.006.

[49] L. Sun, D. Ma, H. Tang, A review of recent trends in wireless power transfer technology and its applications in electric vehicle wireless charging, Renew. Sustain. Energy Rev. 91 (2018) 490–503 https://doi.org/10.1016/j.rser.2018.04.016.

[50] L. Wang, Y. Xu, J. Xu, Realization of wireless charging in intelligent greenhouse with orthogonal coil system uniform magnetic field, Comput. Electron. Agric. 175 (2020) 1–19 https://doi.org/10.1016/j.compag.2020.105524.

[51] W. Huang, X. Qu, S. Yin, M. Zubair, C. Guo, X. Xiong, et al. Long-distance adiabatic wireless energy transfer via multiple coils coupling, Results Phys. 19 (2020) 1–17 https://doi.org/10.1016/j.rinp.2020.103478.

[52] S.Y. Choi, B.W. Gu, S.Y. Jeong, C.T. Rim, Advances in wireless power transfer systems for roadway-powered electric vehicles, IEEE J. Emerg. Sel. Top. Power Electron. 3 (2015) 18–36 https://doi.org/10.1109/JESTPE.2014.2343674.

[53] J. Hu, Y. Zhang, M. Sun, D. Piedra, N. Chowdhury, T. Palacios, Materials and processing issues in vertical GaN power electronics, Mater. Sci. Semicond. Process. 78 (2018) 75–84 https://doi.org/10.1016/j.mssp.2017.09.033.

[54] J. Lutz, J. Franke, Reliability and reliability investigation of wide-bandgap power devices, Microelectron. Reliab. 88–90 (2018) 550–556 https://doi.org/10.1016/j.microrel.2018.07.001.

[55] J.G. Bolger, F.A. Kirsten, L.S. Ng, Inductive power coupling for an electric highway system, 28th IEEE Veh. Technol. Conf., IEEE, 1978, pp. 137–144 https://doi.org/10.1109/VTC.1978.1622522.

[56] A. Esser, Contactless charging and communication for electric vehicles, IEEE Ind. Appl. Mag. 1 (1995) 4–11 https://doi.org/10.1109/2943.469997.

[57] Y. Park, O.C. Onar, B. Ozpineci, Potential cybersecurity issues of fast charging stations with quantitative severity analysis, 2019 IEEE CyberPELS, CyberPELS 2019, Institute of Electrical and Electronics Engineers Inc., 2019 https://doi.org/10.1109/CyberPELS.2019.8925069.

[58] A.P. Sample, D.A. Meyer, J.R. Smith, Analysis, experimental results, and range adaptation of magnetically coupled resonators for wireless power transfer, IEEE Trans. Ind. Electron. 58 (2011) 544–554 https://doi.org/10.1109/TIE.2010.2046002.

[59] E. Asa, J. Pries, V. Galigekere, S. Mukherjee, O.C. Onar, G.J. Su, et al. A novel AC to AC wireless power transfer system for EV charging applications, Conf. Proc.—IEEE Appl. Power Electron. Conf. Expo.—APEC, Institute of Electrical and Electronics Engineers Inc., 2020, pp. 1685–1690 https://doi.org/10.1109/APEC39645.2020.9124602.

[60] Technical Specification, IEC TS 61980-3:2019, Electric vehicle wireless power transfer (WPT) systems - Part 3: Specific requirements for the magnetic field wireless power transfer systems, IEC Webstore. https://webstore.iec.ch/publication/27435, 2019-06-13, (accessed 15.11.20).

[61] J.M. González-González, A. Triviño-Cabrera, J.A. Aguado, Design and validation of a control algorithm for a SAE J2954-compliant wireless charger to guarantee the operational electrical constraints, Energies 11 (2018) 604 https://doi.org/10.3390/en11030604.

[62] V.P. Galigekere, J. Pries, O.C. Onar, G.J. Su, S. Anwar, R. Wiles, et al. Design and implementation of an optimized 100 kW stationary wireless charging system for EV battery recharging, 2018 IEEE Energy Convers. Congr. Expo. ECCE 2018, Institute of Electrical and Electronics Engineers Inc., 2018, pp. 3587–3592 https://doi.org/10.1109/ECCE.2018.8557590.

[63] M. Mohammad, J. Pries, O. Onar, V.P. Galigekere, G.J. Su, S. Anwar, et al. Design of an EMF suppressing magnetic shield for a 100-kW DD-coil wireless charging system for electric vehicles, Conf. Proc.—IEEE Appl. Power Electron. Conf. Expo.—APEC, Institute of

Electrical and Electronics Engineers Inc., 2019, pp. 1521–1527 https://doi.org/10.1109/APEC.2019.8722084.

[64] U.D. Kavimandan, V.P. Galigekere, O. Onar, B. Ozpineci, S.M. Mahajan, A control scheme to mitigate the dead-time effects in a wireless power transfer system, Conf. Proc.—IEEE Appl. Power Electron. Conf. Expo.—APEC, Institute of Electrical and Electronics Engineers Inc., 2020, pp. 3172–3179 https://doi.org/10.1109/APEC39645.2020.9124590.

[65] V.P. Galigekere, R. Zeng, J. Pries, O. Onar, G.J. Sui, Direct envelope modeling of load-resonant inverter for wireless power transfer applications, Conf. Proc.—IEEE Appl. Power Electron. Conf. Expo.—APEC, Institute of Electrical and Electronics Engineers Inc., 2020, pp. 3195–3199 https://doi.org/10.1109/APEC39645.2020.9124589.

[66] M. Mohammad, S. Anwar, O. Onar, J. Pries, V.P. Galigekere, E. Asa, et al. Sensitivity analysis of an LCC-LCC compensated 20-kW bidirectional wireless charging system for medium-duty vehicles, ITEC 2019-2019 IEEE Transp. Electrif. Conf. Expo, Institute of Electrical and Electronics Engineers Inc., 2019 https://doi.org/10.1109/ITEC.2019.8790620.

[67] O.C. Onar, S.L. Campbell, L.E. Seiber, C.P. White, M. Chinthavali, Vehicular integration of wireless power transfer systems and hardware interoperability case studies, ECCE 2016—IEEE Energy Convers. Congr. Expo. Proc., Institute of Electrical and Electronics Engineers Inc., 2016 https://doi.org/10.1109/ECCE.2016.7855553.

[68] G.J. Su, O.C. Onar, J. Pries, V.P. Galigekere, Variable duty control of three-phase voltage source inverter for wireless power transfer systems, 2019 IEEE Energy Convers. Congr. Expo. ECCE 2019, Institute of Electrical and Electronics Engineers Inc., 2019, pp. 2118–2124 https://doi.org/10.1109/ECCE.2019.8912565.

[69] J. Pries, V.P. Galigekere, O.C. Onar, G.J. Su, R. Wiles, L. Seiber, et al. Coil power density optimization and trade-off study for a 100 kW electric vehicle IPT wireless charging system, 2018 IEEE Energy Convers. Congr. Expo. ECCE 2018, Institute of Electrical and Electronics Engineers Inc., 2018, pp. 1196–1201 https://doi.org/10.1109/ECCE.2018.8557490.

[70] J. Pries, V.P.N. Galigekere, O.C. Onar, G.J. Su, A 50-kW three-phase wireless power transfer system using bipolar windings and series resonant networks for rotating magnetic fields, IEEE Trans. Power Electron. 35 (2020) 4500–4517 https://doi.org/10.1109/TPEL.2019.2942065.

[71] A. Foote, O.C. Onar, S. Debnath, J. Pries, V.P. Galigekere, B. Ozpineci, System design of dynamic wireless power transfer for automated highways, ITEC 2019-2019 IEEE Transp. Electrif. Conf. Expo, Institute of Electrical and Electronics Engineers Inc., 2019 https://doi.org/10.1109/ITEC.2019.8790623.

[72] J.H. Kim, B.S. Lee, J.H. Lee, S.H. Lee, C.B. Park, S.M. Jung, et al. Development of 1-MW inductive power transfer system for a high-speed train, IEEE Trans. Ind. Electron. 62 (2015) 6242–6250 https://doi.org/10.1109/TIE.2015.2417122.

[73] M.M. Esfahani, O. Mohammed, Real-time distribution of en-route electric vehicles for optimal operation of unbalanced hybrid AC/DC microgrids, eTransportation 1 (2019) 100007 https://doi.org/10.1016/j.etran.2019.100007.

[74] P. Wang, D. Wang, C. Zhu, Y. Yang, H.M. Abdullah, M.A. Mohamed, Stochastic management of hybrid AC/DC microgrids considering electric vehicles charging demands, Energy Rep. 6 (2020) 1338–1352 https://doi.org/10.1016/j.egyr.2020.05.019.

[75] G. Sharma, V.K. Sood, M.S. Alam, S.M. Shariff, Comparison of common DC and AC bus architectures for EV fast charging stations and impact on power quality, eTransportation 5 (2020) 100066 https://doi.org/10.1016/j.etran.2020.100066.

[76] A.K. Thakur, R. Prabakaran, M.R. Elkadeem, S.W. Sharshir, M. Arıcı, C. Wang, et al. A state of art review and future viewpoint on advance cooling techniques for lithium–ion battery system of electric vehicles, J. Energy Storage 32 (2020) 1–15 https://doi.org/10.1016/j.est.2020.101771.

[77] H. Ambrose, A. Kendall, M. Lozano, S. Wachche, L. Fulton, Trends in life cycle greenhouse gas emissions of future light duty electric vehicles, Transp. Res. D Transp. Environ. 81 (2020) 1–17 https://doi.org/10.1016/j.trd.2020.102287.

[78] M. Huda, M. Aziz, K. Tokimatsu, The future of electric vehicles to grid integration in Indonesia, Energy Procedia 158 (2019) 4592–4597 https://doi.org/10.1016/j.egypro.2019.01.749.

[79] T. Gnann, S. Funke, N. Jakobsson, P. Plötz, F. Sprei, A. Bennehag, Fast charging infrastructure for electric vehicles: today's situation and future needs, Transp. Res. D Transp. Environ. 62 (2018) 314–329 https://doi.org/10.1016/j.trd.2018.03.004.

[80] D. Keiner, M. Ram, L.D.S.N.S. Barbosa, D. Bogdanov, C. Breyer, Cost optimal self-consumption of PV prosumers with stationary batteries, heat pumps, thermal energy storage and electric vehicles across the world up to 2050, Sol. Energy 185 (2019) 406–423 https://doi.org/10.1016/j.solener.2019.04.081.

[81] T. Winther, H. Westskog, H. Sæle, Like having an electric car on the roof: domesticating PV solar panels in Norway, Energy Sustain. Dev. 47 (2018) 84–93 https://doi.org/10.1016/j.esd.2018.09.006.

[82] K.M. Buresh, M.D. Apperley, M.J. Booysen, Three shades of green: perspectives on at-work charging of electric vehicles using photovoltaic carports, Energy Sustain. Dev. 57 (2020) 132–140 https://doi.org/10.1016/j.esd.2020.05.007.

[83] J. Zhang, C. Liu, R. Yuan, T. Li, K. Li, B. Li, et al. Design scheme for fast charging station for electric vehicles with distributed photovoltaic power generation, Glob. Energy Interconnect. 2 (2019) 150–159 https://doi.org/10.1016/j.gloei.2019.07.003.

[84] U. Langenmayr, W. Wang, P. Jochem, Unit commitment of photovoltaic-battery systems: an advanced approach considering uncertainties from load, electric vehicles, and photovoltaic, Appl. Energy 280 (2020) 1–14 https://doi.org/10.1016/j.apenergy.2020.115972.

[85] H. ur Rehman, T. Korvola, R. Abdurafikov, T. Laakko, A. Hasan, F. Reda, Data analysis of a monitored building using machine learning and optimization of integrated photovoltaic panel, battery and electric vehicles in a Central European climatic condition, Energy Convers. Manage. 221 (2020) 1–11 https://doi.org/10.1016/j.enconman.2020.113206.

[86] R.K. Dwibedi, R. Jayaprakash, T. Siva, N.P. Gopinath, Hybrid electric vehicle using photovoltaic panel and chemical battery, Mater. Today Proc. 33 (7) (2020) 4713–4718 https://doi.org/10.1016/j.matpr.2020.08.351.

[87] A. AbuElrub, F. Hamed, O. Saadeh, Microgrid integrated electric vehicle charging algorithm with photovoltaic generation, J. Energy Storage 32 (2020) 1–10 https://doi.org/10.1016/j.est.2020.101858.

[88] K. Kouka, A. Masmoudi, A. Abdelkafi, L. Krichen, Dynamic energy management of an electric vehicle charging station using photovoltaic power, Sustain. Energy Grids Networks 24 (2020) 1–14 https://doi.org/10.1016/j.segan.2020.100402.

[89] K. Seddig, P. Jochem, W. Fichtner, Two-stage stochastic optimization for cost-minimal charging of electric vehicles at public charging stations with photovoltaics, Appl. Energy 242 (2019) 769–781 https://doi.org/10.1016/j.apenergy.2019.03.036.

[90] H.S. Salama, I. Vokony, Comparison of different electric vehicle integration approaches in presence of photovoltaic and superconducting magnetic energy storage systems, J. Cleaner Prod. 260 (2020) 1–16 https://doi.org/10.1016/j.jclepro.2020.121099.

Chapter 12

Simulation, Design, and Application of Hybrid Energy Storage System With Hybrid Power Generation System

1 Hybrid power generation system and hybrid energy storage system

The importance of renewable energy resources (RES) has been gradually increasing and its share inenergy demand is expanding. The sea and ocean are natural sources of energy with wind energy, wave energy and tidal energy. It would be rational to take advantage of the potential of each unit of this natural energy source. In this chapterthe prototype of high potential power generation system has been produced to take advantage of offshore wind and marine current energy. Thushybrid power generation system (HPGS) was established using offshore wind and marine current energy. The sea and ocean energy types have their own unique power generation characteristics like the other RES. By its nature, this power generation does not show continuity and stability depending on the weather conditions.

In addition, an energy storage unit is needed to ensure the continuity of these energy types that are intermittent and unstable due to the nature of RES. Battery technologies that are widely used today have high-energy density. The ultracapacitor storage technology provides high-power density. Thus a hybrid energy storage unit is created with two storage units. The hybrid energy storage system (HESS) consisting of battery and ultracapacitor is added to the HPGS. The purpose of this chapter is to create anHPGS from offshore wind and marine current energies and to integrate the HESS consisting of battery and ultracapacitor. Also, with smart energy management algorithm, it is to meet the demand and to produce quality electricity [1–9].

2 Simulation studies of HPGS and HESS

The MATLAB/Simulink block diagram of the HPGS and HESS structure, controlled by the smart energy management algorithm, is given in Fig. 12.1.

Solar Hybrid Systems. http://dx.doi.org/10.1016/B978-0-323-88499-0.00012-4

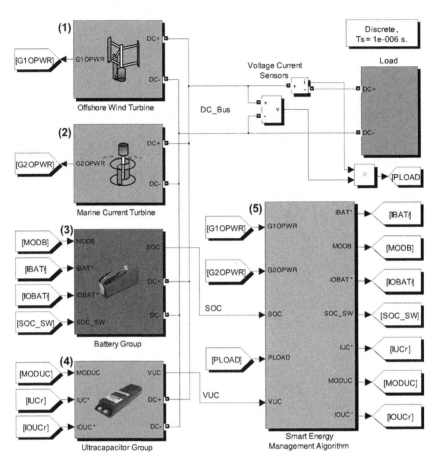

FIGURE 12.1 **HPGS and HESS MATLAB/Simulink block diagram.**

The smart energy management algorithm that includes offshore wind generator, marine current generator, reducers, direct current (DC)/DC boost converters, bidirectional DC/DC converters, battery, and ultracapacitor is used in this simulation study. A permanent magnet generator is simulated in offshore wind and marine current turbines. System parameters used in simulation are given in Table 12.1.

2.1 Offshore wind generator and control algorithm

The block diagram of the offshore wind generator, DC/DC boost converter, and proportional integral (PI) DC bus voltage control algorithm is given in Fig. 12.2. The PI block controls the output voltage from 12 to 24 V thanks to the DC/DC boost converter. The voltage value produced by the offshore wind generator varies depending on the wind speed. Through the DC/DC boost converter, the output voltage is kept at 24 V against changes in the input voltage.

TABLE 12.1 System parameters used in simulation.

Parameters		Value
Offshore wind and marine current generator	Nominal rated power	20 W
	Nominal speed	800 rpm
	Open circuit voltage	14 V
	Maximum current	2 A
	Efficiency	92 %
Battery group (lithium-ion)	Battery voltage	14.2 V
	Battery current capacity	3 Ah
	Battery power capacity	33 Wh
Ultracapacitor group	Ultracapacitor voltage	13.5 V
	Ultracapacitor capacity	2 F
	Ultracapacitor power capacity	50.5 MWh
Load group	Ohmic load	0.2–1.25 Ω

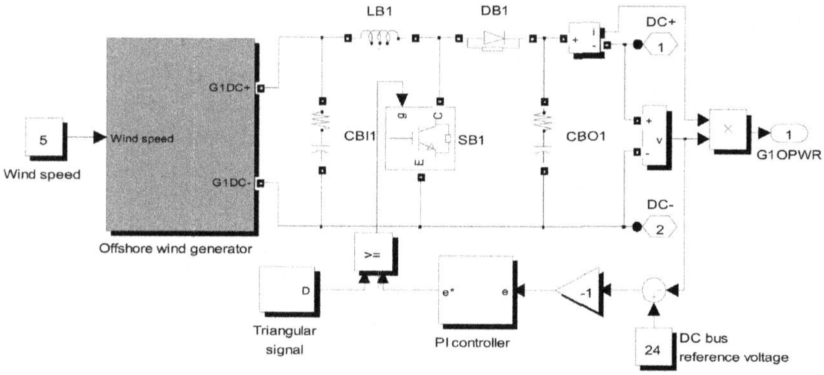

FIGURE 12.2   Offshore wind generator and DC/DC boost converter block diagram.

The current from the offshore wind generator and the DC/DC boost converter is transferred to the DC bus. It meets the load demand, which is the primary task of the system depending on the control of the smart energy management algorithm. At the same time, it provides charging of the battery and ultracapacitor group.

A permanent magnet AC generator is used as an offshore wind generator. This generator has 800 rpm speed, 12 V, and 20 W. The permanent magnet AC generator model, reducer, rpm–m/s conversion units are given in Fig. 12.3.

The permanent magnet AC generator that will be used in the system produces three-phase voltage. The three-phase voltage produced is rectified with a three-phase full-bridge rectifier.

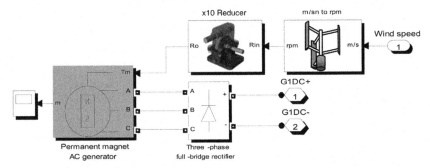

FIGURE 12.3   **Generator, reducer, and full-bridge rectifier block diagram.**

In order for the permanent magnet AC generator to produce power, the angular velocity (w) parameter must be entered. Depending on the angular velocity parameter, the voltage value is obtained from the generator varies. The wind speed entered in m/s applied to the offshore Darrieus wind turbine blades is converted into rpm to calculate the number of revolutions in the turbine shaft. The calculated rpm parameter is applied to the input of the reducer and transferred to the generator shaft after the multiplier conversion is made. In the simulation study, reducer efficiency was entered as multiplier. The block diagrams of rpm conversion and 1:10 reducer conversion ratios are given in Fig. 12.4.

The tests in simulation studies were carried out with a maximum wind speed of 15 m/s. By considering this wind speed and generator speed, the gearbox conversion rate was calculated as 1:10. The conversion of the wind speed entered in m/s applied to the Darrieus wind turbine blades into rpm is given in Eq. (12.1). Reducer conversion rate and efficiency calculation are also given in Eq. (12.2).

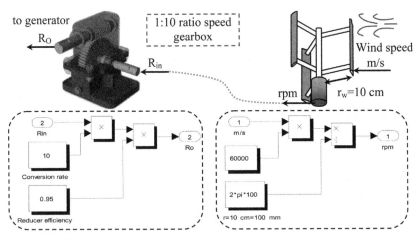

FIGURE 12.4   **Darrieus turbine reducer conversion rate block diagram.**

$$RPM = \frac{60,000 \cdot V_{RH}}{2 \cdot \pi \cdot r} \qquad (12.1)$$

$$Ro = \frac{R_{in} \cdot 10 \cdot 95}{100} \qquad (12.2)$$

where r is the Darrieus wind turbine radius, V_{RH} is the wind speed, R_{in} is the reducer input speed, and R_O is the reducer output speed.

2.2   Marine current generator and control algorithm

Just like offshore wind energy conversion, the marine current energy has the same generator, DC/DC boost converter, and anrpm reducer converter unit. The block diagrams of the marine current generator, DC/DC boost circuit, and control algorithm are given in Fig. 12.5. Unlike the offshore wind energy conversion in the marine current generator, the gearbox conversion rate is different. Although the marine current of the Savonius turbine blade speed is low, the force acting on the blades is waterand the density of the power produced is approximately 800 times higher than the air (the flow density of the water is 1000 kg/m^3 whereas the flow density of the wind is 1.223 kg/m^3). Since the offshore turbine generator will be used in the marine current generator, the reducer output speed must be 800 rpm. Therefore, the gearbox conversion ratio is chosen as 1:60. The block diagram of conversion of Savonius turbine speed in rpm to rpm and 1:60 reducer conversion ratiosis given in Fig. 12.6.

2.3   DC/DC boost converter structure

A DC/DC boost converter is used to increase the energy produced from offshore wind and marine current turbines to DC bus voltage in the system. DC/DC boost converters increase the low DC input voltage to 24 V at the output level. A typical DC/DC boost converter structure is given in Fig. 12.7. The switch

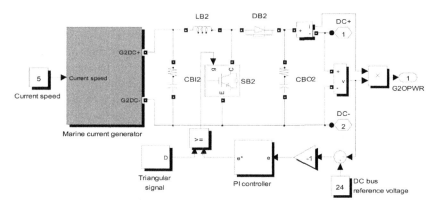

FIGURE 12.5   Marine current generator and DC/DC boost converter block diagram.

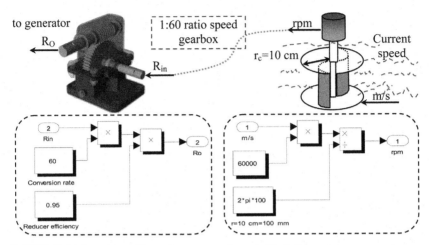

FIGURE 12.6 Savonius turbine reducer conversion rate block diagram.

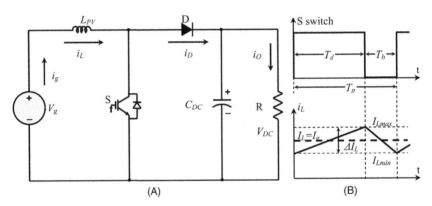

(A)   (B)

FIGURE 12.7 DC/DC boost converter structure.

in the DC/DC boost converter is in transmission in the first range. Inductance is fed by the V_g source. The current passing through the inductance increases linearly and the energy level of the inductance increases. In this range, the capacitor takes over the supply of the load. With the interruption of the switch signal, the diode D is transmitted through the electromotive force produced by the powered inductance. In the second interval the diode is in transmission. The load is fed by the energy accumulated in the V_g source and inductance. The current passing through the inductance decreases linearly and the energy level of the inductance decreases. Also, in this circuit, power elements are exposed to the V_o output voltage.

In the DC/DC boost converter structure, f_p is the switching frequency, T_d is the S switch time whileon-state, T_b is the S switch time while off-state, T_p is the working period, V_g is the DC input voltage, V_{DC} is DC output voltage, I_g is DC input current, I_o is the DC output current, and ΔI_L is the inductance current fluctuation. In the stable regime the equation of the positive and negative fields of the inductance voltage, the average output voltage DC is obtained by Eq. (12.3). Eq. (12.4) gives the occupancy rate equality [10–12].

$$V_{DC} = \frac{1}{1-\lambda} V_g$$
(12.3)

$$\lambda = \frac{T_d}{T_p}$$
(12.4)

From the equality of the powers at the input–output level, the input current is obtained by Eq. (12.5). The amount of fluctuation in the inductance current is given in Eq. (12.6).

$$I_g = \frac{1}{1-\lambda} I_o$$
(12.5)

$$\Delta I_L = \lambda(1-\lambda)\frac{V_{DC}}{f_p L_{PV}}$$
(12.6)

In the DC/DC booster converter the capacitor completely assumes the output current in the first range. The capacitor discharged in the first range fills in the second range. Since the amount of increase and decrease of the capacitor voltage in the steady state is equal, thefluctuation is calculated as in the following equation [13]:

$$\Delta V_{DC} = \frac{\lambda I_o}{f_p C_{DC}}$$
(12.7)

The required coil and capacitor values were calculated by using a DC/DC boost converter circuit (Eqs.12.3–12.7). The values calculated in the designed converter have been tested in the simulation environment, and their compatibility with the system has been ensured. Since the power values produced for HPGS on both systems are the same, the DC/DC boost converter parameters on the two systems are the same. Table 12.2 gives the DC/DC boost converter parameters as used in the HPGS.

2.4   Battery group and bidirectional DC/DC converter control

6S (6 cell) lithium ion battery was used as the battery group. The battery group DC, which is the HESS unit, is connected to the DC bus via a bidirectional DC/DC converter. Three lithium ion batteries with 3.7 V and 1050 mAh are used in the battery storage unit. Batteries are connected in series in order to obtain 12 V in the battery group. Thus the total energy capacity obtained from the battery

TABLE 12.2 DC/DC boost converter parameters used in HPGS.

Parameters	Value
T_p	100 µs
f_p	10 kHz
I_o	5 A
I_g	10 A
V_g	10–20 V
V_{DC}	24 V
L_{PV}	>90 µH
C_{DC}	750 µF

group is 12 Wh. The block diagram model of the battery group, bidirectional DC/DC converter, and controller is given in Fig. 12.8.

The switching element on the converter enables the structure to be operated in a boost or buck mode. Pulse width modulation (PWM) must be applied to the SB3 switch, and the SB4 switch must be in the open circuit position to enable the bidirectional DC/DC converter to operate in buck mode. This control

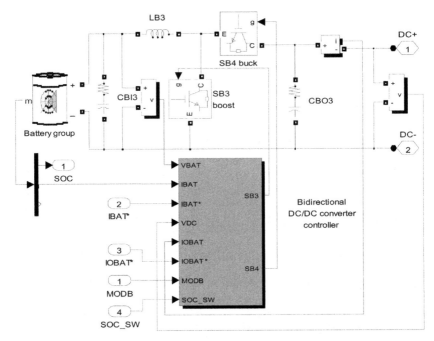

FIGURE 12.8 **Battery group and bidirectional DC/DC converter block diagram.**

algorithm works depending on the energy management algorithm. When the information coming to MODB selection 7 is 1, the control algorithm enables the system to go into a buck mode. The SOC_SW information, which is the eighth entry in the block diagram, controls the situation where the battery group does not need to work (charge/discharge). When 0 is received, it turns off the switch SB3 and SB4 and ensures that the battery group stops the energy flow from the DC. PWM must be applied to the SB4 switch in the block diagram and the SB3 switch must be in the off position in order to make the bidirectional DC/DC converter work in boost mode. The block diagram of the control unit of the bidirectional DC/DC converter structure used in the battery group is given in Fig. 12.9.

Current and voltage control is performed during charge and discharge with the battery group buck and boost mode control given in Fig. 12.10. The cascade PI controller is used in buck and boost mode control. In the buck mode the first

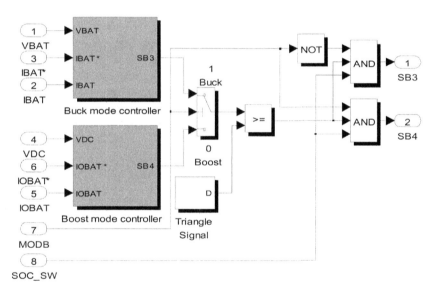

FIGURE 12.9   Battery group bidirectional DC/DC converter control block diagram.

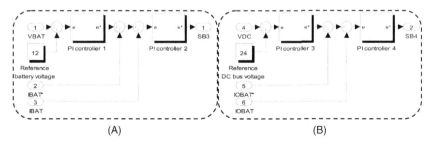

FIGURE 12.10   Battery group mode control block diagram (A) Buckmode controller, (B) boost-mode controller.

PI controller controls 12 V that the battery group should charge. In the buck mode the second cascade PI controller controls the charge current of the battery group. It determines the charge current of the battery group with the current value calculated in the smart energy management algorithm. The buck mode controller is given in Fig. 12.10A. By making cascade PI controller in the boost mode control unit, DC bus voltage is kept constant at 24 V together with the discharge current of the battery group.

The PI controller produces output by comparing the measured and reference current/voltage values. The switching signal is produced by comparing the PI output with a 10-kHz triangular signal. The discharge current value of the battery group is calculated based on the amount of load demanded in the smart energy management algorithm. The boost mode controller is given in Fig. 12.10B.

2.5 Ultracapacitor group and bidirectional DC/DC converter control

A MATLAB/Simulink simulation model of ultracapacitor group,bidirectional DC/DC converter, and control unit is given in Fig. 12.11. As in the battery group, DC/DC bidirectional converter structure is used to ensure the charge/discharge of the ultracapacitor group. Five pieces of Maxwell brand with 2.7 V 5 F ultracapacitor modules were simulated.Ultracapacitor modules were connected in series, and 13.5 V was obtained. The total energy capacity obtained from the ultracapacitor group is 50.5 MWh.

Compared to the battery group, the energy density of the ultracapacitor group is quite low. Since the ultracapacitor group has a much higher power

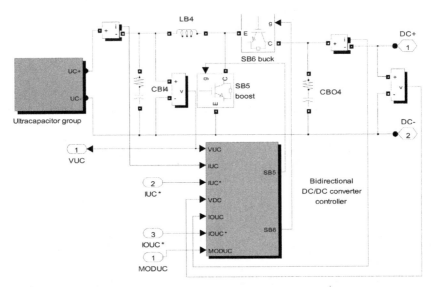

FIGURE 12.11 Ultracapacitor group and bidirectional DC/DC converter block diagram.

density than the battery group, it can provide much larger power support in a shorter time period. In contrast, the battery group has higher energy density, and it can store higher energy compared to the ultracapacitor group and provide energy for a longer period of time. Due to the complementary features of each other, the instant power requirement is met with the ultracapacitor energy storage unit and the battery energy storage unit. Thus anHESS is created with the battery group and the ultracapacitor group.

The Simulink block diagram of the control unit of the bidirectional DC/DC converter structure of the ultracapacitor group is given in Fig. 12.12. Although similar to the control unit of the battery group, it differs depending on the purpose of the ultracapacitor in the system. It calculates the energy amount of the smart energy management algorithm in the system and continuously controls the charge level of the ultracapacitor group. By sending MODUC information to the control unit of the ultracapacitor group, it enables the transducer to operate in buck (MODUC 1 information) or boost (MODUC 0 information) mode. The $I_{UC}*$ reference current value required to charge the ultracapacitor group while operating the controller in buck mode is determined by the smart energy management algorithm.

Due to its structural feature, the ultracapacitor group can draw very high currents during charging. This high charge current can cause voltage drops in the DC bus, resulting ina stable operation of the system. Therefore, ultracapacitor charging current is determined as maximum 5 A. During the operation of the ultracapacitor group in the boost mode, the reference discharge current value is calculated by the $I_{OUC}*$in the smart energy management algorithm based on the amount of load demanded instantly. In the instantpower requirement of the load group is on the demand side, the ultracapacitor group operates in discharge mode. The ultracapacitor group prevents voltage collapse and fluctuations in the

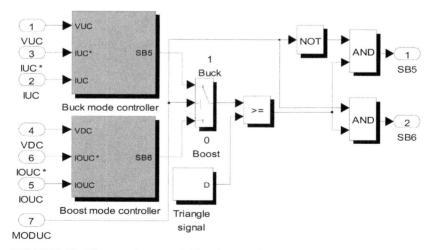

FIGURE 12.12   Ultracapacitor group bidirectional DC/DC converter control block diagram.

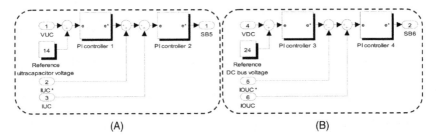

FIGURE 12.13 Ultracapacitor group mode control block diagram (A) Buckmode controller, (B) boostmode controller.

DC bus. On the load demand side, it provides quality, uninterrupted, and continuous energy. It ensures a stable operation of the system and prevents faults.

The ultracapacitor group buck and boost mode control block diagram is given in Fig. 12.13. As in the control of the battery group, the cascade PI controller was used in the buck and boost mode control. The charge level of the ultracapacitor group is controlled at the level of 24 V, which is the DC bus voltage during the discharge.

2.6 Bidirectional DC/DC converter control structure

Bidirectional DC/DC converter structure is used in the battery and ultracapacitor group. The bidirectional DC/DC converter structure used in HESS is given in Fig. 12.14A. The voltage to the DC bus is set at 24 V in HPGS and HESS. Voltage level ofbattery and ultracapacitor groups is 12 V. It is not suitable to use only a buck or boost DC/DC converter in the system. For this purpose, a bidirectional DC/DC converter is used, which can provide charge and discharge in battery and ultracapacitor group control. The bidirectional DC/DC converter unit undertakes to increase the voltage level of the battery and ultracapacitor group to 24 V. In this case, as in Fig. 12.14B, PWM is applied to SB5 switch, and the SB6 switch is the passive state. Likewise, it provides battery and ultracapacitor groups to be charged with excess energy produced from offshore wind and marine current energy. In this case, as shown in Fig. 12.14B, PWM is applied to SB6 switch, and the SB5 switch is the passive state. A coil, two

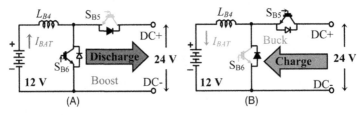

FIGURE 12.14 Bidirectional DC/DC converter structure.

semiconductor switching elements and capacitors are used in the bidirectional DC/DC converter structure.

The capacitors in the bidirectional DC/DC converter structure have been used to prevent voltage fluctuations at the input and output. Capacitor values calculated in a DC/DC boost converter are used in this type of converter in the same way. The capacitor value to be used in the bidirectional DC/DC converter is 750 µF. The coil required for the bidirectional DC/DC converter design has been calculated as given in the following equation[14]:

$$L_{B4}f_p = \frac{2V_{BAT\&UC}^{2}}{P_{BD}} \tag{12.8}$$

where L_{B4} is the bidirectional DC/DC converter coil, f_p is the switching frequency, $V_{BAT\&UC}$ is the bidirectional DC/DC converter input voltage, V_{DC} is the bidirectional DC/DC converter output voltage, and P_{BD} is the bidirectional DC/DC converter output power. Bidirectional DC/DC converter parameters used in battery and ultracapacitor group are given in Table 12.3.

2.7   Smart energy management algorithm

The smart energy management algorithm Simulink block diagram used in HPGS and HESS is given in Fig.12.17. HPGS and HESS have many possible working cases. All possible working cases in the system determine the power/current parameters of the sources and loads in the system. When the possible working cases in the system are considered, a total of nine situations appear as in Table 12.4.

The smart energy management algorithm controls power flows in the system by making decisions based on possible working cases. Energy produced from offshore wind and marine current energy is defined as total energy. This energy is called P_{TOT} and is given in Eq.(12.9). The primary task of HPGS and HESS

TABLE 12.3 Bidirectional DC/DC converter parameters used in HESS.

Parameters	Value
$V_{BAT\&UC}$	10–20 V
V_{DC}	24 V
P_{BD}	30 W
f_p	10 kHz
L_{B4}	>90 µH
C_1	750 µF
C_2	750 µF

TABLE 12.4 HPGS and HESS possible working cases.

Cases	P_{TOT} and P_{LOAD} compare	Battery state ofcharge (SOC)	Ultracapacitorstate of charge (V_{UC})
1	$P_{TOT} \approx P_{LOAD}$	Low	Low
2	$P_{TOT} \approx P_{LOAD}$	High	High
3	$P_{TOT} > P_{LOAD}$	Low	Low
4	$P_{TOT} > P_{LOAD}$	High	High
5	$P_{TOT} < P_{LOAD}$	Low	Low
6	$P_{TOT} < P_{LAOD}$	High	High
7	$P_{TOT} = 0$	Low	Low
8	$P_{TOT} = 0$	High	High
9	$P_{LOAD} = 0$	Low	Low

is to continually provide the load demand. The smart energy management algorithm looks at the difference between generation/demand and keeps the battery and ultracapacitor group charged.

$$P_{TOT} = P_{G1OPWR} + P_{G2OPWR} \tag{12.9}$$

The smart energy management algorithm calculates the current amounts of the units only by looking at their power values. It also provides a control of all converter units in the system. Smartenergy management algorithm continuously scans the system (scanning cycle is chosen as 1 μs in simulation study). HPGS looks at the generated power rating and load demand power and ensures the stable operation of the system. When the energy is not produced, HESS provides the power requirement of the load. Thus intermittent and unstable energy production, the biggest problem of renewable energy sources, is eliminated with HESS.

A current calculation control block diagram of the battery group is given in Fig. 12.15 on the block (1). The control of the battery group takes place by determining the charge or discharge of the battery group by looking at the P_{DIF} power. P_{DIF} power is calculated in Eq. (12.11). This information is sent to the bidirectional DC/DC converter control unit, which controls the battery group with the MODB signal. If P_{DIF} power is greater than zero ($P_{DIF} > 0$), and this energy is used to charge the battery and ultracapacitor group. A value of 1 is generated with the MODB information signal. If the obtained difference power is less than 0 ($P_{DIF} < 0$), the battery group is discharged. The 0 information is given to MODB signal. The reference current value (I_{BAT}^*) that the battery group should charge is calculated in Eq. (12.12). The power required by the

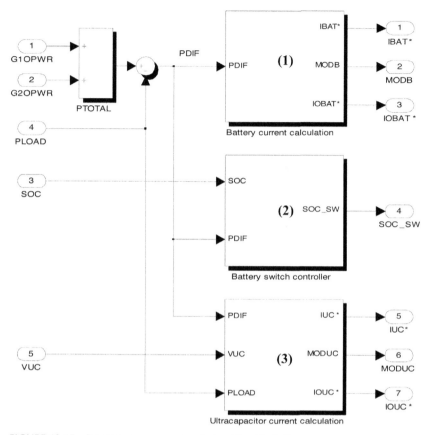

FIGURE 12.15 Smart energy management algorithm block diagram.

load request is calculated in Eq. (12.13) and sends the discharge current reference ($I_{OBAT}{}^*$) to the control unit.

$$P_{DIF} = P_{TOT} - P_{LOAD} \tag{12.10}$$

$$I_{BAT}{}^* = \frac{P_{DIF}}{V_{BAT\_CHAR}} \tag{12.11}$$

$$I_{OBAT}{}^* = \frac{P_{DIF}}{V_{DC}} \tag{12.12}$$

The switch information control block diagram of the battery group in block
(2) of the smart energy management algorithm block is given in Fig. 12.16.

This control block prevents the battery from being over charge/discharge
by controlling the battery state of charge (SOC). Thus it is aimed to extend the
cycle life of the battery group by working in safe operation areas. The smart

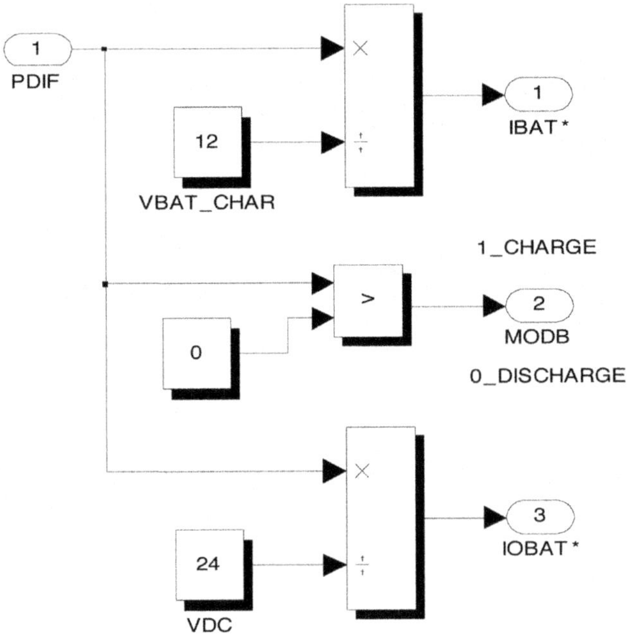

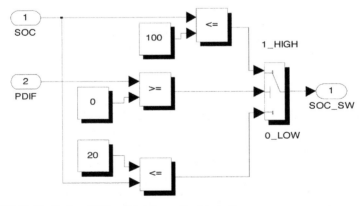

FIGURE 12.16 Battery current calculation block diagram.

energy management algorithm sends the 0 information to the bidirectional DC/DC converter switches when the SOC drops below 20 %. Likewise, when the SOC value is 100 %, it stops the battery group charging. It provides information to the bidirectional DC/DC converter circuit of the battery group via SOC_SW information. As represented in the graph in Fig. 12.17, the SOC_SW information becomes 1 when the battery group is between 20 % and 100 %, and 0 in other cases.

FIGURE 12.17 Battery SOC switch information block diagram.

The control block diagram determining the current calculation and mode status information of the ultracapacitor group in block (3) of the smart energy management algorithm block is given in Fig. 12.18.

In this control block, there is a unit that decides to charge/discharge the ultracapacitor group. This information is decided based on the V_{UC} value and the amount of power demanded by the load. Since the ultracapacitor group responds to suddenly high powers, it must be constantly fueled. This maximum power value is set to 10 W in HPGS and HESS. When this demand power is high, the 0 value from MODUC information is sent to the circuit of the ultracapacitor group. Thus sudden power demand is met without a drop and fluctuation in DC bus voltage. The ultracapacitor group is provided to be continuously charged in a band between 10 and 14 V in charge mode. The reason for this band voltage is that the ultracapacitor group self-discharges.

Due to the characteristic of the ultracapacitor group, it draws a very high charging current from the DC bus during the first charge. This high current

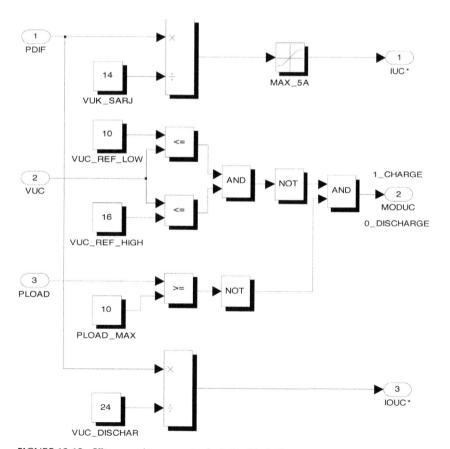

FIGURE 12.18 Ultracapacitor current calculation block diagram.

value can cause a ripple and sag in the DC bus. Therefore, the charging current during the first charge is limited to 5 A. The charge current reference of the ultracapacitor group I_{UC}^* is calculated in Eq. (12.13). The 1 value is sent from MODUC information. When there is a sudden load demand, the ultracapacitor group switches to discharge mode. The reference discharge current value I_{OUC}^* calculated by Eq. (12.14) is generated.

$$I_{UC}^* = \frac{P_{DIF}}{V_{UC\_DISCHAR}}$$ (12.13)

$$I_{OUC}^* = \frac{P_{DIF}}{V_{UC\_DISCHAR}}$$ (12.14)

2.8 Simulation results for Case 1

The $P_{TOT} \approx P_{LOAD}$, P_{BAT}, and P_{UC}SOCs are low in Case 1. The power flow diagram of Case 1 is given in Fig. 12.19. In this case, the total amount of power produced from offshore wind and marine current energy is approximately equal to the power demanded by the load.

In this case, there is no need for battery and ultracapacitor group, depending on the smart energy management algorithm; they will be activated when needed. The SOC is low for HESSand is examined in Case 1. Since HPGS and HESS are modeled in detail in each unit, the total simulation time is chosen as 2 seconds.

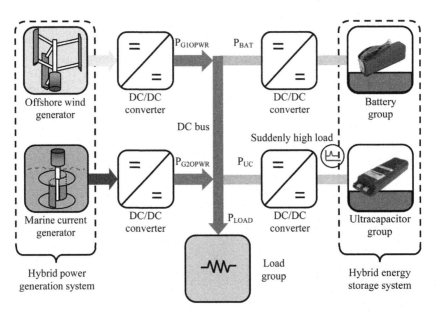

FIGURE 12.19 Power flow diagrams for Case 1.

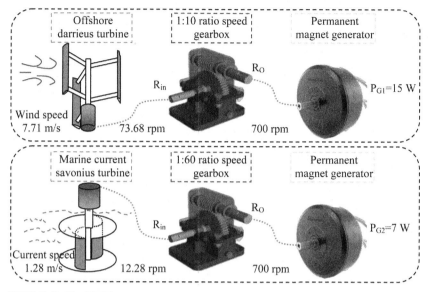

FIGURE 12.20   HPGS parameters for Case 1.

The power value obtained from the generator as a result of the reducer input and output of offshore wind speed of 7.71 m/s is 15 W. Likewise, when the marine current is 1.28 m/s, the power value obtained from the generator as a the result of the reducer input and output is 7 W. Simulation parameter values for Case 1 are given in Fig. 12.20.

The value of the energy produced from the offshore wind generator in the DC/DC boost circuit output current, for Case 1 is 0.6 A in Fig. 12.21.

Since the DC bus voltage is the same for all DC/DC converter outputs, it is given only in Case 1. The power produced from the offshore wind generator is given in Fig. 12.22 and this power is 15 W. The energy generated from the marine current generator in the DC/DC boost circuit output current is 0.38 A. This

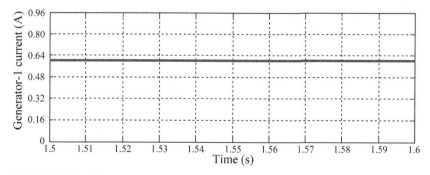

FIGURE 12.21   Offshore wind generator circuit output current for Case 1.

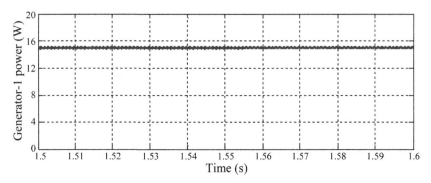

FIGURE 12.22   Offshore wind generator circuit output power for Case 1.

current result is given in Fig. 12.23. The total amount of power generated from the marine current generator is 7 W and is given in Fig. 12.24.

The DC bus voltage in HPGS and HESS is controlled to 24 V. The DC bus voltage is kept constant in all possible working cases. The DC bus voltage value 24 V is given in Fig. 12.25. The battery group current is given in Fig. 12.26 and

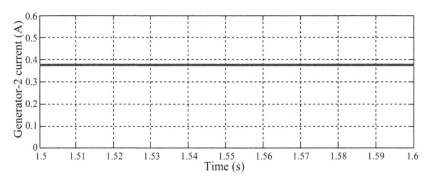

FIGURE 12.23   Marine current generator circuit output current for Case 1.

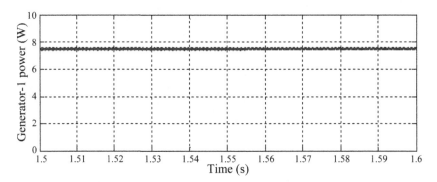

FIGURE 12.24   Marine current generator circuit output power for Case 1.

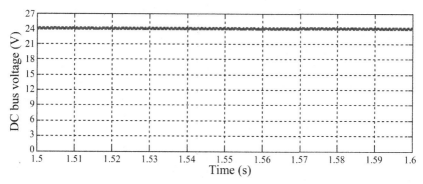

FIGURE 12.25 DC bus voltage for Case 1.

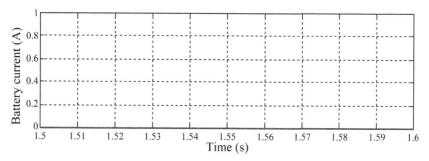

FIGURE 12.26 Battery group current for Case 1.

the current value is 0 A since the battery group is not worked. In this case, battery SOC is seen as 25 % in Fig. 12.27. The current value drawn by HPGS and HESS connected load group is given in Fig. 12.28. The load power is 22.5 W in Case 1.

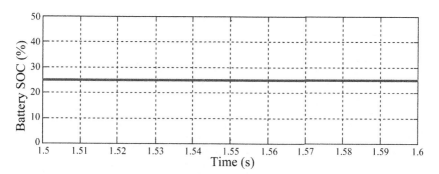

FIGURE 12.27 Battery group SOC for Case 1.

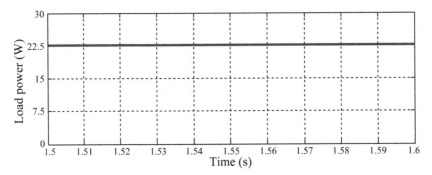

FIGURE 12.28 Load group power for Case 1.

2.9 Simulation results for Case 2

The $P_{TOT} \approx P_{LOAD}$, P_{BAT}, and P_{UC}SOCs are high in Case 1. The power flow diagram of Case 2 is given in Fig. 12.29. Unlike the previous Case 1, this situation is the battery and ultracapacitor group high SOC. Total power produced from offshore wind and marine current generators is 25 W. This power value is approximately equal to the value demanded by the load group. In Case 2 the battery and ultracapacitor group has high SOC, which will prevent voltage fluctuations in the DC bus in the sudden high power demand.

Power generation parameters of HPGS are given in detail in Fig. 12.30. In Case 2 the offshore Darrieus turbine rotates at a speed of 8.81 m/s, and20 W

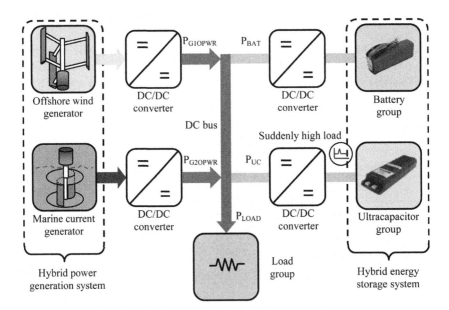

FIGURE 12.29 Power flow diagrams for Case 2.

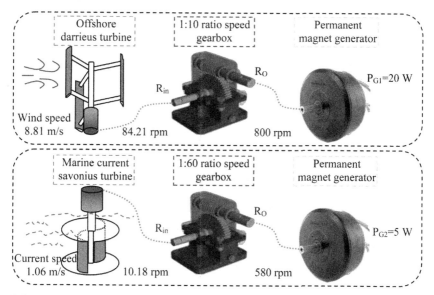

FIGURE 12.30 HPGS parameters for Case 2.

electrical power is generated from the generator output. Marine current Savonius turbine rotates at 1.06 m/s speed and 5 W electrical energy is obtained from the generator with the power transferred with the reducer.

The output current of the DC/DC boost circuit connected to the offshore wind generator is given for Case 2 in Fig. 12.31. The power value produced by the generator is at the maximum level. This power value is 20 W and the power graph is shown in detail in Fig. 12.32. In Fig. 12.33 the current value of the DC/DC boost converter at the output of the marine current generator is given for Case 2. The power value produced from this generator is 5 W and the power graph is shown in detail in Fig. 12.34. In Case 2 the SOC value of the battery

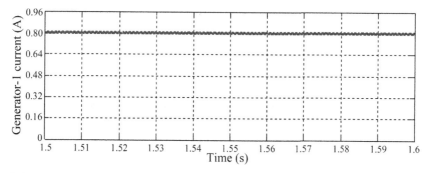

FIGURE 12.31 Offshore wind generator circuit output current for Case 2.

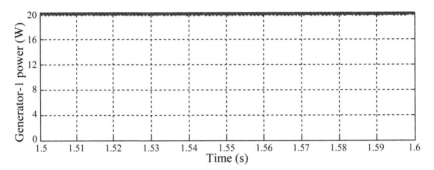

FIGURE 12.32 Offshore wind generator circuit output power for Case 2.

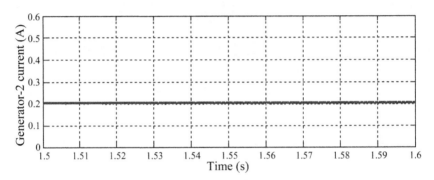

FIGURE 12.33 Marine current generator circuit output current for Case 2.

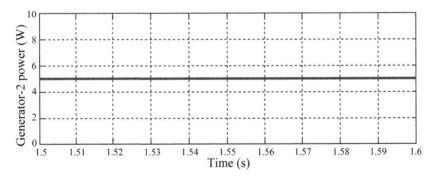

FIGURE 12.34 Marine current generator circuit output power for Case 2.

group is at the level of 80 % and is given in Fig. 12.35. The ultracapacitor group SOC is also high and this voltage value is kept around 12 V with the smart energy management algorithm. The voltage of the ultracapacitor group is given in Fig. 12.36. In Case 2 the power value produced by HPGS is approximately equal to the power demanded by the load. The load power value is given as 25 W in Fig. 12.37.

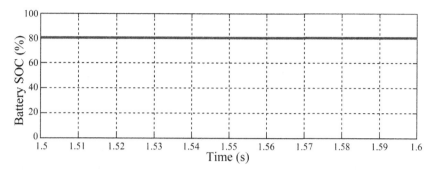

FIGURE 12.35   Battery group current for Case 2.

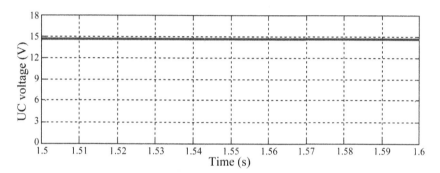

FIGURE 12.36   Ultracapacitor group voltage for Case 2.

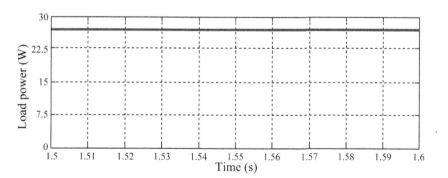

FIGURE 12.37   Load group power for Case 2.

2.10   Simulation results for Case 3

The $P_{TOT} > P_{LOAD}$, P_{BAT}, and P_{UC}SOCs are low in Case 3. The power flow diagram of Case 3 is given in Fig. 12.38. The working case where the power value produced from HPGS is higher than the power value demanded by the load group is examined in this case. The power value produced by HPGS is high and the battery and ultracapacitor group has low SOC. HESS is charged by the

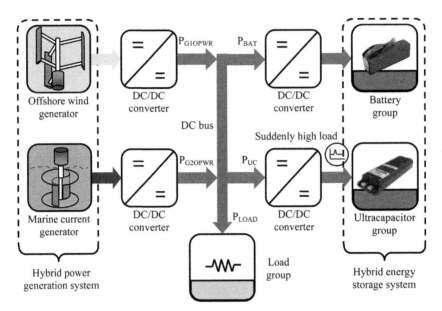

FIGURE 12.38 **Power flow diagrams for Case 3.**

smart energy management algorithm. The smart energy management algorithm calculates the difference in power. It sends the current information that the battery and ultracapacitor need to be charged to the control unit. Thus the system rapidly evaluates dynamic behaviors and ensures stable operation of HPGS and HESS. The simulation parameter values of offshore wind and marine current systems are given in Fig. 12.39. The wind speed of 7.71 m/s coming to the offshore Darrieus turbine is transferred to the generator via the reducer, and 15 W is obtained by the generator. The marine current speed of 1.28 m/s arrival to the Savonius turbine is transferred to the generator via the reducer, and 7 W is produced by the generator.

The offshore wind generator circuit output current is given in Fig. 12.40. The power value obtained from this generator is shown as 15 W in Fig. 12.41.

The marine current generator circuit output current is shown as 0.38 A in Fig. 12.42. The 7 W power value produced by this generator is given in Fig. 12.43. The total power produced by HPGS is 22 W.

Since the total power value generated in Case 2 is greater than the load demand power, the power difference between the load and the P_{TOT} is calculated by the P_{DIF} smart energy management algorithm. This P_{DIF} power value is used to charge the battery and ultracapacitor group.

The case where the ultracapacitor group is completely low is examined. Since the current charging current of the ultracapacitor group is very high, it is limited by the controller to 5 A to prevent DC bus voltage fluctuation. The charging current of the ultracapacitor group is given in Fig. 12.44, and the

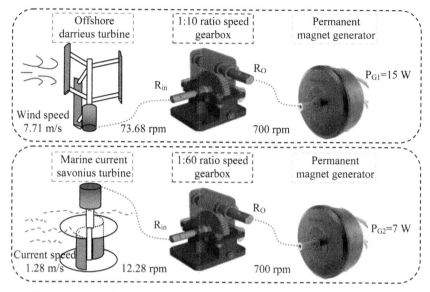

FIGURE 12.39 HPGS parameters for Case 3.

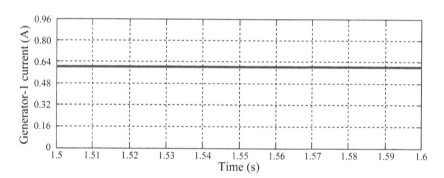

FIGURE 12.40 Offshore wind generator circuit output current for Case 3.

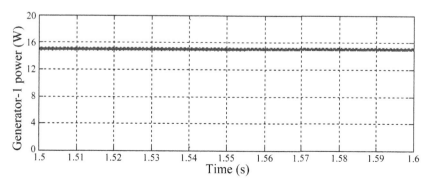

FIGURE 12.41 Offshore wind generator circuit output power for Case 3.

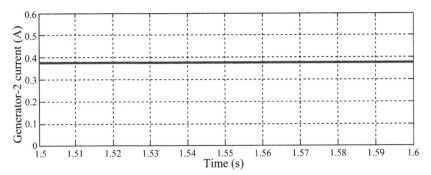

FIGURE 12.42 Marine current generator circuit output current for Case 3.

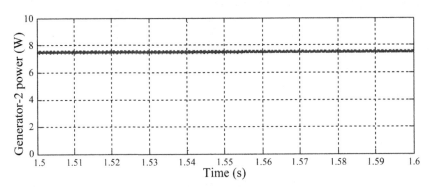

FIGURE 12.43 Marine current generator circuit output power for Case 3.

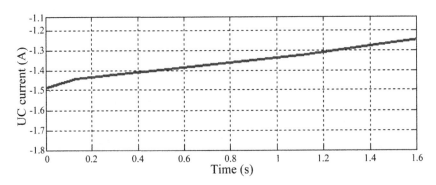

FIGURE 12.44 Ultracapacitor group current for Case 3.

graphic shows that it is charging quickly. Likewise, the voltage of the ultracapacitor group is given in Fig. 12.45, where it increases rapidly.

The smart energy management algorithm calculates the current that the battery group should charge and transmits the reference current information. The charge current value at the input of the battery group is given in Fig. 12.46.

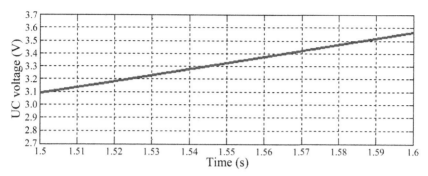

FIGURE 12.45 Ultracapacitor group voltage for Case 3.

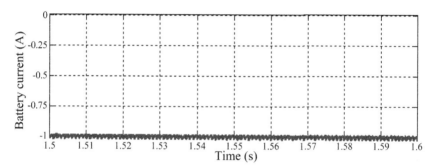

FIGURE 12.46 Battery group current for Case 3.

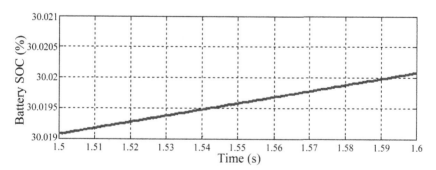

FIGURE 12.47 Battery group SOC for Case 3.

The SOC information of the battery group is shown in detail in Fig. 12.47. In the SOC graphic the scale is given at values close to 30 %. Since the battery group has a high-energy density, a small part of it is observed in 1 second. In Case 2 the power value demanded by the load is 5 W. Load group power for Case 3 is shown in detail in Fig. 12.48.

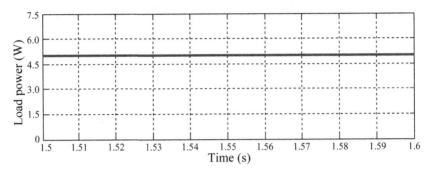

FIGURE 12.48 Load group power for Case 3.

2.11 Simulation results for Case 4

The $P_{TOT} > P_{LOAD}$, P_{BAT}, and P_{UC}SOCs are high in Case 4. The power flow diagram of Case 4 is given in Fig. 12.49. Unlike the previous possible working case, the battery and ultracapacitor group SOC is high in this case. The smart energy management algorithm observes the battery group's SOC value and stops charging when it is 100 %. The voltage level of the ultracapacitor group is constantly monitored and kept at 15 V. Although wind speed and current speed are high, energy is transferred from generators as much as the load demands in the system. Since the generators in HPGS are loaded at different powers, different energy is obtained from the generators.

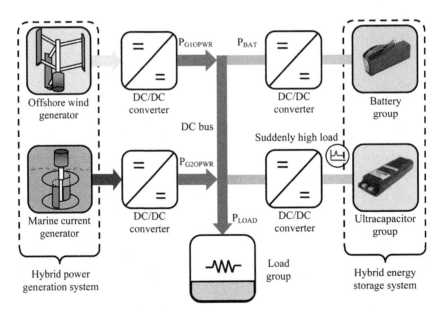

FIGURE 12.49 Power flow diagrams for Case 4.

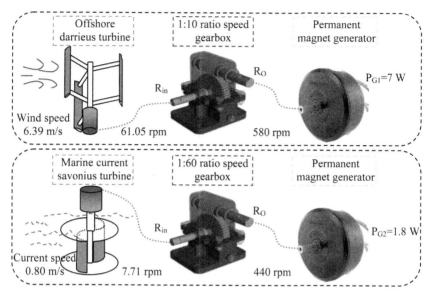

FIGURE 12.50   HPGS parameters for Case 4.

The 6.39 m/s wind speed comes to the offshore wind Darrieus turbine and the values are given in Fig. 12.50. It is transferred to the generator with the reducer and the generator is rotated at a speed of 580 rpm. In this speed the generator can generate 10 W, but since the system has a low power load group, 7 W is produced from the offshore wind generator. The marine current turbine rotates at a flow rate of 0.80 m/s and the generator is rotated at a speed of 440 rpm through the reducer. Likewise, 5 W can be produced from this generator. However, a power of 1.8 W is transmitted due to the load group demand. HPGS is observed to share power equally under possible working cases.

Fig. 12.51 shows the current value of the offshore wind generator circuit output. The amount of power produced by this generator is given in Fig. 12.52

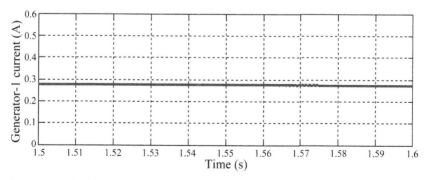

FIGURE 12.51   Offshore wind generator circuit output current for Case 4.

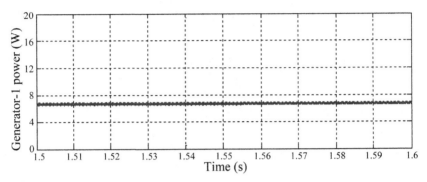

FIGURE 12.52 Offshore wind generator circuit output power for Case 4.

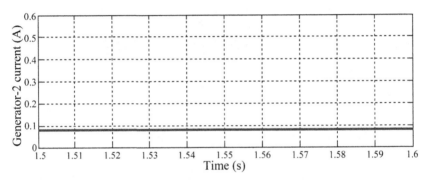

FIGURE 12.53 Marine current generator circuit output current for Case 4.

as 5 W. The graphic of the current in the marine current generator output is given in Fig. 12.53. The power transferred from this generator varies depending on the power demanded by the load group, and this power graph is shown in Fig. 12.54 as 1.8 W.

In Case 4, the battery and the ultracapacitor group are not activated because they have a high SOC level. Fig. 12.55 shows that the battery group is SOC

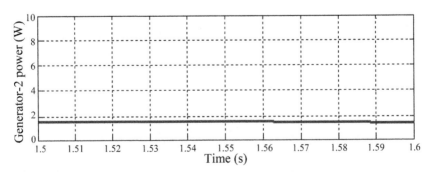

FIGURE 12.54 Marine current generator circuit output power for Case 4.

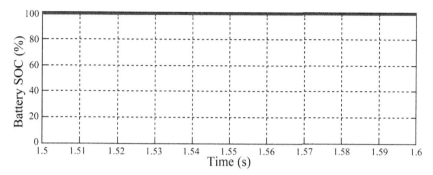

FIGURE 12.55 Battery group SOC for Case 4.

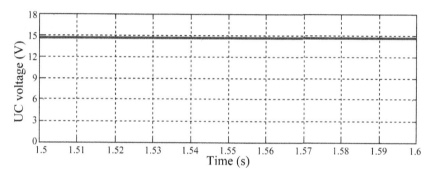

FIGURE 12.56 Ultracapacitor group voltage for Case 4.

100%. Likewise, the voltage indicating the ultracapacitor group SOC is given as 15 V in Fig. 12.56. In this case, if there is a power change of 10 W or more in the load, the smart energy management algorithm activates the battery and ultracapacitor group and meets the load demand. HESS prevents the DC bus voltage fluctuation in the system. The amount of power demand by the load group is shown in Fig. 12.57 as 8 W.

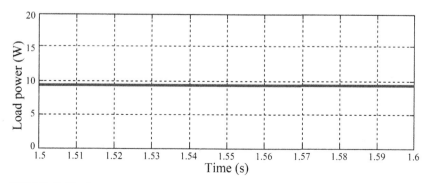

FIGURE 12.57 Load group power for Case 4.

2.12 Simulation results for Case 5

The $P_{TOT} < P_{LOAD}$, P_{BAT}, and P_{UC}SOCs are low in Case 5. The power flow diagram of Case 5 is given in Fig. 12.58. In this case the demand power of the load group is much more than the amount of power produced.

The battery and ultracapacitor SOCsare low. Since the power demanded by the load group is not met from another source, energy will not be provided to the load. If there is power generation in the HPGS, the load demand power will be met. This is the most critical scenario for the system. If the system had supported a grid-connected structure, the energy of the load would be supplied by the grid.

2.13 Simulation results for Case 6

The $P_{TOT} < P_{LOAD}$, P_{BAT}, and P_{UC}SOCs are high in Case 6. The power flow diagram of Case 6 is given in Fig. 12.59.Unlike Case 5, the battery and ultracapacitor group SOCs are high. Also, in this case, there is a need for sudden load demand. In Case 6 the ultracapacitor group supported the battery group and the DC bus fluctuation was prevented. The response of the smart energy management algorithm to the rapidly changing dynamic behavior of the system is examined.

Fig. 12.60 provides HPGS simulation parameter values for Case 6. In this case the offshore wind turbine rotates with a wind speed of 6.39 m/s. The speed of rotation is increased with the reducer and transferred to the generator at a

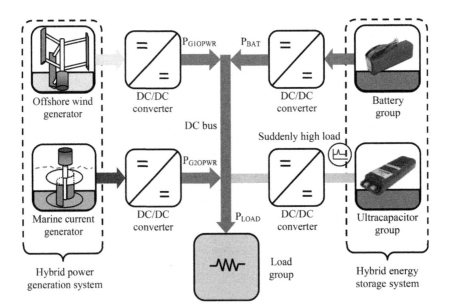

FIGURE 12.58 Power flow diagrams for Case 5.

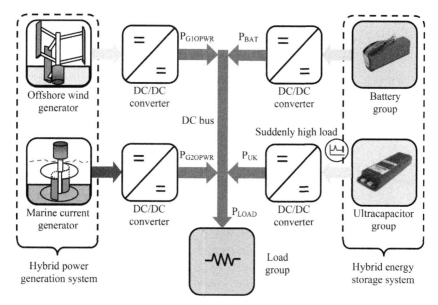

FIGURE 12.59 Power flow diagrams for Case 6.

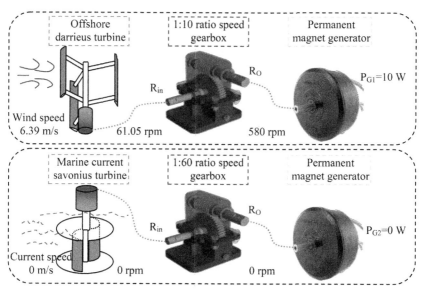

FIGURE 12.60 HPGS parameters for Case 6.

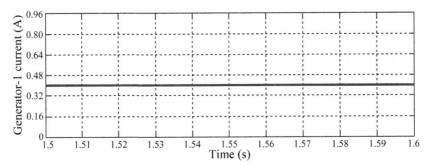

FIGURE 12.61   Offshore wind generator circuit output current for Case 6.

speed of 580 rpmand a power of 10 W is produced by the generator. The marine current turbine is stable and does not provide an energy flow to the system. The offshore wind generator output current graph is given in Fig. 12.61. The power produced from the generator is 10 W and the power graph is shown in Fig. 12.62. The DC/DC boost circuit is passive because no power is generated from the marine current generator.

In Case 6, 10 W is produced from the offshore wind generator, and the power demand by the load group is 20 W. The difference power value P_{DIF-} calculated by the smart energy management algorithm is −10 W. If the difference power is negative, the battery group is activated. The charge and discharge reference current value of the battery group is calculated. During the system operation, a load demand of 13 W is suddenly activated in 1.5 seconds. The smart energy management algorithm detects this sudden load demand and activates the ultracapacitor group. The reference current value that the ultracapacitor group should discharge is calculated and sent to the bidirectional DC/DC converter circuit. Then the ultracapacitor group is disabled. The power demand of the load group is provided by the battery group. Thanks to the smart energy management algorithm and ultracapacitor, the depth of discharge of the battery group is prevented. Thus the battery group cycle life is extended.

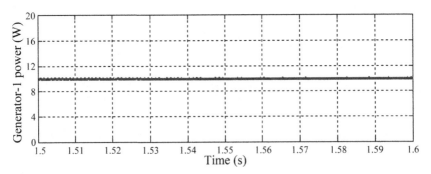

FIGURE 12.62   Offshore wind generator circuit output power for Case 6.

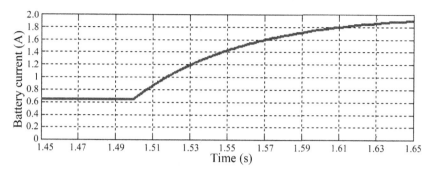

FIGURE 12.63   Battery group current for Case 6.

The battery group current graph in sudden load change working case is given for Case 6 in Fig. 12.63. The battery group is discharged in 1.5 seconds according to the current value calculated by the smart energy management algorithm. The battery SOC value is given in detail in Fig. 12.64. The battery SOC value starts at 90 % and then decreases. The current graph of the ultracapacitor group is given in Fig. 12.65. The sudden load demand in the system is detected by the

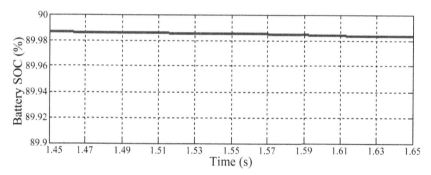

FIGURE 12.64   Battery group SOC for Case 6.

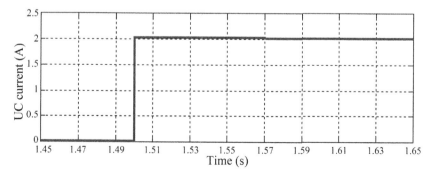

FIGURE 12.65   Ultracapacitor group current for Case 6.

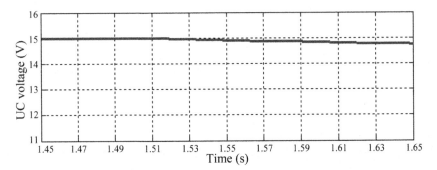

FIGURE 12.66 Ultracapacitor group voltage for Case 6.

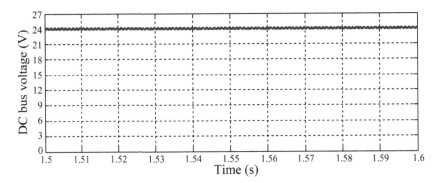

FIGURE 12.67 DC bus voltage for Case 6.

smart energy management algorithm. The ultracapacitor group is discharged with 2 A. The voltage graph of the ultracapacitor group is given in Fig. 12.66. Since the ultracapacitor group has low power energy density, the voltage value decreases very quickly. DC bus voltage is given in Fig. 12.67. Thanks to the ultracapacitor, DC bus voltage fluctuation does not occur.

The load group power first demands 20 W and 12 W additional powers are activated in 1.5 seconds. The system total load power demand is 32 W. The power change graph demand by the load group is given in Fig. 12.68.

2.14 Simulation results for Case 7

The $P_{TOT} = 0$, P_{BAT}, and P_{UC}SOCs are low in Case 7. The power flow diagram of Case 7 is given in Fig. 12.69. In this case, there is no power transfer from HPGS since the offshore wind and marine current turbines do not rotate. The load demand power is met by the battery and ultracapacitor group in the system. In this case the battery group SOC is very low. If the SOC value of the battery group

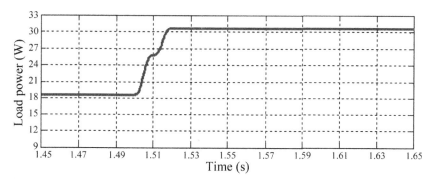

FIGURE 12.68   Load group power for Case 6.

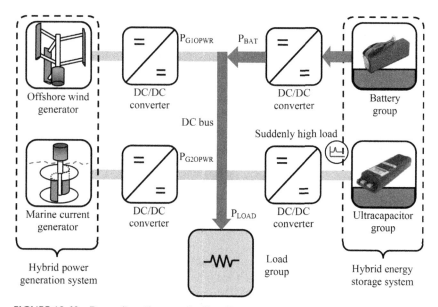

FIGURE 12.69   Power flow diagrams for Case 7.

drops below 20 %, the battery group will be passive. The smart energy management algorithm does not drop below the critical discharge level to protect the health of the battery. Likewise, the ultracapacitor group cannot be charged because it cannot find an energy source from HPGS.

2.15   Simulation results for Case 8

The $P_{TOT} = 0$, P_{BAT}, and P_{UC}SOCs are high in Case 8. The power flow diagram of Case 8 is given in Fig. 12.70. In this case, it does not generate power in both units in HPGS.

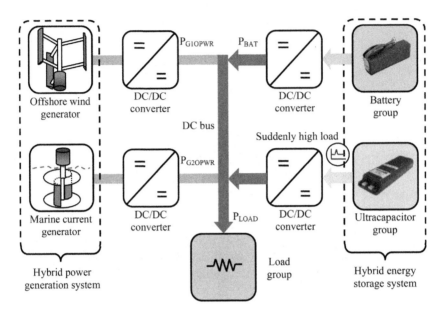

FIGURE 12.70   Power flow diagrams for Case 8.

The battery and ultracapacitor group SOCs in the system are high. The long-time energy demand by the load is provided by the battery group. The ultracapacitor group is activated when there is sudden high load demand.

The input current graph of the battery group is given for Case 8 in Fig. 12.71. The reference discharge current value of the battery group is calculated by the smart energy management algorithm. In this case the SOC value of the battery group is 85 %. Depending on the load demand, the SOC value decreases over time as in Fig. 12.72.

The current value demanded by the load group is shown in Fig. 12.73. The 10 W power value requested by the load group is given in Fig. 12.74.

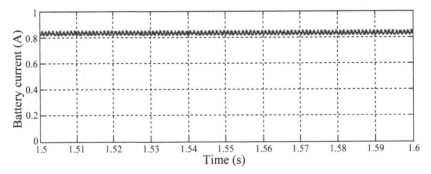

FIGURE 12.71   Battery group current for Case 8.

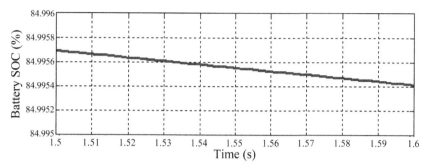

FIGURE 12.72   Battery group SOC for Case 8.

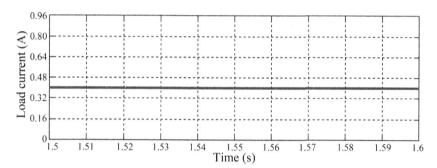

FIGURE 12.73   Load group current for Case 8.

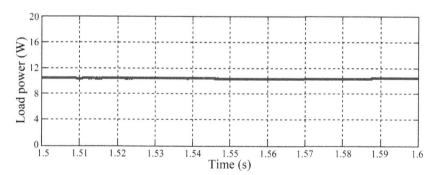

FIGURE 12.74   Load group power for Case 8.

2.16   Simulation results for Case 9

The $P_{LOAD} = 0$, P_{BAT}, and PUCSOCs are low in Case 9. The power flow diagram of Case 9 is given in Fig. 12.75. In this case a small amount of power is produced by HPGS. However, the system does not require any load demand and the power drawn from the system is zero. All of the power produced by HPGS is transferred to HESS. Thus the battery and ultracapacitor group with low SOC

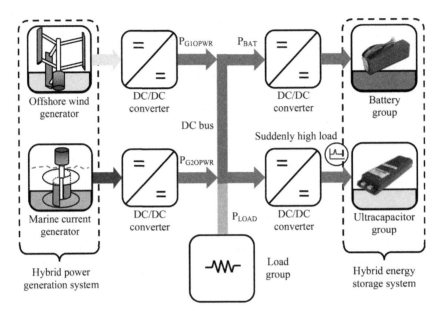

FIGURE 12.75 Power flow diagrams for Case 9.

are charged. Fig. 12.76 provides HPGS simulation parameter values for Case 9. The offshore wind turbine rotates at 4.85 m/s and transfers power mechanically to the gearbox. With the reducer conversion ratio, the 5 W is obtained by rotating the generator to 440 rpm. The marine current turbine rotates at 0.99 m/s, and the generator generates 4 W at 540 rpm.

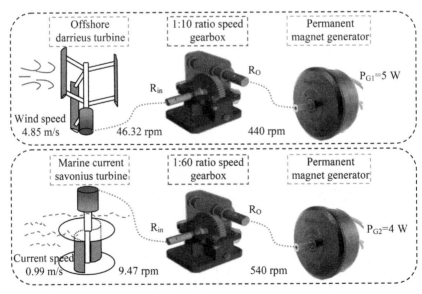

FIGURE 12.76 HPGS parameters for Case 9.

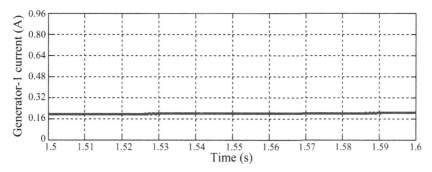

FIGURE 12.77   Offshore wind generator circuit output current for Case 9.

The offshore wind generator output current graph is given in Fig. 12.77. As given in Fig. 12.78, 5 W is transferred to the DC bus. The marine current generator circuit output current is given in Fig. 12.79. The power value produced by this generator is approximately 4 W and is shown in Fig. 12.80.

Since there is no load demand in Case 9, all of the power generated is transferred to HESS. The battery group charges at approximately 0.42 A as given in

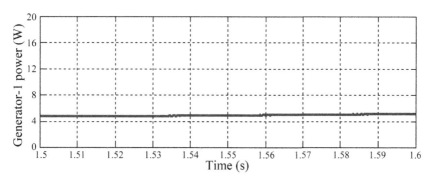

FIGURE 12.78   Offshore wind generator circuit output power for Case 9.

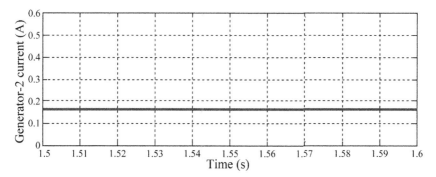

FIGURE 12.79   Marine current generator circuit output current for Case 9.

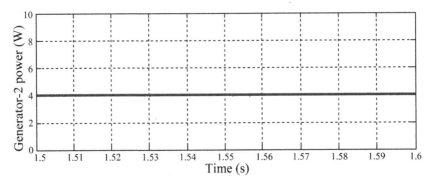

FIGURE 12.80 Marine current generator circuit output power for Case 9.

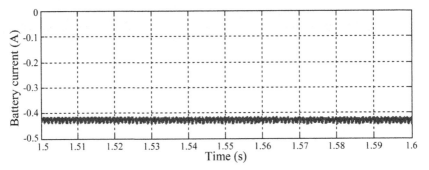

FIGURE 12.81 Battery group current for Case 9.

Fig. 12.81. The SOC value of the battery group is given in Fig. 12.82. The scale in this chart is kept in a small band to better see the SOC value of the battery group. The SOC value of the battery group starts at 30 % and increases gradually.

Fig. 12.83 shows the charge voltage value of the ultracapacitor group for Case 9. Fig. 12.84 shows the charge current graph of the ultracapacitor group. The ultracapacitor group charges very fast compared to the battery group and the charging current goes to zero in a short time.

In this chapter, parameters of all units were calculated for the system. After making the calculations, each system was simulated at the MATLAB/Simulink program, andtest results were obtained. Case transitions were observed considering the nine possible working cases. The importance of HESS was emphasized from the simulation results. With HPGS and HESS structure, energy is stored and primary load demands are supplied. Thusquality energy transfer is realized. The battery and ultracapacitor group were used together in this system. In sudden load changes the ultracapacitor group was activated and the depth ofdischarge of the battery group was prevented. Thus the battery group is expected to cycle life and healthy. Considering the results obtained from the simulation tests, experimental study was started.

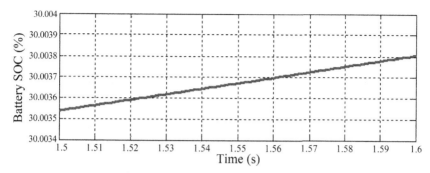

FIGURE 12.82   Battery group SOC for Case 9.

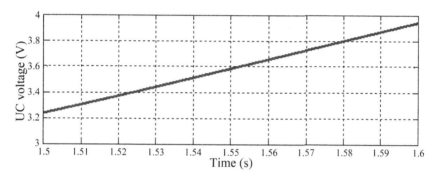

FIGURE 12.83   Ultracapacitor group voltage for Case 9.

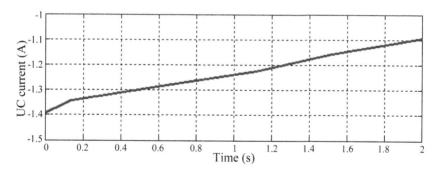

FIGURE 12.84   Ultracapacitor group current for Case 9.

3   Experimental studies of HPGS and HESS

The cards controlling all the subcomponents of HPGS and HESS have been added to the same floating buoy platform. The load group to be used during the experiments is also included in the same system. The drive forces derived from the offshore wind Darrieus turbine and marine current Savonius turbine are transmitted to the generators viareducers. DC electrical energy produced by generators is gathered in DC bus with DC/DC boost converters. The battery and the ultracapacitor group forming the HESS units are connected to the common DC bus with the DC/DC bidirectional converter. The energy taken from the DC bus is transferred to the load. Transducer and control circuits in the system are controlled according to the information received from the sensors. Converters and control circuits in the system arecontrolled according to the information received from the sensors. Depending on the possible working cases of the system, the smart energy management algorithm performs efficient and safe control of all units. The general topology and control diagram of HPGS and HESS is given in Fig. 12.85.

The display of HPGS and HESS with all its subunits is shown in detail in Fig. 12.86. The sensor data in the power converter circuits is important for the smart energy management algorithm. DC bus voltage of the system was determined as 24 V in experimental studies. Battery and ultracapacitor were used in hybrid energy storage units as in simulation studies. When used together,the battery and ultracapacitor are very promising successful systems. Electrical control circuit photo of hybrid power generation and HESS is given in Fig. 12.87. The generator output voltages used in offshore wind and marine current turbines are 12 V. The 1:10 conversion rates gearbox is selected for the offshore wind turbine and the 1:60 conversion rates for the marine current turbine.

The DC/DC boost circuits at the offshore wind and marine current turbine output raise the voltage level to 12–24 V on the DC bus. The 11.1 V voltage level was obtained by using three-piece 3.7 V 1050 mAh lithium ion batteries in the battery group. The 13.5 V was obtained by connecting twopiecesof 2.7 V10 Fultracapacitors in series. There are bidirectional DC/DC converter circuits at the battery and ultracapacitor outputs. Thanks to these circuits, energy storage units are charged and discharged. Resistance groups are used as load

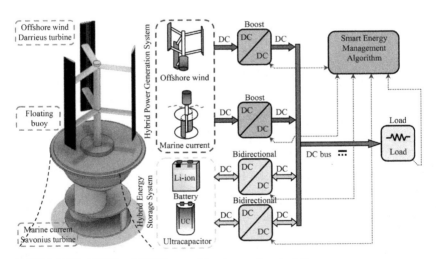

FIGURE 12.85   **HPGS and HESS general topology and control diagram.**

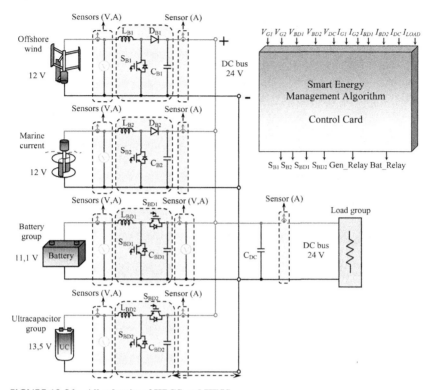

FIGURE 12.86   All subunits of HPGS and HESS.

groups in the system. Load group drawing 10-Wpower was obtained withthese resistors used in the 15 and 30 Ω resistor groups. LTC4150 coulomb counter integration is used to determine the battery group SOC value. Thanks to this coulomb counter integration, the battery group SOC value is sent to the micro-controller. The smart energy management algorithm card is designed to control all units in the system. The PIC 18F4550 microcontroller is used as the main processor in this control card. All current and voltage values in the system are given to this processor as input.

The flow diagram of the smart energy management algorithm used in this chapter study is given in Fig. 12.88. The microcontroller in the system is coded using the MicroC program. Fig. 12.89 gives the graph ofpower values used in experimental studies. The working cases are represented in a daily time frame. In this graph, the power graphs of the offshore wind turbine, marine current turbine, battery, and load group are given respectively. The nine possible work-ing cases have been identified at different working modes. The system dynamic behavior of the smart energy management algorithm developed within the scope of the study is observed in detail with this power graph. The battery group is activated by charging/discharging according to the possible working cases.

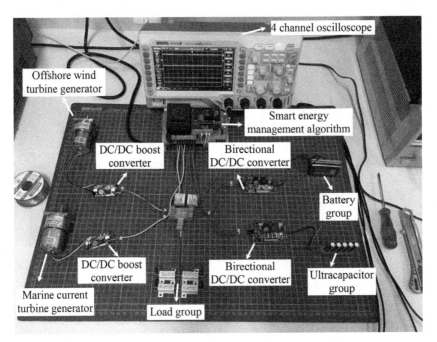

FIGURE 12.87   The photo of HPGS and HESS.

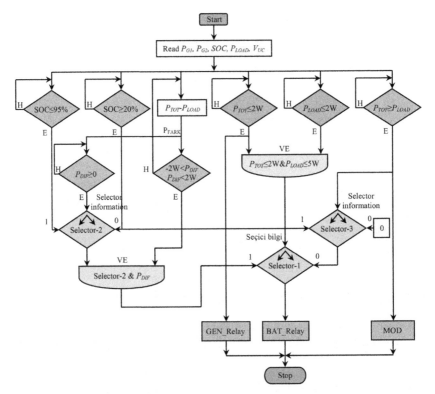

FIGURE 12.88   The flow diagram of the smart energy management algorithm.

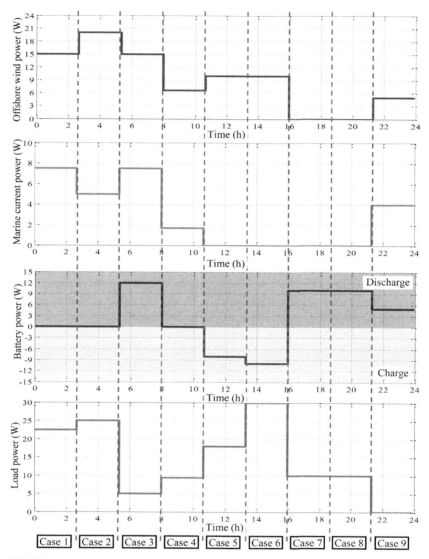

FIGURE 12.89   One-day power graph distribution of working cases.

The dynamic behavior of the system was examined with a daily power graph. In Case 1, 15 W of offshore wind turbine and 7.5 W of marine current turbine are produced. The power of the load group is 22.5 W and the battery group is in a passive state. In Case 2, 20 W of offshore wind turbine and 5 W of marine current turbine are obtained. Since the battery group has a high SOC value, it is passive. The load group is 25 W and its energy is supplied by HPGS. In Case 3 the load group power demands 5 W, when 15 W of the offshore wind turbine and 7.5 W of the current turbine are produced. Since the battery group SOC value is low, it charges with 12 W.

In Case 4, 6.6 W of offshore wind turbine and 1.7 W of marine current turbine are generated, while the load group draws 9.3 W. Since the battery group has a high SOC value, it is passive. There is only battery group SOC difference between Cases 5 and 6. In this case offshore wind turbine 15 W and marine current turbine 0 W are produced. The load group increases from 18 to 30 W and the ultracapacitor group supports the battery group. There is only a difference between battery SOC between Cases 7 and 8. In these two cases the offshore wind turbine and the marine current turbine are passive. The load group demands 10 W and this power is met by the battery group. In Case 9, 5 W of offshore wind turbine and 4 W of marine current turbine are produced. The load group does not demand power, and all power generated is transferred to the energy storage unit.

3.1 Experimental results for Cases 1 and 2

There is only the SOC difference of the energy storage units between Cases 1 and 2. Therefore, a single experimental result was obtained for Cases 1 and 2. While the battery SOC value is low in Case 1, the battery SOC value is high in Case 2. In both cases the power generated from offshore wind and marine current turbines is approximately equal to the load power. In these cases the battery and ultracapacitor are passive. The power flow diagram of the experimental results for Cases 1 and 2 is given in Fig. 12.90.

Offshore wind turbine generator, DC/DC boost output voltages, and currents are given for Cases 1 and 2 in Fig. 12.91. The offshore wind turbine generator output voltage is 13 V, current is 1.23 A, and the power obtained is 16 W. The DC/DC boost circuit at the generator output switches the DC bus to 24 V. DC/DC boost circuit output voltage is 24 V, and the current is 0.62 A. A total of 15 W

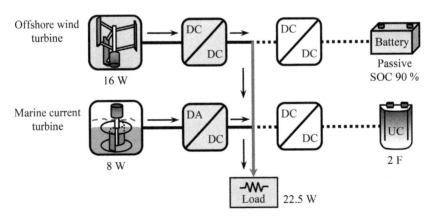

FIGURE 12.90 Power flowchart of experimental results for Cases 1 and 2.

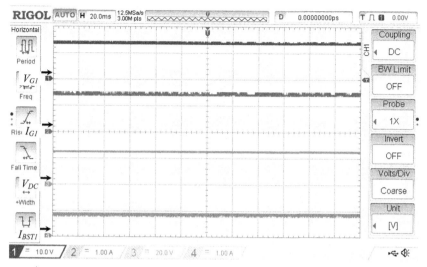

FIGURE 12.91 Offshore wind generator, DC/DC boost voltages, and currents for Cases 1 and 2.

is transferred from the offshore wind turbine generator to the system after DC/DC boost circuit losses.

Marine current turbine generator, DC/DC boost output voltages, and currents are given for Cases 1 and 2 in Fig. 12.92. The marine current turbine generator output voltage is 10 V, the current is 0.8 A, and the power obtained is 8 W.

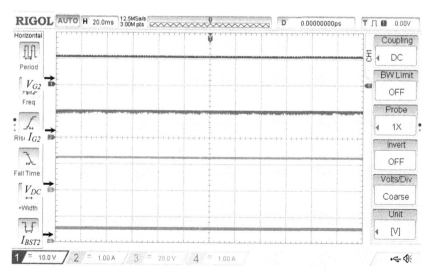

FIGURE 12.92 Marine current generator, DC/DC boost voltages, and currents for Cases 1 and 2.

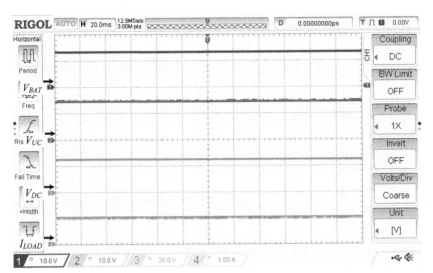

FIGURE 12.93 Battery, ultracapacitor, load voltages, and currents for Cases 1 and 2.

DC/DC boost circuit output voltage is 24 V, current is 0.31 A. A total of 7.5 W is transferred from the marine current turbine generator to the system. Power transfer is provided from both hybrid power generation units in the system.

Battery, ultracapacitor, DC bus voltage, and current drawn by the load are given for Cases 1 and 2, respectively, in Fig. 12.93. Since the load is directly connected to the DC bus, the load voltage is equal to the DC bus voltage. The load group demand is 22.5 W. Since the amount of power produced by HPGS is approximately equal to the power demanded by the load, the battery and ultracapacitor group are inactive. The battery group SOC value is 90 %, battery group voltage value is 12.5 V, and the ultracapacitor group voltage is 13 V.

3.2 Experimental results for Case 3

The power flow diagram of the experimental results for Case 3 is given in Fig. 12.94. In Case 3, 16 and 8 W of power are generated from offshore wind and marine current turbines, respectively. The power demand by the load is 5 W. In this case the battery and ultracapacitor group SOC is low. The battery group with an SOC value of 30 % is charged with 11.6 W. At the same time, the ultra-capacitor group is charged with 5.1 W. After meeting the load demand power, HPGS charges the battery and the ultracapacitor with the remaining power.

Offshore wind turbine generator, DC/DC boost output voltages, and currents are given for Case 3 in Fig. 12.95. The offshore wind turbine generator output voltage is 13 V, the current is 1.23 A, and the power obtained is 16 W. DC/DC boost circuit output voltage is 24 V, current is 0.62 A. A total of 15 W is transferred from the offshore wind turbine generator to the system with a DC/DC boost circuit.

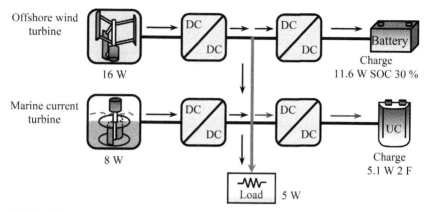

FIGURE 12.94   Power flowchart of experimental results for Case 3.

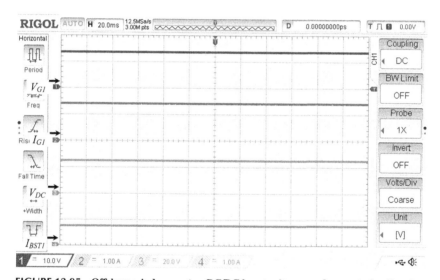

FIGURE 12.95   Offshore wind generator, DC/DC boost voltages, and currents for Case 3.

The current turbine generator, DC/DC boost output voltages, and currents are given for Case 3 in Fig. 12.96. The current turbine generator output voltage is 10 V, the current is 0.8 A, and the power obtained is 8 W. DC/DC boost circuit output voltage is 24 Vand the current is 0.31 A. A total of 7.5 W is transferred from the offshore wind turbine generator to the system. First, the power demanded by the load group is met by the total power provided by HPGS. The remaining power from the load group is transferred to HESS. In this case the load group draws 0.2 A at 24 V.

The input/output current and voltage results of DC/DC bidirectional converter circuit connected to the battery group are given for Case 3 in Fig. 12.97.

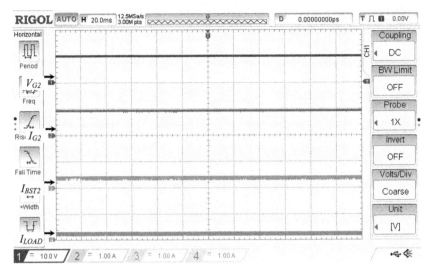

FIGURE 12.96 Marine current generator, DC/DC boost voltages, and currents for Case 3.

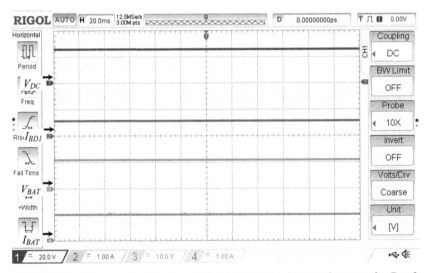

FIGURE 12.97 Battery and bidirectional DC/DC converter voltages and currents for Case 3.

DC/DC bidirectional converter allows the battery to charge. The 24-V DC bus voltage is reduced to the voltage value that the battery should charge. The battery group is charged at 11.2 V. In this case, while 0.5 A current is drawn from the DC bus, the battery group is charged with 1 A and 11.1 W.

The current and voltage results of ultracapacitor and bidirectional DC/DC converter are given for Case 3 in Fig. 12.98. The DC/DC bidirectional converter circuit charges the ultracapacitor with 5 W. Ultracapacitor group voltage is

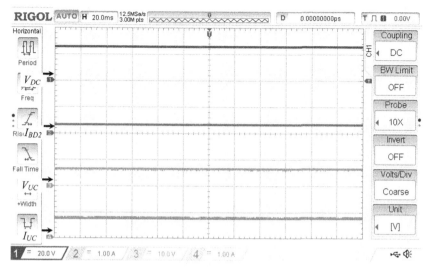

FIGURE 12.98 Ultracapacitor and bidirectional DC/DC converter voltages and currents for Case 3.

6.5 V and current starts from 0.78 A and charges quickly. Ultracapacitor group charges much faster than battery group.

3.3 Experimental results for Case 4

The power flow diagram of the experimental results for Case 4 is given in Fig. 12.99. In this case the offshore wind turbine generator produces 7 W and the marine current turbine generator produces 2 W. HESS units are inactive because the battery and ultracapacitor group SOC are high. A high power is produced from the offshore wind and marine current turbine generator. However, there is

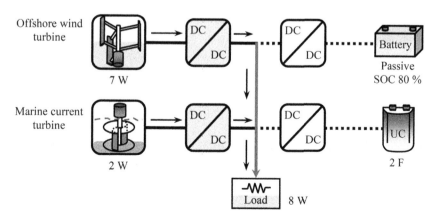

FIGURE 12.99 Power flowchart of experimental results for Case 4.

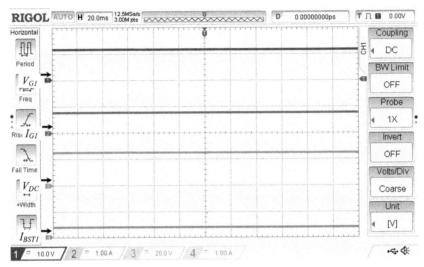

FIGURE 12.100   Offshore wind generator, DC/DC boost voltages, and currents for Case 4.

only 8 W load demand in the system. Even if excess power is generated by the sources, there is only energy flow as much as the load demands. The offshore wind turbine generator, DC/DC boost output voltages, and currents are given for Case 4 in Fig. 12.100. Offshore wind turbine generator voltage is 11 V, current is 0.63 A, and power is 7 W. DC/DC boost converter circuit output voltage is 24 V, current is 0.27 A, and power is 6.5 W.

The marine current turbine generator voltage, current, DC/DC boost output current, and load current are given for Case 4, respectively, in Fig. 12.101. In

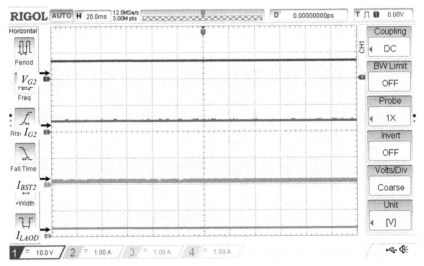

FIGURE 12.101   Marine current generator, DC/DC boost voltages, and currents for Case 4.

this case the generator output voltage is 6 V, current is 0.33 A, and its power is 2 W. DC/DC boost converter output current is 0.06 A. The load demand power only 8 W in this case. Even if HPGS has produced more power, there will be no demand in the system, so produced power will be transferred as much as the load demands power.

3.4   Experimental results for Cases 5 and 6

Cases 5 and 6 are possible working cases where the dynamic behavior of the system is observed. The power flow diagram of the experimental results for Cases 5 and 6 is given in Fig. 12.102. In this case the load group demand power value in the system is increased from 18 to 30 W. In the sudden change of the load, the ultracapacitor group is activated and prevents the battery from thedepth of discharge. Ultracapacitor group helps one to extend thecycle life of battery. Smart energy management algorithm in the system detects sudden load change and activates the ultracapacitor group. The offshore wind turbine generator produces 11 W of power and the load group demands 18 Win Cases 5 and 6. In this case the battery group supports the offshore wind turbine generator. The battery group feeds the load group by discharging with 10.8 W.

The offshore wind turbine generator, DC/DC boost output voltages, and currents are given for Cases 5 and 6 in Fig. 12.103. The offshore wind turbine generator output voltage is 12 V, current is 0.9 A, and the power obtained is 11 W. DC/DC boost circuit output voltage is 24 V, current is 0.41 A. A total of 10 W is transferred from the offshore wind turbine generator to the system with DC/DC boost converter.

Fig. 12.104 shows the results of DC bus, battery group voltage, ultracapacitor, battery group current during sudden load change for Cases 5 and 6. In this

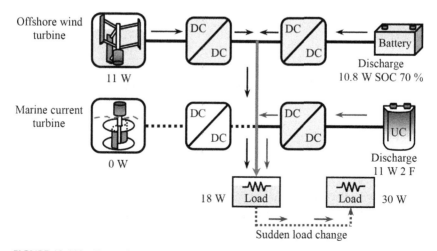

FIGURE 12.102 Power flowchart of experimental results for Cases 5 and 6.

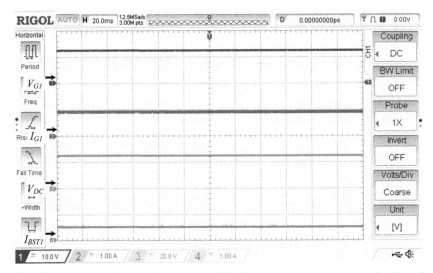

FIGURE 12.103 Offshore wind generator, DC/DC boost voltages, and currents for Cases 5 and 6.

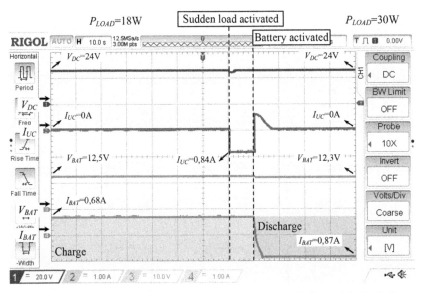

FIGURE 12.104 Battery, ultracapacitor voltages, and currents during sudden load change for Cases 5 and 6.

case the dynamic behavior of the system was investigated when a load group suddenly entered to the system. The DC bus voltage is 24 V, and no fluctuation is observed during sudden load change.

In Fig. 12.105, load voltage and current results are given during the sudden load change for Cases 5 and 6. Thanks to the ultracapacitor group, there is no

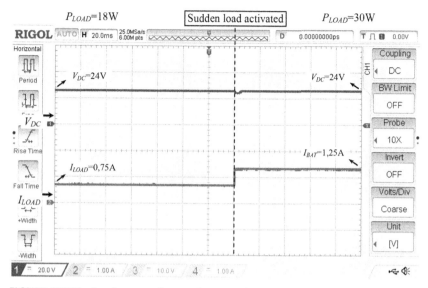

FIGURE 12.105 Load group voltage and current during sudden load change for Cases 5 and 6.

fluctuation in the DC bus voltage. Load group power increases from 18 to 30 W in sudden load transition.

Ultracapacitor group current is normally 0 A. When the load is activated, this case is detected by the smart energy management algorithm. The ultracapacitor group discharges with 0.84 A. The smart energy management algorithm activates the battery group after a sudden load transition. The continuity of demand power is ensured by battery group. After this sudden load transition process is finished, the ultracapacitor group recovers the power it discharges from the DC bus. The ultracapacitor group is charging and its current value decreases to 0 A over time.

The battery group initially discharges with 0.68 A. Discharge current becomes 0.87 A after sudden load transition. While the battery group is discharged with 12.5 V, the battery group continues to discharge at 12.3 V after sudden load increase. In the transition from Case 5 to Case 6, the contribution of HESS is given in detail with experimental results. Ultracapacitor group supports the battery group during sudden load transition and prevents depth of discharge of the battery. DC bus and load fluctuations in the system are prevented.

3.5   Experimental results for Cases 7 and 8

In Cases 7 and 8 no power can be obtained from the offshore wind turbine generator and the marine current turbine generator. The only difference between Case 7 and Case 8 is the battery group SOC value.

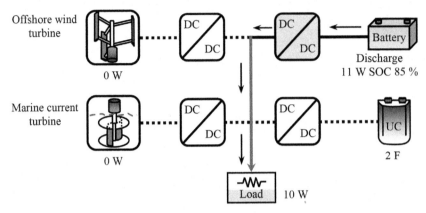

FIGURE 12.106   Power flowchart of experimental results for Cases 7 and 8.

In these cases, when the battery group SOC value drops below 20 %, the battery group is made passive. Excessive discharge of the battery group is prevented. Thus the battery group cycle life can be increased. DC/DC boost circuits connected to generators are in a passive state. The load group connected to the system requires 10 W. In this case the battery group meets the power demanded by the load. The battery group discharged with 11 W, and the battery group SOC value is 85 %. The power flow diagram of the experimental results for Cases 7 and 8 is given in Fig. 12.106.

The battery voltage and current, DC bus voltage, and load current are given respectively for Cases 7 and 8 in Fig. 12.107. Battery group voltage is 12.7 V,

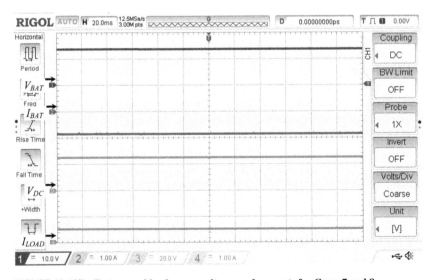

FIGURE 12.107   Battery and load group voltages and currents for Cases 7 and 8.

current is 0.86 A, and discharge power is 11 W. The DC bus voltage is constant at 24 V, and the load draws 0.41 A, demanding a power of 10 W under this voltage.

3.6   Experimental results for Case 9

The power flow diagram of experimental results is given for Case 9 in Fig. 12.108. In this case the offshore wind turbine generator produces 13 W of power and the marine current turbine generator produces 9 W. The load in the system is passive and does not require any power. Battery and ultracapacitor groupSOC is low. Smart energy management algorithm enables these energy storage units to be charged. The battery group SOC value is 20 % and it is charged with 12 W. The ultracapacitor group is charged with 6 W.

The offshore wind turbine generator, DC/DC boost output voltages, and currents are given for Case 9 in Fig. 12.109. The offshore wind turbine generator output voltage is 13 V, the current is 1 A and the power obtained is 13 W. The DC/DC boost circuit at the generator output switches the DC bus to 24 V. DC/DC boost circuit output voltage is 24 V, and current is 0.5 A. A total of 12 W is transferred from the offshore wind turbine generator to the system after DC/DC boost circuit losses.

The marine current turbine generator, DC/DC boost output voltages, and currents are given for Case 9 in Fig. 12.110. Marine current turbine generator output voltage is 10 V, current is 0.9 A, and power is 9 W. The DC/DC boost circuit output current is 0.33 A, and the 8 W transfers power to the DC bus.

The battery group voltage and current and ultracapacitor voltage and current results are given for Case 9 in Fig. 12.111. In this case the battery group is charged with 11 V, 1.1 A current, and 12 W of power. Likewise, the ultracapacitor group is charged. The ultracapacitor charging voltage is 5.6 V, the current is 1.07 A and 6 W.

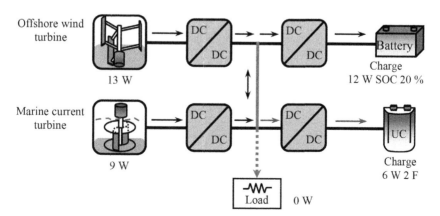

FIGURE 12.108   Power flowchart of experimental results for Case 9.

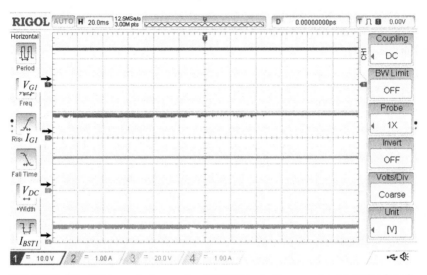

FIGURE 12.109 Offshore wind generator, DC/DC boost voltages, and currents for Case 9.

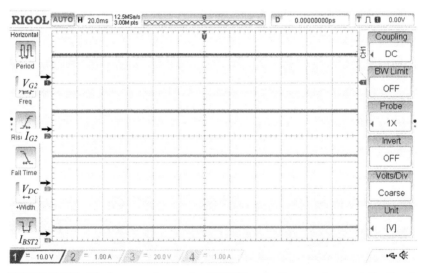

FIGURE 12.110 Marine current generator, DC/DC boost voltages, and currents for Case 9.

Experimental power results of nine possible working cases are given in Table 12.5. In nine possible working cases, offshore wind turbine, marine current turbine, battery group, ultracapacitor group, and power values of the load are given in detail. Thanks to this table, the power analysis of the units in the system can be done comprehensively.

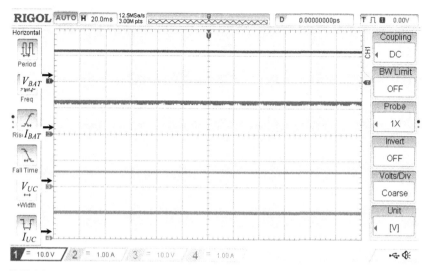

FIGURE 12.111   Battery and ultracapacitor group voltages and currents for Case 9.

TABLE 12.5 HPGS and HESS power values in working cases.

Cases	Offshore wind (W)	Marine current (W)	Battery group status, SOC, power	Ultracapacitor group status, power	Load group (W)
1	16	8	Disable, 20%	Disable, 2 F	22.5
2	20.8	6	Disable, 90%	Disable, 2 F	25
3	16	8	Charging, 30%, 11.6 W	Charging, 5.1 W	5
4	7	2	Disable, 100%	Disable, 2 F	8
5	11	0	Discharging, 45%, 10.8 W	Discharging, 11 W	18
6	11	0	Discharging, 85%, 8 W	Disable, 2 F	30
7	0	0	Discharging, 30%, 11 W	Disable, 2 F	10
8	0	0	Discharging, 95%, 11 W	Disable, 2 F	10
9	13	9	Charging, 20%, 12 W	Charging, 6 W	0

The smart energy management algorithm is coded to the microcontroller and control is provided between the cases. By determining nine possible working cases, offshore wind turbine generator, marine current turbine generator, battery group, ultracapacitor group, and loads were analyzed in detail for each case. The working cases of the circuits were examined in detail with the oscilloscope and the screen image results were given. Transition cases were created to analyze the dynamic behavior of the system. The importance of HESS in these case transitions has been demonstrated by experimental results. The ultracapacitor group has been activated quickly in sudden load transition cases. It has been shown that the depth of discharge of the battery group is prevented by using the battery and the ultracapacitor group together. Ultracapacitor is effective in eliminating the voltage drops experienced by the activation of the loads that draw excess current. The ultracapacitor is activated in the case of overload and thecycle life of the battery is extended.

References

[1] B. Ceran, The concept of use of PV/WT/FC hybrid power generation system for smoothing the energy profile of the consumer, Energy 167 (2019) 853–865, doi: 10.1016/J.ENERGY.2018.11.028.

[2] Y. Han, Y. Sun, J. Wu, An efficient solar/lignite hybrid power generation system based on solar-driven waste heat recovery and energy cascade utilization in lignite pre-drying, Energy Convers. Manage. 205 (2020) 112406, doi: 10.1016/J.ENCONMAN.2019.112406.

[3] A. Mohammadnia, A. Rezania, B.M. Ziapour, F. Sedaghati, L. Rosendahl, Hybrid energy harvesting system to maximize power generation from solar energy, Energy Convers. Manage. 205 (2020) 112352, doi: 10.1016/J.ENCONMAN.2019.112352.

[4] L. Kong, J. Yu, G. Cai, Modeling, control and simulation of a photovoltaic/hydrogen/supercapacitor hybrid power generation system for grid-connected applications, Int. J. Hydrogen Energy 44 (2019) 25129–25144, doi: 10.1016/J.IJHYDENE.2019.05.097.

[5] J. Shi, L. Wang, W.-J. Lee, X. Cheng, X. Zong, Hybrid energy storage system (HESS) optimization enabling very short-term wind power generation scheduling based on output feature extraction, Appl. Energy 256 (2019) 113915, doi: 10.1016/J.APENERGY.2019.113915.

[6] K. Sun, K.-J. Li, J. Pan, Y. Liu, Y. Liu, An optimal combined operation scheme for pumped storage and hybrid wind-photovoltaic complementary power generation system, Appl. Energy 242 (2019) 1155–1163, doi: 10.1016/J.APENERGY.2019.03.171.

[7] M. Mehrpooya, B. Ghorbani, M. Moradi, A novel MCFC hybrid power generation process using solar parabolic dish thermal energy, Int. J. Hydrogen Energy 44 (2019) 8548–8565, doi: 10.1016/J.IJHYDENE.2018.12.014.

[8] M. Nunes Fonseca, E. de Oliveira Pamplona, A.R. de Queiroz, V.E. de Mello Valerio, G. Aquila, S. Ribeiro Silva, Multi-objective optimization applied for designing hybrid power generation systems in isolated networks, Sol. Energy 161 (2018) 207–219, doi: 10.1016/J.SOLENER.2017.12.046.

[9] B. Shi, W. Wu, L. Yan, Size optimization of stand-alone PV/wind/diesel hybrid power generation systems, J. Taiwan Inst. Chem. Eng. 73 (2017) 93–101, doi: 10.1016/J.JTICE.2016.07.047.

[10] S.R. Rex, D.M.M.S.R. Praba, Design of PWM with four transistor comparator for DC–DC boost converters, Microprocess. Microsyst. 72 (2020) 102844, doi: 10.1016/J.MICPRO.2019.07.003.

[11] C.L. Remes, G.R. Gonçalves da Silva, A. Treviso, M.A.J. Coelho, L. Campestrini, Data-driven approach for current control in DC–DC boost converters, IFAC-PapersOnLine 52 (2019) 190–195, doi: 10.1016/J.IFACOL.2019.06.059.

[12] D. Oulad-Abbou, S. Doubabi, A. Rachid, Power switch failures tolerance of a photovoltaic fed three-level boost DC–DC converter, Microelectron. Reliab. 92 (2019) 87–95, doi: 10.1016/J. MICROREL.2018.11.017.

[13] A. Garrigós, D. Marroquí, A. García, J.M. Blanes, R. Gutiérrez, Interleaved, switched-inductor, multi-phase, multi-device DC/DC boost converter for non-isolated and high conversion ratio fuel cell applications, Int. J. Hydrogen Energy 44 (2019) 12783–12792, doi: 10.1016/J. IJHYDENE.2018.11.094.

[14] D.W. Spier, G.G. Oggier, S.A.O. da Silva, Dynamic modeling and analysis of the bidirectional DC–DC boost-buck converter for renewable energy applications, Sustain. Energy Technol. Assess. 34 (2019) 133–145, doi: 10.1016/J.SETA.2019.05.002.

Chapter 13

Examples of Solar Hybrid System Layouts, Design Guidelines, Energy Performance, Economic Concern, and Life Cycle Analyses

1 Solar hybrid system energy performance

1.1 Solar constant and energy generation

The power generation capacity of solar hybrid systems is proportional to the amount of solar irradiation. The sunbeams falling on the earth after being filtered from the earth's atmosphere can be directly converted into electrical energy thanks to solar photovoltaic (PV) panels. The value accepted as the solar constant is 1353 kW/m^2 [1,2]. This value is the maximum amount of electrical energy that can be obtained in m^2 unit area of the sunbeams coming to the earth. The solar constant day/night varies regionally and seasonally. This change is due to the earth's rotation around its axis, its rotation around the sun, and atmospheric effects. Especially climatic changes change the angle of the sunbeams coming to the earth. Also, atmospheric events are factors such as rain, cloud, snow, and fog that affect the sunbeams. As clouds cause the scattering of sunbeams, the energy performance of concentrated solar energy systems especially decreases [3–5].

Another factor affecting energy performance in solar systems is air mass. Air mass is the angle made to the surface of the sunbeam and the solar energy production unit (solar PV panel or concentrated solar systems). This angle varies with solar PV panel positioning. Air mass value of 1.5 is rated as good for solar systems [6,7].

The solar irradiation unit can be measured by pyranometer and gives the electrical power value in m^2 that can be produced by solar PV panels (kW/m^2). Hourly or daily electrical energy amount by the solar PV panel is expressed as kWh.

An electrical energy close to the theoretically calculated value is generated from a solar PV panel calculated and positioned on a sunny day. This situation is called the maximum sun condition. Theoretically expressed value is the solar constant and is 1.353 kW/m^2. However, this power generated in practice can

Solar Hybrid Systems. http://dx.doi.org/10.1016/B978-0-323-88499-0.00013-6

331

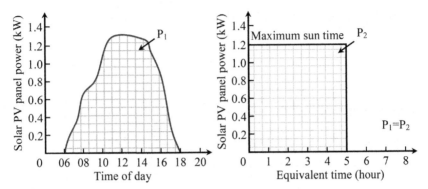

FIGURE 13.1 The cumulative power graph of the total energy value produced from the solar PV panel in 1 day.

never reach this value. The total energy value produced by solar PV panels during the day is variable. The total power generation value can be calculated by taking the cumulative sum of this daily power value [8,9]. The cumulative power graph of the total energy value produced from the solar PV panel in 1 day is given in Fig. 13.1. When the power generation graph is analyzed, the values of P_1 and P_2 are equal. Thus the daily power value that can be taken from a solar PV panel can be calculated on an hourly basis.

The performance of the electrical energy produced by a solar PV panel affects the solar irradiation intensity to the panels, the temperature of the panels, the position angles of the panels, and the shading. Solar irradiation value highly affects the current value to be produced by the solar PV panel. When the solar irradiation value decreases, the current value generated from the solar PV panel decreases. Solar PV panel voltage value is affected at a low rate. Solar irradiation value does not affect the conversion efficiency of solar PV panels, whereas only the current value generated according to the incident irradiation decreases in direct proportion. The temperature highly affects the voltage value of the irradiation to be generated in solar PV panels. Temperature reduces the efficiency of solar PV panel as in each conversion unit [10,11]. A solar PV panel shadowing is caused by natural or environmental factors. In the natural process, factors such as cloud, rain, snow, and fog occur between the sunbeams and the solar PV panel. Environmental factors may include building or tree shading as a result of improper positioning of panels. Other factors can be dust, tree leaves, and bird droppings. In the case of shadowing, power generation in the solar PV panel decreases. If there is solar PV connected in a string, there will be no power flow from the solar PV panel exposed to shading. If it is a solar string with high power capacity, excessive temperature may occur in the solar PV panel exposed to shading. Diodes can be connected to each solar PV panel to solve the problem of power generation in the case of shadowing. Thus damage to the solar PV panels can be avoided in shading condition [12].

1.2 Declination angle, solar altitude angle, tilt angle, solar azimuth angle, and solar zenith angle

Another factor affecting the power generation characteristics of solar PV panels is the positioning angles of the panels. If the solar irradiation were always vertical to the solar PV panels, the power would be continuously produced at maximum power. Sunbeam angles vary according to both latitude and seasons. Since the rotation axis of the earth rotates at an angle of 23.45 degrees with respect to its plane around the sun, the solar irradiation angles for a given location constantly change hourly. Declination is the angle between the sun–earth centerline and the projection of this line on the equatorial plane [13]. The angle of declination varies positively and negatively depending on the seasons. The declination angle is 0 degree only at the spring and autumn equinoxes. The deviation angle is indicated by δ and can be calculated by Eq. (13.1). The +10 is added since the winter equinox occurs before the beginning of each year in Eq. (13.1) and d is number of day.

$$\delta = -23.45° \cdot \cos\left(\frac{360}{365} \cdot (d+10)\right) \tag{13.1}$$

The solar altitude angle is the angle between the sunbeams and the earth's horizontal plane. The solar altitude angle is 0 degree at sunrise and usually 90 degrees when the sun is overhead at noon. The solar altitude angle constantly changes during the day. Solar altitude angle is important for solar PV panel systems. The solar altitude angle varies depending on the location and season where the solar PV panels will be placed [14]. The solar altitude angle is shown by α and is calculated by the following equations:

$$\alpha = 90 + \theta - \delta \tag{13.2}$$

$$\sin(\alpha) = \cos(\theta) = \sin(L) \cdot \sin(\delta) + \cos(L) \cdot \cos(\delta) \cdot \cos(h) \tag{13.3}$$

where θ is the latitude value of the panel location, L is local latitude, and δ is solar declination angle that changes depending on the day of the year.

The angle made by the solar PV panels with the horizontal plane is the tilt angle. The tilt angle is shown with β and thanks to the angle, the sunbeams come vertically to the solar PV panel. Tilt angle is solar PV panel surface angle toward horizontal surface.

The solar azimuth angle is an angle related to the compass direction from which sunlight comes. The solar azimuth angle is the angle value of the sunbeams coming from the south for the Northern Hemisphere and from the north for the Southern Hemisphere. The solar azimuth angle changes during the day, 90 degrees at sunrise and 270 degrees at sunset [15]. The solar azimuth angle is calculated by the following equation:

$$\sin(\gamma) = \frac{\cos(\delta) \cdot \sin(h)}{\cos(\alpha)} \tag{13.4}$$

The solar zenith angle is the angle between the sun and the vertical axis. The zenith angle is similar to the solar altitude angle but is measured from vertical rather than horizontal [16]. Solar zenith angle is calculated by the following equation:

$$\theta = 90° - \alpha \tag{13.5}$$

Knowing the solar azimuth angle will not be sufficient to determine the optimum angle of solar PV panels. Solar PV panel slope angle should be calculated depending on the position of the sun, taking into account all month and seasonal conditions of the year. If all angles related to solar PV panel are calculated, the annual cumulative energy amount to be taken from the solar PV panel will increase. Solar PV panels that will be positioned fixedly for the Northern Hemisphere should be positioned facing south in order to get the best solar irradiation. The direction of the solar PV panels is determined according to the solar azimuth angle for the Southern Hemisphere and Northern Hemisphere. Positive values of solar azimuth angles indicate west direction, while negative values indicate east direction [17]. The depiction of declination angle, solar altitude angle, tilt angle, solar azimuth angle, and solar zenith angle is given in Fig. 13.2.

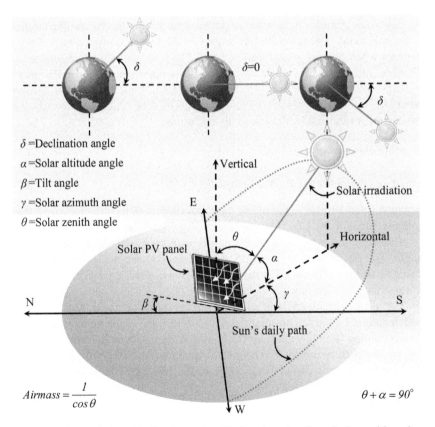

FIGURE 13.2 Depiction of declination, solar altitude, solar azimuth, and solar zenith angles.

2   Solar hybrid system economic concern

A comprehensive feasibility analysis must first be performed in solar hybrid system design. The feasibility study should be analyzed first technically and the second economically. Technically, feasibility studies include the installation area, optimum angles for maximum power, and power sizing. Economically, it is especially the product comparison and initial investment cost that perform the same task. A comprehensive economic analysis includes initial investment cost and cost–benefit analysis processes [18–20].

The return of a solar energy facility to be established should generally be calculated for 25 years. During this period the energy cost demanded from the grid should be analyzed. The facility to be established is expected to amortization for itself as soon as possible. The most effective factor in this amortization period is the grid incentive tariff [21].

Solar PV panel installation often involves high initial costs. While generating electricity from solar PV panels, the source is free of charge. Since there are no moving parts in the solar energy system (for solar PV panels), there are no maintenance costs. All this reduces system operating costs. When considering fossil-based electricity power generation facilities, the installation of solar PV panels is quite simple. Solar PV panel facility installations are encouraged by governments at different rates. The content of this incentive includes variable parameters such as the locality of the material used and the size of the system [22].

An end user who installs a solar PV panel system first wants to meet the energy consumed by her/him from a renewable energy source. If there is excess energy generation by solar PV panels, it sells to the grid. Thus its own consumption and electricity sales system costs will be met. Another cost in this setup phase is the interest rate paid on the borrowed money. Repayment plans should be taken into account when analyzing the amortization of the solar PV system. The purpose of all these economic analyzes is to install the solar PV panel system in the required energy capacity at the right size and at a suitable cost.

Solar energy has an intermittent and noncontinuous structure due to its nature. Therefore energy produced from solar energy systems has the same output characteristic. Due to this feature, it is not possible to provide uninterrupted energy throughout the year with a solar power generation system. All these reasons are also valid for solar thermal power generation systems. In order to overcome this problem, energy storage units containing batteries are used in solar PV panel systems and thermal heat storage units in solar thermal systems. Establishing an energy storage unit will increase the power generation capacity of solar systems. However, this will affect the installation cost. In addition, the maintenance and periodic replacement costs of these energy storage units should also be taken into consideration [23,24].

3   Solar hybrid system life cycle analyses

There are many alternatives to support the energy demanded by the end user with renewable energy sources. Among these alternatives the most suitable from an economic analysis point of view are solar energy systems. One of the most important issues of the solar system is its life cycle. Life cycle analysis takes into account the time value of money. In addition, the cost that will be spent for the system is analyzed in detail. The parameters in the life cycle analysis vary according to the countries, tariff rates, and interest rates. Various criteria are taken into account to make life cycle analysis of a solar system [25].

All costs spent for the solar system are calculated by taking into account the value of money over time. This cost is actually the total cost of the solar system over its lifetime. Life cycle cost can be predicted over a specified period of time. This cost analysis is based on the idea of compensating for future costs. In this analysis the money spent for the system is able to catch the gain as a result of putting the money in the bank with the highest interest. A value factor affecting the life cycle analysis in solar systems is the lifetime savings [26]. This lifetime saving represents the money value to be earned throughout the life cycle of the system.

One of the factors affecting the results of solar systems feasibility analysis is the payback period. The payback period is the time taken to recover the money spent on the solar system. During this period, energy is not taken from the grid. It is thought that all of the energy needed is taken from the solar system. The recovery of the cost spent for solar systems is expressed by the return on investment [27]. The return on investment is generally desired to be as soon as possible. It is calculated by making a table for the life cycle of solar systems and writing the revenues spent and planned to be recovered.

The life cycle analysis also reflects the savings achieved by using solar energy. Conventional power generation systems often require high costs in the initial installation stage. However, solar systems, especially solar PV panels, can be installed at low costs. In addition, source/fuel costs are variable in traditional electricity generation systems and unit costs are gradually increasing. This factor also resets the fuel cost in solar systems. In the life cycle analysis, the initial investment and operating costs are taken into account throughout the solar system operation. When the solar systems reach the end of their life, the parts in the system can be considered as scrap that affects the amortization period [28,29].

The initial investment in solar systems includes equipment cost, engineering cost, employee cost, shipping cost, legal application cost, and area costs. If it is a facility with a solar thermal structure, the initial investment costs include concentrated solar collectors, pump, pipe, trough, insulation material, storage tank, and heat exchangers. As the size of the solar energy system increases, the capacity value increases, but this causes an increase in initial investment costs. It is very important to determine the optimum capacity required for solar systems. In order to find out the total cost of a solar system clearly, all equipment and parts must be identified completely. In this optimum analysis the operating efficiency of the system should also be predicted. This efficiency value is especially important when calculating the system payback [30–32].

4 Solar hybrid system design guidelines

Solar hybrid system design has many parameters that need to be evaluated. First of all, it should be determined that the system to be installed will be on-grid or off-grid structure. After selecting the solar hybrid building type, the system is sizing. One of the critical parameters in solar energy system sizing is the determination of the amount of load. In addition, during the installation stage, the location where the solar energy system will be installed, mounting angles, local solar irradiation data are the parameters to be considered [33].

Solar irradiation data at the installation location are needed to estimate the daily output power data for solar PV panels. In order for solar PV panels to obtain maximum solar irradiation values, optimum angle values should be calculated. These angle values should be adjusted for the floor or roof applications where the installation will be made. The area where the solar PV panels will be located should not be shaded. If the monthly average of the loads in the solar energy system is constant, the solar PV panel power capacity should be determined according to the worst solar irradiation month data.

After determining the location of the solar energy system, the planned power value should be analyzed. This power value should be determined according to the load demand power to be used in the system. Determining the demand power in solar PV panels is an important factor for design parameters. When calculating the load demand power, the daily consumption values of the loads in the system should be determined by calculating. Load demand is important in determining the capacity of the battery group in off-grid systems. According to this load demand, Ah capacity of the battery group will be selected.

Solar PV panels should be designed in a size that can meet all loads during the day. Thus the solar PV system can meet the demand regardless of both the grid and the energy storage unit. The battery capacity directly determines the availability of the system in off-grid solar PV systems. The more battery groups, the longer it supports the energy demand. The large battery group size will increase the total cost of the system. In addition, solar PV panel capacity and battery group should be selected in proportion to each other. If it is desired to charge the battery group in addition to the daily use of power, the solar PV panel capacity should be high. If the energy produced by the solar PV panels only meets the demand power, the battery group will never be charged. In such a case an extra battery group will have been purchased [34–36]. When determining the capacity of the battery group, design should be made according to the load power. The battery group should not be deeply discharged in high power demands. In such a case the cycle life of the battery group decreases and creates costs. For off-grid solar PV systems, the battery group capacity should be selected to meet the total loads for a week.

After determining the demand power in the solar energy system, the PV panel is selected. Today, there are solar PV panels with different technology structures. The most used and preferred PV panel technologies are monocrystalline, split-cell module, polycrystalline, and passivated emitter and rear contact (PERC) module PV panels [37]. In solar energy systems the power range of a solar PV panel module varies between 250 and 400 W at standard temperature conditions of 25 °C per module. Maximum power values stated in catalog values are given under 1100 W/m² solar radiation values. Different solar PV panel manufacturers give different W/m² and efficiency factors. Array can be obtained by connecting solar PV panels in series or in parallel, depending on the demand power. The comparison of power production capacities of monocrystalline, split-cell module, polycrystalline, and PERC module PV panels with the same power values is given in Fig. 13.3. These values are calculated under the same conditions by taking from solar PV panels with the same capacity.

The total efficiency of the system should be considered during the dimensioning and design in solar PV panel systems. Factors affecting total efficiency are cables, terminal connections, power converter efficiency, battery group efficiency, and dust in solar PV panels. The total loss rate in a solar PV panel system is approximately 10 %–20 %. Correct sizing of the solar system will reduce the total cost and increase the total life and performance of the system [38].

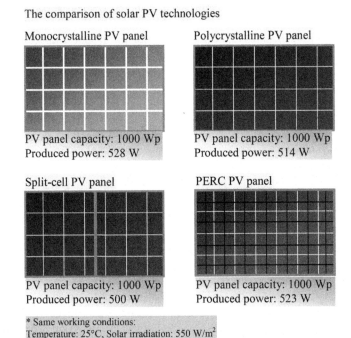

The comparison of solar PV technologies

Monocrystalline PV panel

PV panel capacity: 1000 Wp
Produced power: 528 W

Polycrystalline PV panel

PV panel capacity: 1000 Wp
Produced power: 514 W

Split-cell PV panel

PV panel capacity: 1000 Wp
Produced power: 500 W

PERC PV panel

PV panel capacity: 1000 Wp
Produced power: 523 W

* Same working conditions:
Temperature: 25°C, Solar irradiation: 550 W/m²

FIGURE 13.3 **Comparison of power production capacity of different solar PV panel technologies.**

One of the sizing items in off-grid solar PV systems is the battery group. The number of batteries that can be used in series or in parallel will be determined as a result of the design. One of the factors determining the number of batteries is power capacity. The power capacity will determine the size of the battery group in Ah. In addition, the total efficiency of the batteries should be taken into account when calculating the losses. Temperature factor affects solar PV panels, power converter circuits, and battery group. During seasonal use, attention should be paid to the temperature effect.

In the case of excessive electrical energy in solar PV panel systems, it can be given to the grid. In this case, bidirectional meters are used to measure the power transferred to the grid. The difference between the consumed and transferred power to the grid is calculated with these bidirectional meters. The electricity sales are carried out at the rate of the tariff thanks to the bidirectional meters. In addition, it is necessary to have leakage current fuse and over current/voltage protection equipment together with bidirectional meters. All these equipment affect the total cost of the system [39].

One of the protection requirements in solar PV panel systems is grounding equipment. Grounding equipment provides protection against lightning and possible overvoltage in the system. Grounding is provided by connecting to metal boxes and mounting parts in the system.

Direct current (DC) energy received by solar PV panels is transformed into alternative current (AC) energy with inverters. Inverters are connected to the grid synchronously and ensure the flow of energy. Inverters also work by simulating the grid to feed the loads. Inverters in solar PV panel systems ensure that the frequency and voltage are produced at standard values. Inverters convert the DC energy received in the battery group to AC energy in off-grid structure. Inverters also operate solar PV panels at their maximum power point. Thus inverters ensure that maximum energy is harvested from solar PV panels. The inverter power size should always be higher than the total system power. If high-power motor loads are to be operated in the system, this parameter should be considered in the sizing analysis [40].

One of the important parts of the design phase in solar PV systems is automatic and manual circuit breaker. This protection equipment gathers all cabling operations in the system at a single point. This provides ease of maintenance and fault detection in the system. When there is maintenance or cleaning in the system, the electric energy is turned off with the circuit breakers, and safe intervention is ensured. In addition, the data of the solar PV system can be monitored remotely, facilitating data tracking and analysis. These data tracking systems help one to monitor the voltage, current, and power data of the solar PV panel, load, and battery group via the internet. Thanks to the data tracking system, it can make power generation predictions by looking at the historical data. In addition, this remote monitoring system provides automatic notification in the case of fault [41,42].

There is a battery control circuit for safe charge/discharge of the battery group in solar PV systems. The battery control circuit ensures that the battery group remains fully charged by looking at its state of charge. The battery control circuit works bidirectionally. The battery control circuit also enables solar PV panels to operate at their maximum power point. The battery control circuit ensures a long cycle life of the battery group by controlling the optimum power flow. The input and output parameters (current/voltage) of the battery control circuit must be calculated in detail at the design stage.

5 Solar hybrid system layouts

5.1 On-grid solar PV panel system layouts

As an example, solar PV panel application with a power capacity of 5 kWp has been designed. The on-grid structure is connected to the existing grid in a single phase with low voltage. In the solar PV panel application example, the design was made by accepting the monthly average power consumption value of 4 kW. Annual average power consumption value is 5.5 kWh. The 5-kWp solar PV panel power application installed in Kütahya Dumlupınar University Electrical and Electronics Engineering Department was sized for this location. Solar PV panels are positioned to be mounted on the ground in an open area. For a solar PV panel with a power capacity of 5 kWp, an area of 40 m^2 is required as a result of the slope and shading calculations. Total 20 solar PV panels with a total power capacity of 250 W are used in the solar system. The layout and calculation of solar PV panel is given in Fig. 13.4. Solar PV panels are placed facing south. Simav is located at 39.11° latitude and at 29.01° longitude at Turkey. Tilt angles of solar PV panels are calculated as 41 degrees in average according to this position. Simav's yearly average total sunshine duration is 2690 hours and daily average sunshine duration is 7.37 hours. The power generation coefficient for the Simav region is taken as 1.1. Simav's annual mean solar irradiation is 1773 kWh/m^2. Considering the total efficiency of the whole solar PV panel system as 0.75, the expected annual mean production is approximately 7313 kWh. Simav's daily average temperature is 20.1 °C.

There are 10 solar PV panels connected in serial a single string and two strings connected in parallel in solar PV panel system. Thus 320 V DC bus voltage was obtained in solar PV panel system. The layout and calculation of solar PV panel is given in Fig. 13.4. The expected production capability from the solar PV system can be calculated with the following equation:

$$E_{tot} = E_{PV} \cdot P_{peak} \cdot E_{mai} \cdot \eta_{sys} \tag{13.6}$$

where E_{tot} is expected production capability from the solar PV system, E_{PV} (kWh/kWp) is annual mean production optioned solar PV panels, P_{peak} is peak power produced form solar PV panels, E_{mai} (kWh/m^2) is mean annual irradiation at located (Simav), η is total efficiency of solar PV panel system (inverter, cables, and connections, losses due to shading, and losses due to temperature effect).

Location: Simav, Turkey Latitude: 39.11°, Longitude: 29.01°, Altitude: 830 m
Solar PV panel capacity: 5 kWp
Simav's yearly average total sunshine duration: 2690 h, daily: 7:37 h/min
Simav's annual mean solar irradiation: 1773 kWh/m^2
The expected annual mean production: 7313 kWh
Simav's daily average temperature: 20.1°C
Solar PV system efficiency: 75 %

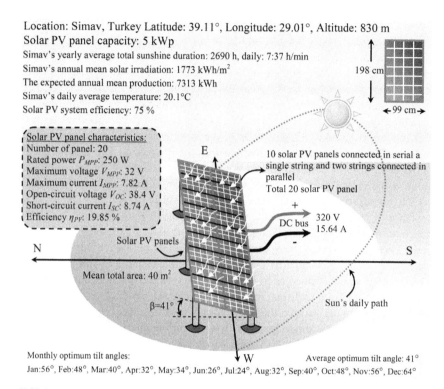

198 cm

←99 cm→

Solar PV panel characteristics:
Number of panel: 20
Rated power P_{MPP}: 250 W
Maximum voltage V_{MPP}: 32 V
Maximum current I_{MPP}: 7.82 A
Open-circuit voltage V_{OC}: 38.4 V
Short-circuit current I_{SC}: 8.74 A
Efficiency η_{PV}: 19.85 %

E

10 solar PV panels connected in serial a
single string and two strings connected in
parallel
Total 20 solar PV panel

+
DC bus   320 V
          15.64 A
-

N   Solar PV panels   S

Mean total area: 40 m^2

β=41°   Sun's daily path

W   Average optimum tilt angle: 41°
Monthly optimum tilt angles:
Jan:56°, Feb:48°, Mar:40°, Apr:32°, May:34°, Jun:26°, Jul:24°, Aug:32°, Sep:40°, Oct:48°, Nov:56°, Dec:64°

FIGURE 13.4   The layout and calculation of solar PV panel.

The minimum and maximum operating temperatures of solar PV panels are −40 °C and +85 °C. The maximum open-circuit voltage of 20 solar PV panels to be used in the system according to the standard conditions has been calculated as 384 V. For the same solar PV panels, the operating voltage value at the maximum power point is 320 V. The operating current of the solar PV panels at the maximum power point is 15.64 A.

The total efficiency of the solar PV system has been calculated to be 75 %. The total derating and efficiency representation of solar PV system is given Fig. 13.5. This total efficiency includes derating of solar PV panels by manufacturer (5 %), derating for dirt over solar PV panels (5 %), derating of solar PV panel over temperature (0.5 %), AC (2 %), and DC (3 %) bus cable loss, and efficiency of the inverter (95 %). Cabling and connection work should be done carefully during system installation in order to have high total efficiency. One of the external factors affecting efficiency is dust for solar PV panels. Solar PV panels should be cleaned periodically with water to eliminate the shadow caused by dust.

The annual and monthly average energy production values of the on-grid solar PV panel system were determined by location using the "Photovoltaic

Geographical Information System" tool offered free of charge on the internet by the European Union Science and Information Service. The annual energy generation value for the location determined in this calculation tool has been calculated as 7313 kWh. The monthly average energy production graph for the region determined in the 5 kWp solar PV system is given in detail in Fig. 13.5. In addition, monocrystalline silicon structure is preferred as PV technology.

The choice of inverters in solar PV systems is one of the important issues. The input current and voltage values of the inverter that will be connected to the solar PV system must be compatible with the PV panels. The input voltage of the inverter to be selected should not be within the limit of the total voltage value of the solar PV panels. Since the open-circuit voltage of solar PV panels is high, attention should be paid to this voltage level.

On-grid inverters must be synchronous with the grid and operate in accordance with standard values. The leakage current values generated by the inverter during the energy transfer to the grid should be low. If high leakage current values are generated by the inverter, it will cause the circuit breaker to which the system is connected to blow. Inverters should also work in solar PV panels continuously at their maximum power point. Depending on the different products of the manufacturer, there are models with two maximum power point tracking methods in a single inverter. The reason for this is to always harvest maximum energy from solar PV panels. Inverters act as isolation between solar PV panel and grid.

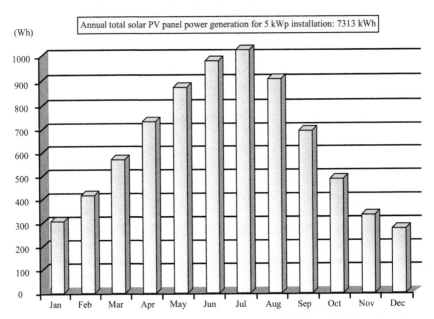

FIGURE 13.5   Monthly average energy production graph of 5-kWp solar PV panel system.

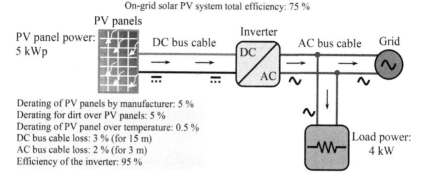

On-grid solar PV system total efficiency: 75 %

PV panels

PV panel power: 5 kWp

Inverter

DC bus cable

AC bus cable   Grid

DC

AC

Derating of PV panels by manufacturer: 5 %
Derating for dirt over PV panels: 5 %
Derating of PV panel over temperature: 0.5 %
DC bus cable loss: 3 % (for 15 m)
AC bus cable loss: 2 % (for 3 m)
Efficiency of the inverter: 95 %

Load power: 4 kW

FIGURE 13.6   **The total derating and efficiency of on-grid solar PV system.**

A single-phase inverter model has been chosen for the solar PV system application example. The operating voltage range is between 200 and 600 V at the maximum power point at the DC input of this inverter. The input power capacity of the inverter is 5.3 kW, and the maximum DC input current is 18 A. The AC voltage on the grid side of the inverter is 230 V, the output power capacity is 5 kW, and the output operating frequency is 50 Hz. The efficiency of the inverter is 95 %. The total derating and efficiency of on-grid solar PV system is given in Fig. 13.6.

The use of multicore cable in the DC bus cable connection in the solar PV panel system installation will prevent voltage drop. It will be appropriate to use single-core cable between the grid and the inverter. One of the most important issues in cable selection is that it should not prevent voltage drop in the system. Voltage drop in cables will affect system efficiency. AC and DC bus cable losses in solar PV system are 5 %. Outer frames and carrier constructions of solar PV panels should be grounded with 2.5 mm^2 yellow–green N07V-K cable. Grounding will provide security against lightning strikes and leakage currents and voltages.

In the solar PV application example, the annual average electrical energy consumption of the load group is 1500 kWh. The annual amount of electricity generated from solar PV panels is 7313 kWh. The selling price of the electricity tariff (including taxes) in the grid to which the solar PV panel is connected is 0.12$/kWh. Profit will be obtained from the electricity distribution company as much as this amount of electricity. The annual gain of this total energy is 7313 kWh × 0.12$/kWh = 877.56$. The total cost of 5 kWp capacity solar PV panel system is 10,000$, including taxes. When the initial investment cost and the total electricity energy cost to be obtained are compared, the amortization period is 10,000$/877.56$ = 11.3 years. After the amortization period, 877.56$ annual profit is obtained according to the electricity tariff prices. The annual amortization and profit graph of the solar PV panel system is given in Fig. 13.7.

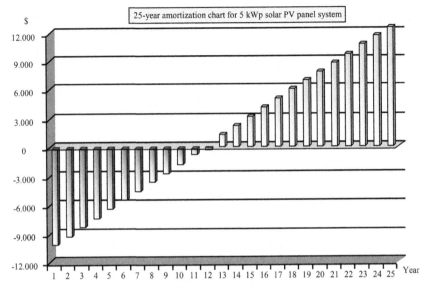

FIGURE 13.7 The annual amortization and profit graph of the solar PV panel system.

5.2 Off-grid solar PV panel system layouts

Battery group is also used in solar PV panel systems. When determining the battery capacity, the load power in the system should be determined. In the solar PV panel example, the total load is 4 kW. First, the load profile of the solar PV system is determined, and the average daily energy demand required is calculated. Power converters are selected according to the load profiles to be used in the solar PV system. The battery group stores electrical energy in order to use it at night and when sunlight is not sufficient. When determining the daily electrical energy need in the solar PV panel system, it is calculated by Eq. (13.7) for DC electricity energy demand and Eq. (13.8) for AC electricity energy demand.

$$\Sigma \left(\frac{\text{Wh}}{\text{week}} \right)^{DC} = \sum_{j=1}^{k} \sum_{i=1}^{n} \left(P_{ij}^{DC} \cdot n_{ij} \right) \cdot \left(\frac{\text{hour}}{\text{day}} \right)_{ij} \cdot \left(\frac{\text{day}}{\text{week}} \right)_{ij} \tag{13.7}$$

$$\Sigma \left(\frac{\text{Wh}}{\text{week}} \right)^{AC} = \sum_{j=1}^{k} \sum_{i=1}^{n} \left(P_{ij}^{AC} \cdot n_{ij} \right) \cdot \left(\frac{\text{hour}}{\text{day}} \right)_{ij} \cdot \left(\frac{\text{day}}{\text{week}} \right)_{ij} \tag{13.8}$$

where i is the number of electrical devices and j is the types of electric devices.

The total electrical energy demand in the solar PV panel system is calculated by the next equation:

$$\Sigma \left(\frac{\text{Wh}}{\text{week}} \right)^{AC} = \frac{\Sigma \left(\dfrac{\text{Wh}}{\text{week}} \right)^{DC}}{\eta_{DC/DC}} + \frac{\Sigma \left(\dfrac{\text{Wh}}{\text{week}} \right)^{AC}}{\eta_{DC/AC}} \tag{13.9}$$

where $\eta_{DC/DC}$ is *DC/DC* converter circuit efficiency and $\eta_{DC/AC}$ is *DC/AC* converter circuit efficiency.

The battery capacity is calculated on the basis of the battery voltage value in the solar PV panel system. The electrical energy requirement demanded in the solar system is calculated in Ah/day with the following equation:

$$\left(\frac{Ah}{day}\right) = \frac{\Sigma\left(\frac{Wh}{week}\right)}{V_{bat}} \cdot \frac{1}{7}\left(\frac{week}{day}\right) \tag{13.10}$$

The voltage level of the battery group should be determined by considering the distance between the battery group and the load and the PV output voltage. As the distance between the load and the battery group increases, the battery voltage should be chosen at the same rate to tolerate the voltage drop. When determining the battery group capacity, the number of days that the loads in the system must meet without receiving energy from solar PV panels should be taken into account. In addition, the number of cloudy days of solar PV panels also affects the battery group capacity. The deep of discharge current demand of the loads in the system determines especially the number of batteries. Since the operating temperature of the battery group affects the voltage levels, attention should be paid to the temperature coefficient in the calculations.

The battery group generally consists of n_S serial and n_P parallel branches in the solar PV system. The number of batteries in the serial branch is usually determined by the voltage level needed for the system. The number of parallel branches in the battery group is calculated with Eq. (13.11). The total efficiency of the battery group has been chosen as 85 %.

$$n_P = \frac{\left(\frac{Ah}{day}\right)\cdot\left(N_{cloudyday}\right)\cdot\left(I_{DoD}\right)\cdot\left(K_{temp}\right)}{\left(n_{serial}\right)\cdot\left(C_{bat}\right)} \tag{13.11}$$

where $N_{cloudyday}$ is the cloudy days of the solar PV panels, I_{DoD} is the deep of discharge current of the battery group, K_{temp} is the temperature coefficient of the battery, and C_{bat} is the total capacity (Ah) of the battery group.

It has been calculated to be compatible with the previous application example for the off-grid solar PV panel system. Total load power in solar PV system is 4 kW, and it is calculated that the load group demands a total daily capacity of 300 Ah. In the solar PV system example, the battery group will be designed to meet the entire load demand for a week. The battery group is designed to meet an average of 20 kWh energy demand for a week. The voltage of the battery group has been selected as 48 V. Lead–acid batteries will be used as battery technology. The capacity selection of the battery group has been selected as

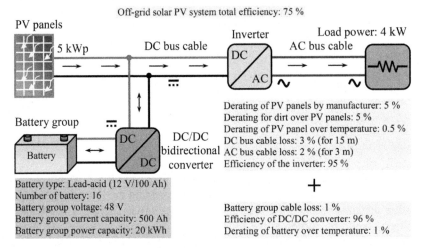

FIGURE 13.8 **The total derating and efficiency off-grid solar PV system.**

500 Ah, which is greater than the calculated demand power value. The battery voltage used in the solar PV panel system is 12 V and the battery capacity is 100 Ah. There are 4 batteries connected in serial a single string and 4 strings connected in parallel and a total of 16 batteries are used. The total derating and efficiency off-grid solar PV system is given in Fig. 13.8.

References

[1] D. Dirnberger, G. Blackburn, B. Müller, C. Reise, On the impact of solar spectral irradiance on the yield of different PV technologies, Sol. Energy Mater. Sol. Cells 132 (2015) 431–442 https://doi.org/10.1016/j.solmat.2014.09.034.

[2] D. Waterworth, A. Armstrong, Southerly winds increase the electricity generated by solar photovoltaic systems, Sol. Energy 202 (2020) 123–135 https://doi.org/10.1016/j.solener.2020.03.085.

[3] R. Kumar, C.S. Rajoria, A. Sharma, S. Suhag, Design and simulation of standalone solar PV system using PVsyst software: a case study, Mater. Today Proc. (2020) https://doi.org/10.1016/j.matpr.2020.08.785. In press.

[4] S. Verma, S. Mishra, S. Chowdhury, A. Gaur, S. Mohapatra, A. Soni, et al. Solar PV powered water pumping system – a review, Mater. Today Proc. (2020) https://doi.org/10.1016/j.matpr.2020.09.434.

[5] H.J. Han, M.U. Mehmood, R. Ahmed, Y. Kim, S. Dutton, S.H. Lim, et al. An advanced lighting system combining solar and an artificial light source for constant illumination and energy saving in buildings, Energy Build 203 (2019) 109404 https://doi.org/10.1016/j.enbuild.2019.109404.

[6] C.A. Gueymard, A reevaluation of the solar constant based on a 42-year total solar irradiance time series and a reconciliation of spaceborne observations, Sol. Energy 168 (2018) 2–9 https://doi.org/10.1016/j.solener.2018.04.001.

[7] B. Bora, R. Kumar, O.S. Sastry, B. Prasad, S. Mondal, A.K. Tripathi, Energy rating estimation of PV module technologies for different climatic conditions, Sol. Energy 174 (2018) 901–911 https://doi.org/10.1016/j.solener.2018.09.069.

[8] A.K. Tripathi, S. Ray, M. Aruna, S. Prasad, Evaluation of solar PV panel performance under humid atmosphere, Mater. Today Proc. (2020) https://doi.org/10.1016/j.matpr.2020.08.775. In press.

[9] M.D. Udayakumar, G. Anushree, J. Sathyaraj, A. Manjunathan, The impact of advanced technological developments on solar PV value chain, Mater. Today Proc. (2020) https://doi.org/10.1016/j.matpr.2020.09.588. In press.

[10] L. Guan, S. Liu, J. Chu, R. Zhang, Y. Chen, S. Li, et al. A novel algorithm for estimating the relative rotation angle of solar azimuth through single-pixel rings from polar coordinate transformation for imaging polarization navigation sensors, Optik (Stuttg) 178 (2019) 868–878 https://doi.org/10.1016/j.ijleo.2018.10.080.

[11] A.Z. Hafez, A. Soliman, K.A. El-Metwally, I.M. Ismail, Tilt and azimuth angles in solar energy applications – a review, Renew. Sustain. Energy Rev. 77 (2017) 147–168 https://doi.org/10.1016/j.rser.2017.03.131.

[12] Z. Jin, K. Xu, Y. Zhang, X. Xiao, J. Zhou, E. Long, Installation optimization on the tilt and azimuth angles of the solar heating collectors for high altitude towns in Western Sichuan, Procedia Eng. 205 (2017) 2995–3002 https://doi.org/10.1016/j.proeng.2017.10.225.

[13] S. Soulayman, Comments on solar azimuth angle, Renew. Energy 123 (2018) 294–300 https://doi.org/10.1016/j.renene.2018.02.063.

[14] K.A. Balmes, Q. Fu, The diurnally-averaged aerosol direct radiative effect and the use of the daytime-mean and insolation-weighted-mean solar zenith angles, J. Quant. Spectrosc. Radiat. Transf. 257 (2020) 107363 https://doi.org/10.1016/j.jqsrt.2020.107363.

[15] V. Poulek, T. Matuška, M. Libra, E. Kachalouski, J. Sedláček, Influence of increased temperature on energy production of roof integrated PV panels, Energy Build. 166 (2018) 418–425 https://doi.org/10.1016/j.enbuild.2018.01.063.

[16] N.A.S. Elminshawy, M. El Ghandour, H.M. Gad, D.G. El-Damhogi, K. El-Nahhas, M.F. Addas, The performance of a buried heat exchanger system for PV panel cooling under elevated air temperatures, Geothermics 82 (2019) 7–15 https://doi.org/10.1016/j.geothermics.2019.05.012.

[17] R.A. Rafael, D.L.F. Pottie, L.H. Leonardo, B.J. Cardoso Filho, M.P. Porto, A directional-spectral approach to estimate temperature of outdoor PV panels, Sol. Energy 183 (2019) 782–790 https://doi.org/10.1016/j.solener.2019.03.049.

[18] B. Fina, H. Auer, W. Friedl, Cost-optimal economic potential of shared rooftop PV in energy communities: evidence from Austria, Renew. Energy 152 (2020) 217–228 https://doi.org/10.1016/j.renene.2020.01.031.

[19] P. Pal, V. Mukherjee, P. Kumar, M. Elizabeth Makhatha, Viability analysis of direct current (DC) standalone hybrid photovoltaic (PV)/hydrogen fuel cell (HFC) energy system: a techno-economic approach, Mater. Today Proc. (2020) https://doi.org/10.1016/j.matpr.2020.10.405. In press.

[20] A. Datas, A. Ramos, C. del Cañizo, Techno-economic analysis of solar PV power-to-heat-to-power storage and trigeneration in the residential sector, Appl. Energy 256 (2019) 113935 https://doi.org/10.1016/j.apenergy.2019.113935.

[21] B.R. Lukanov, E.M. Krieger, Distributed solar and environmental justice: exploring the demographic and socio-economic trends of residential PV adoption in California, Energy Policy 134 (2019) 110935 https://doi.org/10.1016/j.enpol.2019.110935.

[22] G.A. Thopil, C.E. Sachse, J. Lalk, M.S. Thopil, Techno-economic performance comparison of crystalline and thin film PV panels under varying meteorological conditions: a high solar resource southern hemisphere case, Appl. Energy 275 (2020) 115041 https://doi.org/10.1016/j.apenergy.2020.115041.

[23] A. Allouhi, Solar PV integration in commercial buildings for self-consumption based on life-cycle economic/environmental multi-objective optimization, J. Clean. Prod. 270 (2020) 122375 https://doi.org/10.1016/j.jclepro.2020.122375.

[24] P. Mirzania, N. Balta-Ozkan, A. Ford, An innovative viable model for community-owned solar PV projects without FIT: comprehensive techno-economic assessment, Energy Policy 146 (2020) 111727 https://doi.org/10.1016/j.enpol.2020.111727.

[25] M.Y. Ali, M. Hassan, M.A. Rahman, A.A. Kafy, I. Ara, A. Javed, et al. Life cycle energy and cost analysis of small scale biogas plant and solar PV system in rural areas of Bangladesh, Energy Procedia 160 (2019) 277–284 https://doi.org/10.1016/j.egypro.2019.02.147.

[26] A. Sagani, J. Mihelis, V. Dedoussis, Techno-economic analysis and life-cycle environmental impacts of small-scale building-integrated PV systems in Greece, Energy Build 139 (2017) 277–290 https://doi.org/10.1016/j.enbuild.2017.01.022.

[27] I. Celik, Z. Song, A.B. Phillips, M.J. Heben, D. Apul, Life cycle analysis of metals in emerging photovoltaic (PV) technologies: a modeling approach to estimate use phase leaching, J. Clean. Prod. 186 (2018) 632–639 https://doi.org/10.1016/j.jclepro.2018.03.063.

[28] C. Luerssen, O. Gandhi, T. Reindl, C. Sekhar, D. Cheong, Life cycle cost analysis (LCCA) of PV-powered cooling systems with thermal energy and battery storage for off-grid applications, Appl. Energy 273 (2020) 115145 https://doi.org/10.1016/j.apenergy.2020.115145.

[29] J. Ling-Chin, O. Heidrich, A.P. Roskilly, Life cycle assessment (LCA) – from analysing methodology development to introducing an LCA framework for marine photovoltaic (PV) systems, Renew. Sustain. Energy Rev. 59 (2016) 352–378 https://doi.org/10.1016/j.rser.2015.12.058.

[30] H.G. Ozcan, H. Gunerhan, N. Yildirim, A. Hepbasli, A comprehensive evaluation of PV electricity production methods and life cycle energy-cost assessment of a particular system, J. Clean. Prod. 238 (2019) 117883 https://doi.org/10.1016/j.jclepro.2019.117883.

[31] O.C. Akinsipe, D. Moya, P. Kaparaju, Design and economic analysis of an off-grid solar PV in Jos-Nigeria, J. Clean. Prod. 287 (2020) 125055 https://doi.org/10.1016/j.jclepro.2020.125055.

[32] G. Li, Q. Xuan, G. Pei, Y. Su, Y. Lu, J. Ji, Life-cycle assessment of a low-concentration PV module for building south wall integration in China, Appl. Energy 215 (2018) 174–185 https://doi.org/10.1016/j.apenergy.2018.02.005.

[33] S. Yadav, S.K. Panda, Thermal performance of BIPV system by considering periodic nature of insolation and optimum tilt-angle of PV panel, Renew. Energy 150 (2020) 136–146 https://doi.org/10.1016/j.renene.2019.12.133.

[34] Y. Zhang, H. Chen, Y. Du, Lightning protection design of solar photovoltaic systems: methodology and guidelines, Electr. Power Syst. Res. 174 (2019) 105877 https://doi.org/10.1016/j.epsr.2019.105877.

[35] M. Aghaei, N.M. Kumar, A. Eskandari, H. Ahmed, A.K.V. de Oliveira, S.S. Chopra, Solar PV systems design and monitoring, Photovolt. Sol. Energy Convers. 1 (2020) 117–145 https://doi.org/10.1016/b978-0-12-819610-6.00005-3.

[36] S. Sreenath, K. Sudhakar, A.F. Yusop, E. Solomin, I.M. Kirpichnikova, Solar PV energy system in Malaysian airport: glare analysis, general design and performance assessment, Energy Rep. 6 (2020) 698–712 https://doi.org/10.1016/j.egyr.2020.03.015.

[37] E. Taveres-Cachat, F. Goia, Exploring the impact of problem formulation in numerical optimization: a case study of the design of PV integrated shading systems, Build. Environ. 188 (2020) 107422 https://doi.org/10.1016/j.buildenv.2020.107422.

[38] R. Stropnik, U. Stritih, Increasing the efficiency of PV panel with the use of PCM, Renew. Energy 97 (2016) 671–679 https://doi.org/10.1016/j.renene.2016.06.011.

[39] F.M. Zaihidee, S. Mekhilef, M. Seyedmahmoudian, B. Horan, Dust as an unalterable deteriorative factor affecting PV panel's efficiency: why and how, Renew. Sustain. Energy Rev. 65 (2016) 1267–1278 https://doi.org/10.1016/j.rser.2016.06.068.

[40] E.Y.T. Chen, Y. Chen, B. Guo, H. Liang, Effects of surface morphological parameters on cleaning efficiency of PV panels, Sol. Energy 194 (2019) 840–847 https://doi.org/10.1016/j.solener.2019.10.087.

[41] M. Nethra, B. Kalidasan, Earth tube heat exchanger design for efficiency enhancement of PV panel, Mater. Today Proc. (2020) https://doi.org/10.1016/j.matpr.2020.02.387. In press.

[42] D. Wang, H. Wang, X. Zhang, Mission profile-oriented configuration of PV panels for lifetime and cost-efficiency of PV inverters, Microelectron. Reliab. 114 (2020) 113944 https://doi.org/10.1016/j.microrel.2020.113944.

Index

Note: Page numbers followed by "f" indicate figures, "t" indicate tables.

Printed in the United States
by Baker & Taylor Publisher Services